ALLE · ZEIT · WACH · 1842

SPRINGER STUDY EDITION

Monique Mainguet

Desertification

Natural Background and Human Mismanagement

Second Edition

With 85 Figures

Springer-Verlag
Berlin Heidelberg New York London
Paris Tokyo Hong Kong Barcelona
Budapest

Professor Dr. MONIQUE MAINGUET

Université de Reims
Laboratoire de Géographie Physique Zonale
57, Rue Pierre Taittinger
51100 Reims, France

Cover picture:
Djida, Vallée du serpent. Red dune system, interspersed with trees, very degraded after the drought 1968–1985. This photograph was taken in July 1985, just after the first rain. The trees, appearing almost lifeless, show some foliage among dry branches. The soil apposes a red, almost reactivated dune (background), and an interdunal depression with little, discontinuous and broken crusting (foreground).
(Copyright by M. Mainguet)

The first edition was published as Vol. 9 of the series „Springer Series in Physical Environment".

ISBN 3-540-57746-7 2. Aufl. Springer-Verlag Berlin Heidelberg New York
ISBN 0-387-57746-7 2nd ed. Springer-Verlag New York Berlin Heidelberg

ISBN 3-540-52519-X 1. Aufl. Springer-Verlag Berlin Heidelberg New York
ISBN 0-387-52519-X 1st ed. Springer-Verlag New York Heidelberg Berlin

Library of Congress Cataloging-in-Publication Data. Mainguet, Monique, 1937- Desertification : natural background and human mismanagement / Monique Mainguet. – 2nd ed. p. cm. „Springer study edition"–Prelim. p. Includes bibliographical references and indexes. ISBN 3-540-57746-7 (Berlin : acid-free). – ISBN 0-387-57746-7 (New York : acid-free) 1. Desertification. I. Title. GB611.M24 1994 551.41'5–dc20

Printed in Germany

Typesetting: International Typesetters Inc., Makati, Philippines
32/3130-543210 – Printed on acid-free paper

“Development is a gradual process of learning.” *(Eicher 1986)*

“The relationship between people and nature in most of Africa has been extractive: people took crops, and wood, and put nothing back, but gave nature time and space to restore herself. That relationship depended on an abundance of land and forests. But as population density grew, that abundance could no longer be taken for granted. Nature is no longer given time to restore herself. Her capital of resources is being depleted, and the whole system grinds gradually downhill.” *(Harrison 1987)*

“I see a fulfilment of the great Law of all worlds, that while the wisdom of Man thinks it is working one thing, the wisdom of Nature constrains it to work another, an quite a different and far better thing.” *(Abbott 1884)*

Preface

The different phenomena called *desertification* have attracted most attention in those countries where their impact on human life is most severe. Great human misery, malnutrition, starvation, and the undermining of the ability of these drylands to support life have most often been described in the Sahel of Africa and in western Rajastan (India). Are they in fact caused by one or several aspects of what is called desertification?

When I agreed to write this book on *Desertification: Natural Background and Human Mismanagement,* I was well aware that there is an extensive and growing body of literature on the subject. The only printout of desertification bibliography of the Office of Arid Land Studies, University of Arizona, at the beginning of the 1980's was almost 30 m long! So why another book? Was it because I felt:

1. It was necessary to offer further shades of meaning to the concept of desertification. Is desertification an existing phenomenon? In which ecosystem does land degradation reach the level of desertification?
2. How much land degradation is due to natural causes and how much to man's activities?
3. At what level can the solutions to land degradation be applied? Do seeds of hope exist?

I also agreed to write this book because I think that my fieldwork in the last three decades in the severely drought-prone countries of Africa and in almost all the deserts of our planet has provided me with valuable first-hand experience and information.

I use the world "experience" as it is defined by Claude Bernard: "Dans la langue française le mot expérience au singulier signifie … l'instruction acquise par l'usage de la vie".

The concept of desertification is relative in connection with human activities. Without people there is no desertification.

1. People are the producers of desertification. Why has there been the temptation to maximize the natural causes an to minimize the human responsibilities?
2. Desertification is mainly a function of mistaken human optimism about rainfall; it can be due to the assumption that good rains will

always fall. However, in a year of poor rains, agricultural methods suitable only for years of good rains can lead to crop failure, lack of adequate cover, wind erosion, and desertification.

3. Desertification can be considered as a useful alarm signal from the environment to indicate when human activities have reached the threshold of tolerance.

This book is not an attempt to synthetize all the available scientific information on desertification, but a selection of confirmed information highlighting the problem.

I shall also try to identify those policies which could prevent desertification and those which can improve land degradation which has not reached the point of being irreversible.

MONIQUE MAINGUET

Acknowledgements

When I decided to write this book in English, I knew that I was taking on a mountainous task. It might well have become too frustrating without the help I have had from many sources too numerous to mention individually. I must, however, express my gratitude to four people in particular.

Mrs. Mary Peel agreed to read through the first draft and amend any mistakes in my English. In the outcome, this resulted in much more than that. Because of her familiarity with much of the basic material acquired over 40 years from her late husband Ronald Peel, himself an arid lands specialist, she was able to correct an question my meaning in a way which made me clarify my thinking an the aim and development of my arguments on numerous points.

W. Richards also agreed to read the book and to offer his criticisms and advice an the plan and content. He made me aware that I had run away from dealing adequately with all the complexities of the African drylands.

Marie-Christine Chemin responded patiently to my demands on her for the organizing of the material at the final stages of the first edition.

Harold Dregne and Philipe Duchaufour, with immense kindness an efficiency, made numerous remarks, precious for the preparation of the second edition. Also, my colleague and friend René Létolle helped me unfailingly, thanks to his great knowledge in chemical, physical and hydrologic phenomena, to improve the first edition.

Finally, my debt to my husband Serge is immeasurable.

MONIQUE MAINGUET

Contents

Chapter 1 What is Desertification?: Definitions and Evolution of the Concept

1.1 Definitions of Desertification

As commonly used, the word *desertification* means an environmental crisis which produces desert-like conditions in any ecosystem; for different media, the meaning of the word is not the same: some use it for *desertion* (rural depopulation) due to economic difficulties.

El-Baz (1988), quoting Budge, writes: "The word desert originated as an ancient Egyptian hieroglyph pronounced tesert, meaning a place that was forsaken or left behind" and later:

"From this came the Latin verb *deserere*, to abandon. From the latter came *desertum*, a waste place or wilderness, and *desertus*, meaning abandoned or relinquished. This in itself implies that the desert had been a better place. In it, there was life -- in some places teeming life. There was much vegetation, grasses and trees, many animals and human beings. Then something happened, and the place became a wasteland; it was *deserted*."

According to Glantz' inventory (pers. comm.), the word desertification has more than 100 definitions: a testimony to the complexity of the phenomenon.

Deserts, as normally conceived, were during times past the result of deep climatic changes. Since the continual increase of human populations there has been an accelerated rate of degradation and production of desert-like conditions.

Earth-forming processes and man-made processes have two different time scales. Desertification has a much more immediate time scale and has variously been described as a series of conditions, as a process, as the ultimate stage of degradation, or as the encroaching desert, which latter description has gained acceptance over the past 30 years, and which I hope this book will totally disprove.

The analysis of the definitions reveals four main meanings. Because of its fourfold meaning, the word has high ambiguity.

First meaning. Desert areas are considered as the starting areas from where desertification begins, with the corollary of desert encroachment. This meaning has often been adopted by governments, national bodies, and mass media (as mentioned later in this book), even though the UNCOD[1] *Round-Up, Plan of Action, and Resolution* already stated the contrary in the year 1978:

"Deserts themselves are not the sources from which desertification springs. Except for hot winds, the deserts themselves supply none of the essential impetus for the processes described. Desertification breaks out, usually at times of drought stress, in areas of naturally vulnerable land subject to pressures of land use. These degraded patches, like a skin disease, link up to carry the process over extended areas. It is generally incorrect to envision the process as an advance of the desert frontier engulfing usable land on its perimeter: the advancing sand dune is, in fact, a very special and localized case.

[1]UNCOD, United Nations Conference on Desertification (1977) Nairobi (Kenya).

Desertification, as a patchy destruction that may be far removed from any nebulous front line, is a more subtle and insidious process."

Four years before UNCOD, Rapp (1974) avoided the mistaken theory of "encroaching desert',' insisting on: "the spread of desert-like conditions in arid or semi-arid areas due to man's action or, climatic change." By desert-like conditions Rapp means climatic conditions and desert-like mechanisms of evolution of the landscape: those of major water and wind erosion on bare surfaces.

The second meaning from Mabbutt (1978a) involves a continuum of change and means all the intermediate steps between a sound and an irreversibly degraded landscape: "change in the character of land to a more desertic condition" involving "the impoverishment of ecosystems as evidenced in reduced biological productivity and accelerated deterioration of soils and in an associated impoverishment of dependent human livelihood systems".

The same ideas of gradual impoverishment and change in ecosystems were launched by Garduno (1977):

"Desertification is the impoverishment of arid, semi-arid, and some subhumid ecosystems by the impact of human activities. It is the process of change in these ecosystems that leads to reduced productivity of desirable plants, alterations in the biomass and diversity of life forms, accelerated soil degradation, and increased hazards for human occupancy."

Third meaning. "An intricate process of land degradation", Street (1987). Street insisted on the role of processes in land degradation.

The same destructive processes are mentioned by Kates et al. (1977):

"It involves destructive processes in which the productive base deteriorates and the social system is imperilled. Unlike drought, which is usually a short-term diminution of available moisture, the physical processes involved in desertification are long-term, chronic, and pervasive".

One further complexity needs noting — the difference between short-term climatic causes and long-term physical processes.

Meaning four. The ultimate nonproductive and desert-like stage of deteriorated environment. "Desertification is the name given to the processes whereby such ecosystems lose this capacity to revive orto repair themselves" (Hare 1985b).

Hare introduces the idea of natural irreversible deterioration.

Already introduced by Kates et al. (1977), the idea of the impoverishment of social system is further developed by Warren and Maizels (1977). They introduced the theory of a sustained economical impact: decline in the yield.

"A simple graphic meaning of the world desertification is the development of desert-like landscape in areas which were once green. Its practical meaning ... is a sustained decline in the yield of useful crops from a dry area accompanying certain kinds of environmental change, both natural and man-induced."

The explanation is ambiguous and much confusion has resulted between the first meanings, which concern processes in space and time, and the last, which is a state.

More precise is the term *land degradation* to describe the results of deteriorating processes and the word *desertification* to define in dry ecosystems an environment which has turned into a desert.

The words *dry ecosystems* or *drylands* cover a complex set of ecosystems.

Ecozones	Called in Africa	Rainfall (mm/year)
Arid	Saharan	<50–150
Semi-arid	Sahelian	150–600
Subhumid (with a marked dry season)	Sudanian	600–1200

All the intermediary combinations can exist: Saharo-Sahelian, Sahelo-Sudanian, etc.

The word desertification must be used in a limited zone only: etymologically, it means the change-over when dry-subhumid or semi-arid ecosystems become arid or hyperarid-like landscapes. This evolution shows itself in two ways: (1) deteriorated vegetative cover, (2) truncated top soil or soil, irreversible in human terms or within reasonable economical limitations.

This spatial limitation of desertification will be retained here, and used within the dry or seasonally dry ecozones or more simply drylands, which can be defined by their lack or scarcity of rain. According to Mabbut (1978): the term dryland has been used to designate those areas subject to periodic drought stress which are most liable to desertification. Hydrologists consider the 700 mm isohyet as the lower limit of permanent runoff. This value can also be considered as the hydrological limit of drylands. In the local rivers of drylands, water is available only during the short rainy season, the runoff is ephemeral, accidental, intermittent, or seasonal, according to the length of the rainy season.

After reviewing a large number of authoritative definitions of desertification, the definition proposed in 1977 by the United Nations Conference on Desertification (UNCOD) seems the most comprehensive (Resolution no. 7).

"Desertification is the diminution or destruction of the biological potential of the land, and can lead ultimately to desert-like conditions. It is an aspect of the widespread deterioration of ecosystems, and has diminished or destroyed the biological potential, i.e., plant and animal production, for multiple use purposes at a time when increased productivity is needed to support growing populations in quest of development."

Ten years later, Ahmad and Kassas (1987), in the introduction to their book *Desertification: Financial Support for the Biosphere*, offer the following refinements.

1. In the Preface

- "the scale and urgency of desertification as presented to UNCOD (1977) and addressed by the PACD[2] are confirmed; desertification has continued to spread and intensify despite efforts undertaken since 1977;

[2]PACD, Programme Activity Center to Combat Desertification.

— land irretrievably lost through desertification continues at about 6 million ha annually, and land reduced to zero or negative economic productivity is more than 20 million ha annually;
— areas of productive land affected by at least moderate desertification include 3100 million ha of rangeland, 335 million ha of rainfed croplands and 40 million ha of irrigated land;
— rural populations in areas severely affected by desertification number 135 million (1984), compared with 57 million in 1977."

2. In the Introduction

a) "Desertification is a process of environmental degradation by which productive land is made nonproductive and desert-like. The total area of once productive land lost in recent decades is estimated at over 9 million km^2, the present annual rate of loss being of the order of 60 000 km^2. The significance of this figure is indicated by the fact that the world's present food-producing land area is about 13 million km^2.

b) Desertification is a global environmental problem, which directly affects some 100 countries in the world and indirectly affects the remainder. Its seriousness is accentuated by the fact that food production needs to be substantially increased to provide for the growing population of the world.

c) Desertification and, in particular, its acceleration in recent decades, is essentially due to interactions between societal systems and fragile ecosystems. These include actions of man, particularly the misuse and over-use of land resources. There is no evidence that desertification is caused by large-scale climatic change.

d) The effort to combat desertification should be integrated into national development plans and priorities. The shortage of resources (capital, knowledge, and trained manpower) represents a major obstacle to action for halting desertification and reclaiming desertified lands. Desertification is thus particularly overwhelming in the least-developed countries."

These two definitions are the most complete to date; but they still do not clearly distinguish between **causes, mechanisms, manifestations**, and **impact of desertification**. Perhaps a better distinction of these four components of desertification could be given by the following definition, which I now propose:

"Desertification, revealed by drought, is *caused* by human activities in which the carrying capacity of land is exceeded; it proceeds by exacerbated natural or man-induced *mechanisms*, and is made manifest by intricate steps of vegetation and soil deterioration which *result*, in human terms, in an irreversible decrease or destruction of the biological potential of the land and its ability to support population."

In the above definition the economic dimension is not mentioned. This difficulty is well expressed by Nelson (1988):

"To cover our own reluctant use of the word desertification, and to start to uncover the issues, we offer a definition:
Desertification is a process of sustained land (soil and vegetation) degradation in arid, semi-arid and dry subhumid areas, caused at least partly by man. It reduces productive potential to an extent which can neither be readily reversed by removing the cause nor easily reclaimed without substantial investment."

Most definitions of desertification contain explicit reference to drought as the basic cause, when drought in fact only plays the role of revealer and intensifier of the phenomenon.

Hare (1985a) has most clearly described drought as an intensifier of desertification: "Climatic drought is the failure of expected rains: it is hence a shock to any economy that is already over-stressing its food system and natural resources".

Human activities are the determining factor at all stages of desertification: its causes, the exacerbated mechanisms of deterioration, and its impacts, the most tragic of which is famine. Underlying all desertification are demographic forces.

As Hugo (1984) commented on the relationship between famine and demographic processes, it is somewhat surprising that our understanding of the relationships between desertification and demographic explosion is so extremely limited. No well-developed theory yet provides a conceptual framework linking population growth and desertification.

In addition to the relationship between population growth and desertification, maladjustment of environmental management policies increases environmental deterioration and desertification. We lack methods of analyzing these effects and the scattered literature has no case studies differentiating between causes, processes, and impacts and thereby identifying the levels at which solutions may be found.

In this connection, particular attention must be paid to the level of the actual physical processes of soil degradation; this is the key to the problem. In the short term, these mechanisms are the appropriate level for measures against desertification to become applicable and effective: since, with reasonable financial and human support, processes such as wind erosion, water erosion, salinization, and waterlogging may be reduced and controlled. In contrast, reducing explosive demographic growth or changing unadapted traditional community practices require longer-term action.

The battle against desertification requires overlapping short-term physical and longer-term social strategies.

The first aim of this book is to analyze the status in the late 1980's of progress in understanding and combating desertification.

The attempts to understand desertification in the second part of the book seek the causes in time and space of the phenomenon and by so doing, try to make a more realistic assessment of it and to classify the solutions within clear boundaries, and by case studies.

The book was initially planned to emphasize the physical and chemical processes of land deterioration, but it soon became apparent that the socio-economic factors are so closely interwoven with the physical factors that they could not be neglected.

I agree with Baker (1982) when he writes: "It would be quite mistaken to believe that desertification is a process which can be arrested by physical or technical means alone; quite the contrary. The evidence is overwhelming that selective or symptomatic treatment of this sort has accentuated the problem quite seriously."

The attempt to manage desertification has to integrate natural and human processes; a complex relationship is inherent in the fact of desertification. In the final chapter desertification management with its complex interconnections will be analyzed. Desertification is a geographical dryland phenomenon characterized by the principe de connexité of Brunhes (1956): "Les faits de la réalité géographique sont étroitement liés entre eux et doivent être étudiés dans leurs multiples connexions. L'idée du tout terrestre."

This principle of interconnection is applicable to desertification, which has the complexity of a real world situation and requires the integration of its ecological, political, and socio-economical components.

1.2 The Evolution of the Concept of Desertification

"The word *concept*, used with the German meaning of Begriff, according to Kant, signifies a general but not an absolute idea" (Larousse du XXe siècle 1929; 1st ed.).

1.2.1 The Age of Awareness

Desertification was first referred to by Aubreville (1949), who observed "Ce sont de vrais déserts qui naissent aujourd'hui sous nos yeux, dans des pays où il tombe annuellement de 700 à plus de 1500 mm de pluies". "(There are real deserts being born today, under our very eyes, in areas where the annual rainfall is between 700 and 1500 mm").

This sentence of Aubreville is neither premonitory nor intuitive: it stems directly from field observations and it directs our attention to dry-subhumid areas. The work done in our laboratory (Laboratoire de Géographie Physique Zonale, University of Rheims, France) in collaboration with the French Institut Géographique National on the assessment of desertification between 1952 and 57 to 1987 comes to the same conclusion: the greatest danger is not in the Saharo-Sahelian ecozones, but in the subhumid Sudanian ecozone. Aubreville's text (1949) uses the word desertification to include both soil and vegetative deterioration caused, in part at least, by human activities.

The word desertification was created a long time after the alarm had been sounded on environmental vulnerability. One of the most touching outcries is in Lowdermilk's paper (1935) *Man-Made Deserts*, which begins: "The history of civilizations is a record of struggles against the progressive desiccation of civilized lands. The more ancient the civilization, the drier and more wasted, usually, are the supporting countries."

"In fact, so devastating seems the occupation of man that, with a few striking exceptions, a desert or near-desert condition is often associated with his long habitation of a region. Two major factors are believed to account for the growth of man-made deserts. In the first place, semi-arid to semi-humid regions proved the most favorable sites for the early development of human culture. Such areas, however, stand in a condition of delicate ecological balance between humid and true desert climates." "In the second place, processes of soil erosion are accelerated by the exposure of soil surfaces hitherto protected by complete mantles of vegetation, whether grass or forest, by heavy grazing and cultivation. It is only within the past decade that experimental studies of these processes have been made. So enormous have been the differences in soil wastage and superficial runoff of rain waters from bared sloping lands, as compared with similar surfaces protected by a complete coverage of vegetation, that new light is thrown on the problem of the decadence of former civilizations."

The term desiccation is here used for:

1. The result of a set of actions which transform more or less irreversibly a landscape in areas which did not previously have these characteristics.
2. The man-made desert; induction of desert or near-desert conditions.
3. "Disturbance of the coverage of vegetation and soils", resulting in "man-made deserts".
4. The processes of soil erosion which are accelerated by the exposure of soil surfaces hitherto protected by a vegetative cover.

In another sentence Lowdermilk insists on the fact that the death of a civilization is not the result of climatic changes but of mismanagement

"Recently, the archeologists have turned back the pages of history, not merely centuries, but thousands of years. Their post-mortems on buried civilizations suggest that it has been the hand of man, more than climatic change, which has reduced once rich and populous regions to desolation and poverty. After a long struggle, a civilization either died or its people migrated to more productive regions. Many ancient civilizations, once revelling in a golden age of prosperity, are crumbling in ruins or lie buried in sands and debris, largely caused by the destructive treatment of the lands on which they were dependent for sustenance. If modern peoples are to escape a similar fate by man-induced impoverishment and the desiccation of their lands, it would seem well to take a measure of these destructive processes and forces, and by intelligent land planning and land use provide for the sustained productivity of agricultural lands and the protection of grass lands and forests for food, textiles, raw materials and continued water supply."

Jacks (1939), in a very clear-sighted paper on human-accelerated erosion, writes:

"To gain control over the soil is the greatest achievement of which mankind is capable. The organization of civilized societies is founded upon the measures taken to wrest control of the soil from wild Nature, and not until complete control has passed into human hands can a stable superstructure of what we call civilization be erected on the land."

The author notes that geological erosion or denudation, which is natural erosion, "is a beneficial process without which the world would have died long ago". And he adds: "The same process, accelerated by human management, has become one of the most vicious and destructive forces that has ever been released by man." And in another definitive sentence:

"Soil erosion is altering the course of world history more radically than any war or revolution. Erosion is humbling mighty nations, re-shaping their domestic and external policies and once and for all it has barred the way to the El Dorado that a few years ago seemed almost within reach."

The soil scientists were probably the first to become aware to the potential danger of land degradation. Aubreville wrote his 1949 text shortly after the Second World War at a time of food shortage and severe drought in Europe.

Although Aubreville, first and foremost a field scientist, based his paper solely on field observations, it was forgotten for almost 30 years, until the 1980's.

1.2.2 The Age of Wrong Perception or Misconception

The 1970's and the first half of the 1980's can be considered the age of wrong perception of desertification as a result of several preconceptions:

1. The desire to find a world-wide unique concept of desertification.
2. The belief that the problem of desertification can be resolved by a universal panacea viz. satellite imagery. There was as a corollary the belief that a general view must replace investigations into every corner of the planet. Analysis of areas with apparent damage, on the scale of a continent or a country, cannot provide reliable knowledge of degradation, unless related to local observations on the ground and to realistic local statistics.
3. The belief that drought, desertification, and famine are directly linked led finally to panic and the conviction that the problem of desertification had to be solved urgently. The psychology of urgency resulted in the decision to hold the 1977 UN Conference on desertification in Nairobi. Some 500 delegates from 94 countries met from 29 August to 9 September to discuss the problem of desertification.

"The simplistic fears of a few years ago are now replaced by a well-founded sense of danger" reads the introduction of the UNCOD report. So to the notion of urgency was added the notion of danger and of the extent of the problem.

In the same report appeared the first worldwide statistics linking the countries, the surfaces threatened by desertification, and the estimate of threatened population (Table 1).

"What has also become clear is that desertification is not a problem that concerns just a few countries. Based on climatic data, more than a third of the earth's surface is desert or semi-desert and more than 15 percent of the world's population live in these areas.
If we go by data on the nature of soil and vegetation, the total area is some 43 percent of the earth's land surface. The difference is accounted for by the estimated extent of man-made deserts (9,115,000 square kilometres), an area larger than Brazil.
Further, some 30 million square kilometres (19 per cent+ of the earth's land surface) are threatened with desertification, and this threatened area is distributed among more than two-thirds of the world's 150 countries."

The UN statistics (Table 2) are now being questioned by some scientists (Forse 1989). Speaking of desertification, Forse writes in the *New Scientist*:

"The word and the image have sustained a decade-long effort by the United Nations Environment Programme (UNEP) to combat the spread of deserts. Now an increasing number of scientists are arguing that the image is a mirage and the efforts have been largely misdirected. Countries such as Mali, on the southern fringes of the Sahara desert, have used international aid to plant millions of trees as a barrier against the encroaching sands. But is the desert really moving?"

Table 1. Estimates of populations in areas undergoing severe desertificatioin. (UNEP UNCOD 1978)

(In thousands)

Region	Total population	Urban-based	Cropping-based	Animal-based	Area (km^2)
Mediterranean Basin	9820	2995	5900	925	1 320 000
		31%	60%	9%	
Sub-Saharan Africa	16 165	3072	6014	7079	6 850 000
		19%	37%	44%	
Asia and the Pacific	28 482	7740	14 311	6431	4 361 000
		27%	54%	19%	
Americas	24 079	7683	13 417	2979	17 545 000
		32%	56%	12%	
Total	78 546	21 490	39 642	17 414	30 076 000
		27%	51%	22%	

A review of the action against desertification was held in 1984. The conference was preceded by a study based on (1) a questionnaire of 27 pages distributed in 1982, (2) regional assessments by UNSO (United Nations Sahara-Sahelian Office), and UN regional commissions, and (3) case studies. The answers to the questionnaire were not very satisfactory and only 20 responses were used for the report. The other replies were too patchy and imprecise, mainly because the questionnaire was too complicated. Often the data it asked for did not exist.

According to Tolba, Director of UNEP: "35% of the world's land surface is at risk. . . each year 21 million ha are reduced to near or complete uselessness." These figures are criticized by Nelson (1988), of the World Bank, and by Forse (1989). The increase of 35% of land affected and of 35% of people affected by desertification between 1977 and 1984 can only be explained by the inclusion in 1984 of the subhumid ecozone, which was excluded in 1977.

1.2.3 The Age of Doubt

That 10 years after UNCOD signs of future control of desertification or land degradation can be seen, may be due to some improvements in the understanding of the concept of desertification.

Nelson (1988) says:

"The problem of desertification has been poorly characterized by public statements, books, magazines, and sometimes scientific articles. It has been poorly characterized in five ways. First, the impression has been conveyed that the extent of the problem of desertification is well known, when in fact the evidence is extraordinarily scanty. Second, the degree to which there is professional agreement among scientists and practitioners on the extent, causes, and solutions has been overestimated. Third, the extent of desertification as an irreversible state has probably been exaggerated,

Table 2. UNEP estimation of annual land degradation, costs and benefits of corrective measures

(1)	(2)	(3)	(4)	(5)	(6)	(7)	(8)	(9)	(10)
Type of land[a]	Annual rate of land degradation (per 1000 ha)[b]	Estimated value			Gain	Estimated cost of salvage[d] ha^{-1}	Total (2 × 7) million	Net gain ha^{-1} (5)–(7)	Total net benefits (2) × (9) million
		if not salvaged ($\$/ha^{-1}$)[c]	if salvaged	(4)–(3) ha^{-1}	(2) × (5) Total million				
				$	$	$	$	$	$
Irrigated	125[e]	200	2000	1800	225	850 (250–2000)	106	950	119
Range	3200	2	20	18	58	10 (1–50)	32	8	26
Rain-fed crop	2500	50	450	400	1000	100 (50–150)	250	300	750
Total	5825				1283		388		895

[a] Arid and semi-arid lands only.
[b] Annual rate of land degradation is based on annual rate of change of classes of land to more degraded conditions. The degree of degradation from higher to lower classes of land has been converted to more limited areas assumed to be deteriorating from land yielding highest net return (if salvaged) to land at the point of going out of production (if not salvaged).
[c] In view of difficulties in quantifying social values, these estimates are rough conservative approximations of orders of magnitude of capitalized values. Values are calculated using an assumed net income at half of gross income divided by an assumed opportunity cost of 10%, with a slight adjustment for rangelands to reflect lower opportunity costs. If social factors are included, the values would be substantially higher.
[d] Figures within parentheses give ranges of salvage costs. It follows from footnote[b] that cost of salvage is the maximum, equivalent to the cost of reclamation or restoration of practically completely desertified land. Because desertification is a continuous process, the more prudent course of action would be to begin corrective investment as soon as practicable and initially to lands which offer the highest returns to ensure continued maximum production.
[e] Due to waterlogging, salinization and, to a lesser extent, alkalinization.

although it is correct to classify it as a serious problem. Fourth, the image created has too often been of inexorably advancing sands, as opposed to more subtle, more complex, pulsating deteriorations, sometimes with reversals, but at least, with substantial periodic remissions, radiating out from centers of excessive population pressure. Fifth, the availability of profitable technologies to combat the problem has been overestimated because the gap between what is socially profitable and what is perceived as privately profitable has been underestimated."

According to Warren and Agnew (1988b) the main areas of ignorance about desertification are:

1. Ignorance of the extent of desertification:

"An example of a misleading statistic is the claim by UNDCPAC[3] (1987) that 35% of Earth's land surface is threatened with desertification. Yet this is the area that is arid and at least half of that area is very arid: by UNDCPAC's own acknowledgment, this zone is not in danger -- about half of it is too arid for any form of agriculture."

2. Ignorance about climatic data. Neither near-future weather trends nor wide long-term fluctuations can be predicted with confidence. The general paucity of the data from developing countries and the disagreement between scientists over the causal mechanisms of climatic change are combined to increase the doubt and weaken the validity of the forecasting programs.
3. Ignorance of the causes and the manifestations of desertification. It is difficult to decide what the real causes of land degradation are. Sandford (1976) proposed what he considers the four classic causes:
   a) unadapted social and economic structures,
   b) uncontrollable climatic events,
   c) human fallibility of the pastoralists, agriculturalists, donors, planners, leading to the lack of appropriate technologies,
   d) human and animal population growth.
4. How can the fluctuations of the vegetation zones be interpreted? According to Dregne and Tucker (1988), in the Sahel the biomass shifts seasonally by 5 to 6 km towards the south in the dry season, and the same distance back to the north in the rainy season each year. But they note that "a permanent shift could only be established with perhaps 30 to 40 years of observation by meteorological satellites and ground studies."
5. The ignorance in establishing trends and speeds, because trends and speeds require sequential surveys. Is desertification progressing or regressing? How can resilience or recoverability of soils be assessed? When can one speak of irreversible degradation?

   Naveh (in Finkel 1986) points out how the semi-arid Mediterranean uplands have suffered long and severe land degradation because of long-lasting severe human pressures. However, these pressures did not lead to irreversible degradation. The author demonstrates the regenerative and protective power of the vegetative cover, the potential for soil protection and

[3]UNDCPAC, United Nations Desertification Control Programme Activity Center.

soil building by the vegetation, and the obvious resilience of this environment.

Toulmin (1988) shows how rapid and astonishing is the recovery of an apparently irreversibly degraded area following a good rainy season in Niger. After the rainy season, grasslands barren and dry for 20 or 30 years looked like green velvet.

Thanks to good rainfall, and in spite of plagues of locusts the western Sahel shifted from a deficit of 1 million tons of cereals in 1987, to an excess in 1988 of 1 million tons (Cilss 1989a). But this surplus cannot hide three other worrying difficulties:

a) the problem of stocking, and how to capitalize on the surplus,
b) the discrepancy between the excess production of millet, sorghum, and maize and the lack of specific requirement for growing imported cereals, wheat and rice,
c) the countries still in deficit: the Cape Verde Islands, Senegal, Mauritania, and Chad.

6. Poor information on range conditions. As few analyses of the state of *range* in the 1950's exist, comparisons with the 1980's are often not possible. Sandford (1976) highlights the contradiction between increasing herds and decreasing quality of range conditions. The same observations are valid for crop yield: increasing cultivated areas, but less yield.
7. The data necessary for land classification are not available in most of Africa. Thus, it is difficult to assess the progress of soil erosion whether by wind or water. It is almost impossible to decide whether the state of erosion results from the immediate past few years or from longer fluctuations. According to the observation of 1988, after the good rainy season the reaction of vegetation reflects mainly the good rainfall and not the poor dry run of 1983 to 1984. The resilience of the vegetation had not been destroyed.

In a discussion on the ability of the dry environment to recover, the fundamental question is: can all degraded areas recover if given enough rain, or can degradation remove soil to such an extent that recovery is not possible? This question remains without an answer at the present level of knowledge, because the threshold of recovery varies according to physical and human changes in time and space.

1.2.4 The Myth of the Encroaching Desert

Harold Dregne found that the earlier reference to "desert encroachment" was by Frank Cana (1915).

Dregne and Tucker (1988), in their paper with the provocative title of *Desert Encroachment*, relate the history of this idea.

Bovill (1921) is the creator of the concept of the "encroaching Sahara." Bovill's paper is based on observation of the river Senegal, which dried up, and on dry wells in Senegal and Sokoto (North Nigeria).

In a paper to the Royal Geographical Society, Stebbing (1938), a British forester, gained widespread attention, by observing in the 1930's the mixed degradation of deciduous forest and savannas in the Sahel north of Nigeria, south of Niger, and east of Mali. He quoted a French political officer as saying that the Saharan margin had advanced toward the south at a rate of 1 km per year for 300 years. This extension of the desert was the result of human activities. All these assertions are anecdotal.

But 2 years previously, an American soil scientist, Lowdermilk (1935), had written:

> . . ."Comparatively slight disturbances of the coverage of vegetation and soils, such as are brought about by human occupation for grazing and cultivation, are sufficient to extend the borders of the desert far beyond the natural true desert into more humid climates. . . ."

In spite of severe criticism of Stebbing's "encroaching desert" theory, the assertion continued and reached its peak when Lamprey (1975) attempted to quantify the rate of advance of the desert margin.

In 1975, Lamprey, based on aerial field investigations, affirmed that the southern limit of the Sahara was advancing at the rate of 5.5 km per year. According to Hellden (1984), his assertion was based on the comparison of the southern margin of the desert in western Sudan at two different periods. The first was in 1958 according to the map of Harrison and Jacks. This map was plotted on a vegetation map and based on a climatic map achieved with a very poor network of meteorological stations. The second period was based on an aerial survey conducted by Lamprey himself in 1975[4].The conclusion was that in 17 years the desert edge has encroached by 90 to 100 km. Lamprey's conclusions have been strongly challenged in Hellden (1984, 1988), Olsson (1985), Dregne and Tucker (1988), Ahlcrona (1988), Warren and Agnew (1988a), and Nelson (1988).

In the 1980's (as is now the case with the destruction of the ozone layer), desertification became an international or national political topic.

Warren and Agnew have given three good examples of this use:

> "On the 14th of March, 1986, Vice-President Bush was being urged to give aid to the Sudan because desertification was advancing at 9 km per annum."

> On the 11th of September, 1986, in a debate in the European Parliament on the subject of aid to Africa, Winifred Ewing (Scottish Nationalist) declared that aid must go to the Sahel, because the desert was advancing at 8 km a year. (Agence France Presse).

> On August the 4th, 1984, the late President Kountche of Niger urged his countrymen to join "the fight against the advancing Sahara" in order to avoid the humiliation and disgrace of desertification. He used the occasion to crack down on merchants who stole food-aid and slack civil servants, and to sack 30 traffic police. On April the 15th, 1985 in announcing even more draconian measures against desertification, he shelved plans to liberalize the domestic political system "in the face of the more pressing problem of how to feed the population....... 'We cannot talk politics on an empty stomach,' he said. (He) called on Niger citizens to step up their fight against the advancing Sahara desert. . . .". (Agence France Presse)

[4]According to Dregne (1992), "Lamprey made his observations while travelling with a team on the ground."

The team of the University of Lund was the first to ring the alarm bell of doubt. Based on a very solid methodology, including repeated field investigations, sound and diachronic use of satellite images, and a reliable background in geography and botany, Hellden (1984), Olsson (1985), and Ahlcrona (1988) demonstrated that Lamprey had not made a distinction between short-term drought effects and long-term trends. According to the Swedish scientists, drought has a severe impact on crop yields and vegetative cover. They found plenty of evidence of low long-term damage to the natural resources. Ahlcrona observed on satellite images from the Central Sudan that there had been little change in the albedo index (reflectance from the vegetative cover) or in the vegetation index NDVI (Normalized Difference Vegetation Index — Colwell 1974). Her fieldwork demonstrates that behind the absence of quantitative change of vegetation a qualitative change was hidden: good pastures were being replaced by the unpalatable species *Calotropis procera*. Ahlcrona's conclusion, confirming Hellden 1984, is that there is degradation but no change to more desert-like conditions with less vegetation.

Using other precise measurements Olsson (1985) writes:

"... no woody species has been eradicated from the area, no ecological zones have shifted southwards, and the boundaries between different vegetation associations appear to be the same now as they were 80 years ago. A gradual move toward a more useless woody species composition can be traced in limited areas under a heavy population pressure. On the other hand, there are also examples of soil and vegetation recovery within the area which clearly demonstrate the dynamics of the countryside."

With some irony, Warren and Agnew (1988a) noted: "How could one measure the advance of the desert edge, when moreover that edge itself is ill-defined (to say nothing of the poverty of the data)?".

In 1988 Hellden reconfirms his doubts of Lamprey's assessment and concludes his paper very energetically with a summary of its observations:

"It was NOT possible to verify that:

- desertification takes place through a systematic Sahara desert boundary encroachment
- any major and systematic sand dune complex encroachment takes place
- desertification takes place through a systematic expansion of desertified village/water hole perimeters
- variations in crop yield and food production indicated desertification or that these variations are caused by a systematic expansion of cultivation into marginal and vulnerable lands
- there has been significant northward transgression of the cultivation limit during the past 100 years.

There was no creation of long-lasting desert-like conditions during the 1962-1979 period in the area corresponding to the magnitude described by many authors. There was, however, a severe drought impact on crop yield during the Sahelian drought 1964–1974. The drought effects on the natural vegetation productivity was most probably of the same magnitude. The impact of the Sahelian drought was short-lasting followed by a fast land productivity recovery."

One has to place the late 1930's concern about soil erosion and the encroaching Sahara in the context of the Great American Dust Bowl in the USA a few decades before. Both had as great an effect on science at the time as the ozone hole has had on global climatic change studies today.

1.2.5 The Dawning of a New Realism

After the age of doubt, can we really speak of the age of a new realism?

We do not know if the drought period is finished in Africa even with the rainfall becoming better since 1985. But we know from different scientists:

1. That the apparently long-lasting desert-like conditions which appeared in the Sahel from 1968 to 1985 in the Sahelian ecozones in fact represented degradation rather than desertification.
2. For instance, Toulmin, at the end of 1988, in Mali after the rainy season observed a remarkable recovery of some badly degraded areas, and
3. in July 1985 I had a very interesting experience during a visit to Mali. The "brousse tigrée" west of Sokolo (Mali, 14°45N) was looking dead, the trees had split trunks, falling bark, broken branches, and no leaves (which is normal for deciduous trees at the end of the dry season). The vegetation was so desolate that on the map of degradation indicators (vegetation and soil) it was decided to mark it as "area with dead trees". But one week after the first few rains on these skeletal trees, beautiful green leaves were springing miraculously out of the little twigs. Our mapping was wrong; we had observed with eyes accustomed to trees of the temperate zone. The forest was not dead even after the run of low rainfall in 1983 and 1984 and the climatic fluctuations which had begun at the end of the 1960's.

The theory of the encroaching desert, which has now been scientifically rejected, is still a fixed idea in the minds of governments, donors, and journalists; this must change.

A certain scepticism has inspired new methods of evaluation of the problem of desertification such as:

1. Remote sensing supported by verifications on the ground for a better perception of the extent and processes of degradation.
2. Monitoring systems in the affected countries: in Ethiopia, the ILCA range land monitoring system; in Jodhpur (Radjastan, India) CAZRI (Central Arid Zone Research Institute); in Kenya, the Department of Resources Survey and Remote Sensing (formerly KREMU), working with UNEP in assessment of desertification in part of the Marsabit District, in northern Kenya, in the lake Baringo basin, and in an area containing the Kora National Reserve and surrounding marginal farmlands.
3. Awareness that desertification is a set of processes, each of which singly or in combination varies locally from one place to another.
4. The necessity to transfer studies to the village experimental level is also becoming commonly accepted. On this scale a serious inventory of the natural resources and human data must be made. When solutions to problems are proposed, measurements and analyses of their outcomes must be undertaken at the village and country level.

1.3 General Conclusion, Chapter 1

The word *desertification*, created four decades ago, became a trap which ambushed scientists, planners, donor countries, governments of the affected countries, and the mass media!

It was a trap because of the uncertain dimensions of the problem, because of the erroneous attempt to define different realities with the same word, and also because it became a political alibi or an "institutional fact" for the donor countries and the assisted nations (Thompson et al. 1986, quoted by Warren and Agnew).

For the donors the objective is to maintain the institutions already working or simply to have a clear conscience, and for the assisted nations it is to attract a permanent flow of money; but this does not mean that dryland problems do not exist; nor does it mean that after a decade of failures, successes cannot occur. The strategies must change because we know now why we have failed: that is by believing that one model of solutions can be applied for a whole ecozone. We know now that multidisciplinary solutions must be envisaged on very local scales.

We are also aware that the extent of the problem must be redefined; this will be the subject of the second part of this book. In Chapter 3 desertification in several continents will be analyzed.

In this book "desertification" is the ultimate step of land degradation: irreversibly sterile land, meaning irreversible in human terms and within practicable economic limitations. (Please see new Chapter 7 at end of book.)

We are able to distinguish the three components of desertification: causes, processes or symptoms, and effects. Can we fight on a reasonable time scale against the causes? Certainly not, when the causes are natural (drought for example). Two or three generations will be necessary to transform socio-economic structures. But the battle can be immediate and efficient on the level of factors of deterioration, and for this reason Chapter 4 of this book analyzes the physical and physiochemical factors leading to land degradation or desertification and Chapter 5 considers remedies.

Chapter 2 "Desertification" or Land Degradation: Location and Dimension in Time, Vulnerability of Soil and Plants in Drylands

2.1 Introduction

After the discussion in Chapter 1 of the existence or nonexistence of desertification, I have great difficulty in accepting the use of the word "desertification" at all, because it so often implies an irreversible condition. I prefer the term "land degradation" and I shall try to prove that the processes of land degradation offer far more hope for recuperation than is often deemed possible. Meanwhile I shall continue to use the word "desertification" because it is the commonly accepted one, but with a restricted meaning of irreversible land degradation.

The danger of potential desertification, of drought-prone areas, necessitates the knowledge of where the more vulnerable areas are and the formulation of global theories about their location.

Even if it has not been scientifically demonstrated that deserts are expanding at their margins, even if the idea of encroachment must now be abandoned, a certain relationship between deserts, their semi-arid and subhumid margins, and vulnerability exists. Our first questions will be: where are the deserts and their margins? Where are the drought-prone areas, the areas with rainfall deficit? Quantitatively the dryness ratio will help to determine what is a rainfall deficit in areas periodically or seasonally dry.

2.2 Location of "Desertification" or Land Degradation, Threatened Areas and Drought-Prone Regions

In the 1920's Martonne and Aufrère (1928) launched the idea of dryland diagonals: les diagonales arides.

Our planet has five main diagonal zones of drylands:

1. In northwest Mexico the Sonoran Desert and its extension in the dry basins of the southwestern United States.
2. The Atacama Desert, running along the western foot of the Andes from south of Ecuador to Central Chile. The belt crosses the Andes at the latitude of Antafogasta and extends eastwards into Patagonia.
3. The vast diagonal running from the Atlantic Ocean to China, including the Sahara, the Arabian Desert, the Rajastan Desert of Pakistan and India, the deserts of Iran and the former USSR, and, in China and Mongolia, Taklamakan and Gobi.

4. The Namib-Kalahari and its surroundings in Southern Africa.
5. Most of the Australian continent.

These belts are divided into:

a) arid zones with rainfall below about 100–150 mm per year, insufficient to support cropping,
b) drylands with rainfall below about 200–250 mm, where vegetation can be sufficient to support pastoralism,
c) semi-arid zones with rainfall between 250–300 and 600 mm per year where cultivation of drought-resistant crops is possible,
d) dry-subhumid zones as defined by Nelson (1988) as with 600–1200 mm of rainfall and a maximum of 6 months growing season. Dryness can be a seasonal or intermittent constraint.

Since the emergence of humankind, climates, and consequently environment, have changed continuously in time and space.

Brooks (1926) distinguished the "solar climate" and the "geographical climate", noting:

"We have discussed the factors influencing the distribution of temperature, namely, winds and ocean currents, the heat received from the sun, the heat transmitted by the earth's atmosphere, and the reflection from cloud surfaces. We may divide these factors into those which are constant along a given line of latitude and give rise to the "solar climate," and those which vary from place to place even in the same latitude, and so give rise to the local distribution of climates. The latter may be termed the "geographical climate," since it depends mainly on the distribution of land and sea and partly also on the relief of the land surface."

This classification can provide a simple and general framework for our purpose, to understand the location of the great deserts and their semi-arid and subhumid fringes and to distinguish between (1) "zonal arid, semi-arid, and subhumid eco-zones", and (2) dry areas located on the leeside of the main mountain ranges in the mid-latitude westerlies. This distinction is highlighted by Alissow's 1954 map, published with slight modification by Suzuki (1981).

Alissow's map (Fig. 1) and Suzuki's explanations lead to a sound theory of the distribution in the world of arid, semi-arid, and subhumid areas.

"One of the most important pieces of information derived from this map is on the mechanism of formation of the great deserts in the world. The formation of a desert can be directly understood through the location of each front. The main rain bringers are the ITC and the polar front. The area where one of them seasonally passes gets rain or snow regularly at least once or twice a year. The area, on the other hand, indicated as TT on the map, does not get any rain through these frontal or convergence zones. This area is located beyond the reach of polar front and ITC. Here we find the world's largest deserts; the Sahara, the Arabian, the Thar, the Kalahari, the Namib, the Australian, the Atacama, and the Californian.

It should be noted that most of these deserts do not extend as far as the east coasts of respective continents. This is due to the meandering of the mid-latitude westerlies. Such east-west difference of climates is not indicated on the map, which may be regarded as the major insufficiency in Alissow's classification.

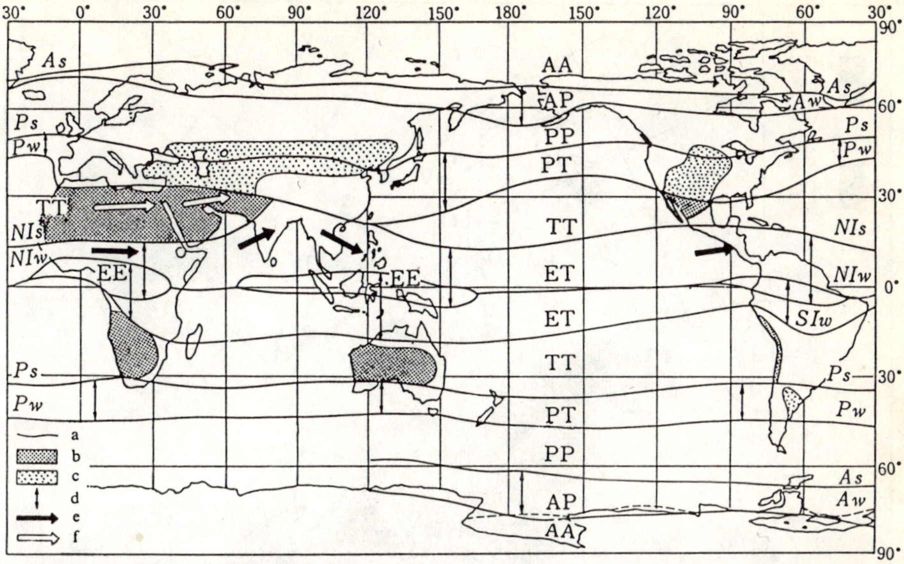

Fig. 1. Dynamic classification of climates (After Alissow 1954) with the distribution of drylands. (After Suzuki 1981). *A* Arctic or Antarctic front; *P* Polar front; *NI*, NITC; *SI*, SITC; *w* location in northern winter; *s* location in northern summer; *A* arctic or antarctic air; P polar air; T tropical air; E equatorial air. Combination of these air masses in summer and winter are shown in figure: *a* Northern or southern limit of front or convergence zone; *b* arid zone beyond the reach of frontal or convergence zone; *c* arid zone due to the leeside location to the westerlies; *d* seasonal migration of frontal or convergence zone; *e* northern summer location of the equatorial westerlies; *f* hypsithermal northern summer location of the equatorial westerlies

"The rest of the big deserts and semiarid areas in the world, e.g., Taklamakan, Gobi, Patagonia and the western part of the USA are explained by a simpler mechanism. They are all formed on the leeside of the mountain ranges in the zone of mid-latitude westerlies." (Suzuki 1981).

Thompson (1975) sums up the five main processes causing aridity (lack of moisture): continentality, cold off-shore currents, topography and dynamic anticyclonic subsidence, and high pressure systems. Deserts are located where one or more of these parameters operates for sufficient time over a sufficient surface.

The map of UNESCO (1977, in Zonn 1986) is another interesting and complementary representation of the distribution of dry or seasonally dry ecozones on our planet. It shows that all the continents have dry ecosystems from hyperarid to subhumid. Thirty percent of the earth has a dry climate (Fig. 2).
"A world Map of Desertification was published in 1977 at a scale of 1:25 million by FAO, UNESCO, and WMO showing the extent and degree of desertification hazards by bioclimatic zones. The degree of hazard was assessed on the basis of severity of climate, sensitivity to degradation of terrain, soils and vegetation cover, and pressure of human and livestock populations. Although the map

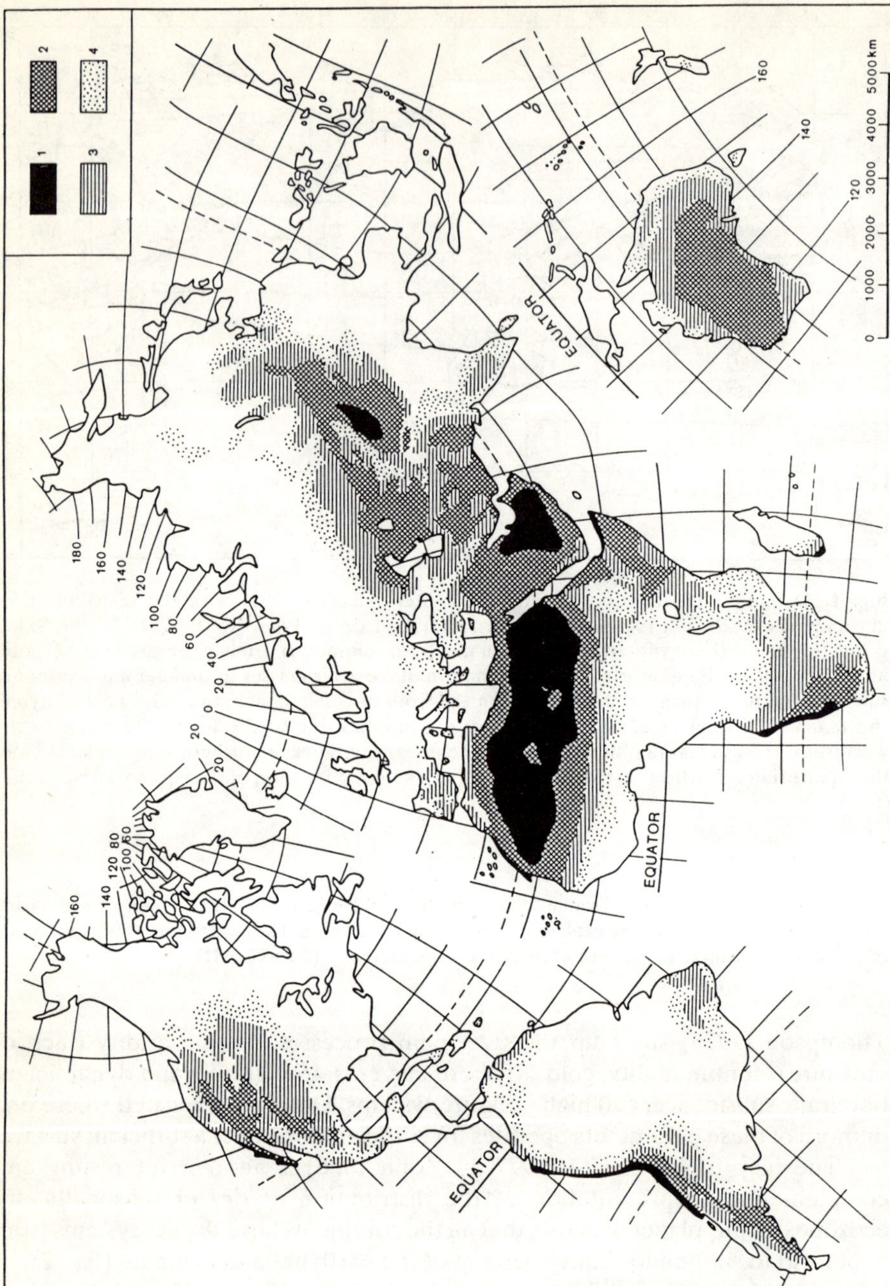
1
2
3
4
0 1000 2000 3000 4000 5000km
EQUATOR
EQUATOR
EQUATOR
180
160
140
120
100
80
60
40
20
0
20
20
160
140
120
100
80
60
160
140
120
160

depicts the threat, rather than the actuality, of desertification, it may be taken as expressive of the magnitude of the problem nevertheless." (Mabbutt 1978b).

The real extent of the problem had to be mapped as scientists agree that the areas threatened by desertification correspond not only to the semi-arid margins but also to the dry- subhumid ecozones on the fringes of these arid areas (UNESCO map, Fig. 2). They are located on the two margins of the Sahara: in the southern Maghreb, Libya, and Egypt; in the southern Sahara in the seven Sahelian countries; in southern Africa around the Namib and the Kalahari desert, in east Africa around the Chalbi desert; in parts of the Saudi Arabian peninsula, in the Middle East; in Rajastan in India and in areas around the Gobi desert, the southern former USSR mainly in Kazakstan, Kirghizia, Turkmenistan, and Uzbekistan, in Mexico, and on the western side of the USA, in South America, and in Australia.

In these areas, when human activity has destroyed the natural resources, harsh climatic conditions make natural rehabilitation difficult and slow, and reclamation needs substantial investment. Here, we encounter for the first time the connection with economic factors (see Nelson's definition in Chapter 1 of this book).

2.3 Dryness Ratio and Drought

2.3.1 Dryness Ratio

The previous chapter has shown as simply as possible the world distribution of the ecosystems where rainfall deficit is endemic. Dryness must be quantified because its effects can be different as a function of its degree.

Budyko (1958) gives a dryness ratio derived from the equation which correlates the hydric (R) and thermic (P) balances of the earth surface:

$$R = H + LE + \text{energy reserve}$$

$$P = N + E + \text{water reserve},$$

where H = thermic turbulent flux towards atmosphere, E = evapotranspiration, LE = thermic turbulent latent flux, N = water excess (run off and infiltration).

Meigs (1953, 1961) prepared for UNESCO a classification of arid climates. He attempts to correlate efficient rainfalls with potential evapotranspiration on a humid surface with a moisture index (Im).

◄

Fig. 2. World distribution of drylands. (UNESCO 1977). *1* hyperarid; *2* arid; *3* semi-arid; *4* subhumid

According to Mather (1974):

$$Im = 100\,(\frac{P}{Ep} - 1),$$

where Ep = potential evapotranspiration

Later, Lettau (1969) simplifies Budyko dryness ratio:

$$D = \frac{R}{LP},$$

where R = annual net radiant energy balance at the Earth's surface[5], P = rainfall annual average, L = temperature required to evaporate a year's rainfall. Hare (1985b) uses Budyko's dryness ratio for west Africa and obtains the following relations (Table 3).

Table 3. Dryness ratio in Africa south of the Sahara. (Hare 1985b)

Dryness ratio value	Vegetation zone	Remarks
>10	True desert (Saharan)	Vegetation confined to oases
7–10	Sahelo-Saharan	Semi-desert scrub much affected by pasturing
2–7	Sahelian and Sahelo-Sudanian	Dry savannas and forests severely affected by over-pasturing and cultivation
1–2	Sudano-Guinean and Sudanian	Moist savannas and dry forests largely cleared
<1	Guinean	Formerly tropical rainforest, now much used by cultivators

Hare also gives a useful map of the areas with dryness ratio between 2 and 10 covering the Sahelo-Sudanian, Sahelian, Sahelo-Saharan, and Saharan ecosystems (Fig. 3).

This map shows that arid and semi-arid ecosystems susceptible to drought exist on all continents but with greater development on the western fringe of Latin America and on the fringes of the diagonal which covers the Sahara, Arabia, the Russian semi-arid ecosystem, and the Chinese dry zones. There is another fringe in South Africa, and another is the whole of central Australia. The gradient of the dryness ratio in west Africa appears distributed according to the latitude from the Guinea Coast to the Sahara. Such a regular pattern consistent with the parallels of latitude is unique on our planet.

[5]The net radiant energy balance at the Earth's surface is composed as follows. Direct sunshine; + scattered solar radiation from the air (× the fraction which is absorbed at the Earth's surface); + infrared radiation from the air to the surface; – infrared radiation emitted by the plants and soil.

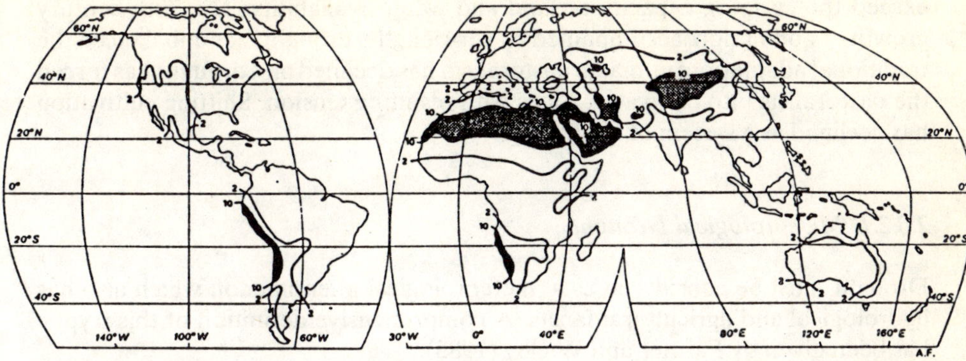

Fig. 3. World distribution of the zone with dryness ratios between *2* (outer limit of arid zone) and *10* (the desert margin). (From data of Henning for the world climate conference of 1979). (Hare 1985b)

2.3.2 Drought

Drought and aridity are clearly related phenomena. Aridity results from a permanent rainfall deficit. Drought is a prolonged departure from normal rainfall. For Farmer and Wigley (1985) "drought is an interval of conditions more arid than the prevailing norm, a temporary aridity."

In Africa, the Sahelian and Sudanian ecosystems are the most frequently drought-affected. They are also the ecozones where drought susceptibility is at a maximum.

To understand why desertification is so commonly related to drought we must analyze drought.

It seems naive to say that rainfall deficit is the primary cause of drought. We shall see that in some cases drought can be dissociated from low rainfall. It is more complicated to explain rainfall deficit and to link it with global climatic change. The only honest answer is that as yet "we don't know". With quiet philosophy, Farmer and Wigley wrote in 1985:

"The most obvious gaps in our knowledge are: we do not know why the present drought exists; we do not know why droughts have occurred in the past; we do not know how long the present drought will last; and we do not know why the main rainfall deficits in the most recent Sahel drought are in August/September, part of the rainy season."

But we are able to explore the protracted droughts and to understand how failure of rainfall, which under natural conditions would have little effect, becomes a disaster under some socio-economic changes.

Drought is to be seen in the light of ecosystem change, including the climatic changes and feedback processes of land and resource use. The demographic explosion in semi-arid and subhumid ecosystems increases accordingly the needs for food, fiber, and energy. The demands made on vegetation, soils, and water exceed the carrying capacity of land and water availability. The 20th century growth of population, accompanied by very lengthy drought, has unbalanced the traditional adaptation; nomadic pastoralism has declined and drought has forced the pastoralists into cultivated zones with resulting tension. Shifting cultivation has declined as a system of livelihood.

2.3.2.1 Meteorological Drought

Drought must be considered as a meteorological phenomenon which also has hydrological and agricultural facets. A comprehensive definition of this tryptic has been given by Farmer and Wigley (1985):

"In general, drought is a protracted departure from normal water availability, a water deficit that exists for a long enough period to cause hardship. More specifically, whether or not drought occurs depends on the type of activity for which water is being used and on the expected or required availability of water."

From the same authors: "Meteorological drought occurs when precipitation is significantly below expectations. Its definition involves only precipitation statistics and depends on the time of year and on location."

The geographical extent of drought was categorized by climatologists into three classes: local, regional, and subcontinental. In central Europe the droughts are generally local, and concern only one country (e.g., France in the 1940's) or a part of a country. The African droughts are on a larger regional scale. The same drought which occurred south of the Sahara and which was called the Sahelian drought severely affected seven countries from Mauritania and Senegal to Mali, Burkina Faso, Niger, Sudan, and Ethiopia.

This vast spreading of drought implies causal phenomena on a large spatial and long-term scale, a topic which has not been studied in any depth in Africa.

A second characteristic is typical of the African droughts: they are a recurring phenomenon but do not show any periodic behavior of useful predictive value. The recurrent droughts vary according to location and time, but even that concept must be modified. Puterbaugh et al. (1983) and Nicholson (1983) found low correlations between the rainfall of west, east, and south Africa, and suggest that there is little possibility of links that might have predictive value. Local controls, but over quite large regions, play a major role in determining African rainfall. Different mechanisms are operating in different parts of Africa without common large-scale factors operating on a longer time scale.

The Sahel is not the only ecozone affected by droughts. The Sahelo-Sudanian and the Sudanian, even the Sudano-Guinean zones, are affected. Nicholson (1980) has shown the normal coefficient of variability of the rainfall for Africa south of the Sahara (Table 4).

Hurault (1972) has demonstrated in Adamaoua (Cameroon) the existence of two dry spells 20 000 B.P. and 1500–1300 B.P.. The numerous charcoal and

Table 4. Ecozones, annual rainfall, variability and rainy season. (Nicholson 1980)

Zone	Mean annual rainfall (mm)	Coefficient[a] of variability (%)	Length of rainy season (months)
Sahelo-Saharan	50–100	50+	1–2
Sahelian	100–400	30–50	2–3
Sudanian	400–1200	20–30	3–5
Sudano-Guinean	1200–1600	15–20	5–8

[a]Standard deviation of annual totals expressed as a percentage of the annual mean.

pottery levels in south Congo are also indicators of previous dry periods in wet tropical and equatorial zones.

All these assertions are vague but qualitatively very valuable. Figure 4 shows four diagrams from Nicholson (1983), showing:

1. during the first 80 years of the 20th century the absence of a general trend in climatic changes for all the dry and wet tropical ecozones;
2. a relative similarity between wet and dry spells for all the ecozones mainly since the 1950's;
3. that the last dry spell was common to all the tropical ecosystems north of the Equator (Figs. 5 and 6).

2.3.2.2 Hydrological Drought

"Hydrological drought occurs when water resources used for industry, for human or animal consumption, or to support agriculture (e.g., by irrigation) reach low levels. Hydrological drought is usually reflected in low levels of rivers or lakes, reservoirs, or groundwater. These levels are determined not only by precipitation, but also by water usage and by evaporation and evapotranspiration." (Farmer and Wigley 1985).

Hydrological droughts are probably those which, thanks to past lake level oscillations and river flow variations, can be the most objectively studied.

The lake level is an indicator of integrated rainfall amount over the lake catchment area, modified by the effect of the coefficient of runoff/infiltration and evaporation. The different types of land use, changes in the agricultural surface, and deforestation modify the coefficient and the evapotranspiration rates. So it is better to use the criterion of lake level as an indicator of trends and not of refined fluctuations. In the early Holocene (10 000 B.P.), most lake levels were high, the highest being between 13 °N. and 2 °S. A second general trend is the lowering of the lakes since 2500 B.P.

The east African lakes, among which are the Kenyan lakes Turkana, Baringo, Bogoria, Nakuru, Elementeita, Naivasha and Magadi, experienced high levels in the early 1960's after heavy rains in the years 1961–1962. The levels of Lake Victoria (Fig. 7) have been known since the beginning of this century. Like the former group, its last high level occurred in the 1960's and since then there

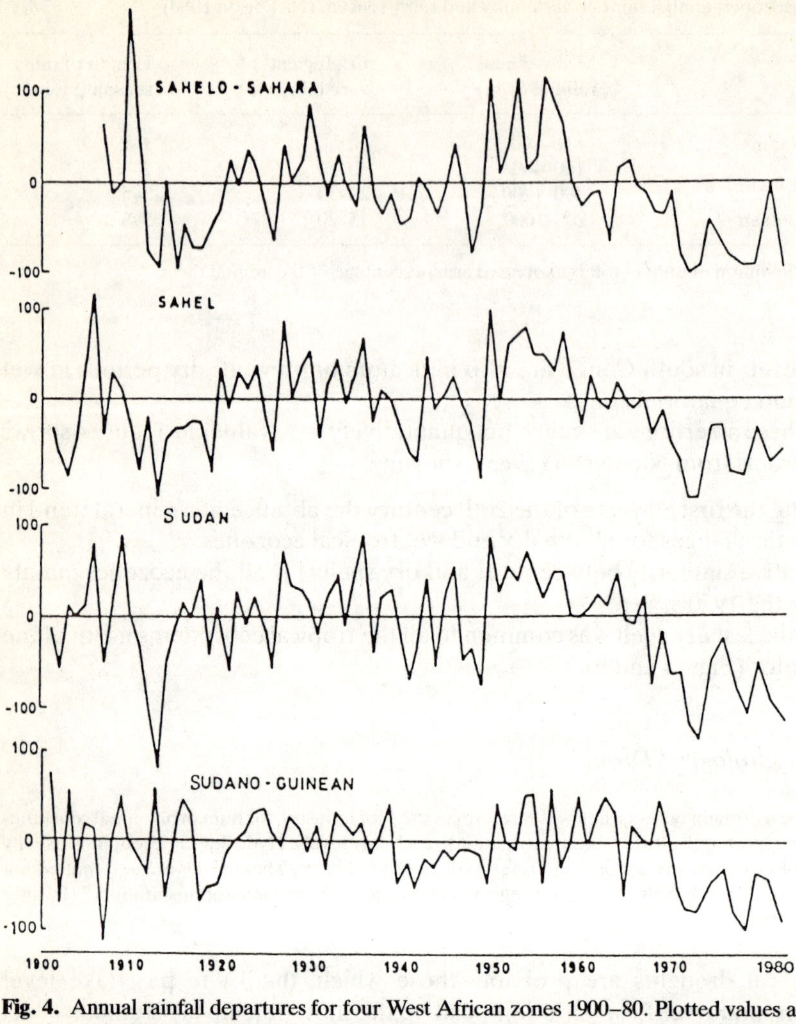

Fig. 4. Annual rainfall departures for four West African zones 1900–80. Plotted values are approximately equivalent to a regionally averaged "percent of standard departure" (Nicholson 1983, in Farmer and Wigley 1985). The rainfall has a discontinuous distribution; therefore, the "averaging" degrades the information and caution is required.

Fig. 5. Normalized area average rainfall variations for two regions in West Africa showing the large spatial extent of the recent decline. (Farmer and Wigley 1985). The *bottom parts* of the individual sections show the number of stations used each year in calculating the average (*one mark* equals ten stations)

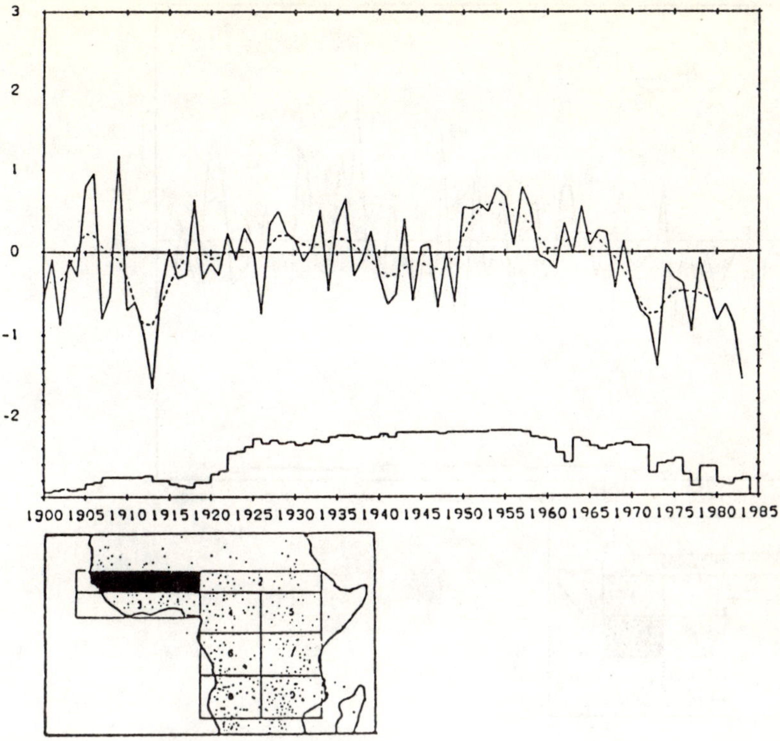
3
2
1
0
-1
-2
1900 1905 1910 1915 1920 1925 1930 1935 1940 1945 1950 1955 1960 1965 1970 1975 1980 1985

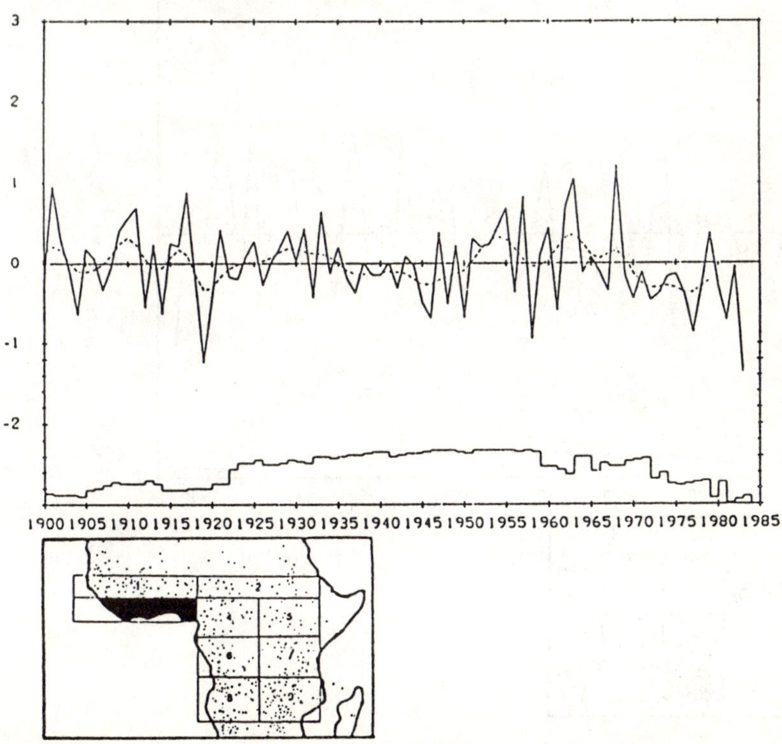
3
2
1
0
-1
-2
1900 1905 1910 1915 1920 1925 1930 1935 1940 1945 1950 1955 1960 1965 1970 1975 1980 1985

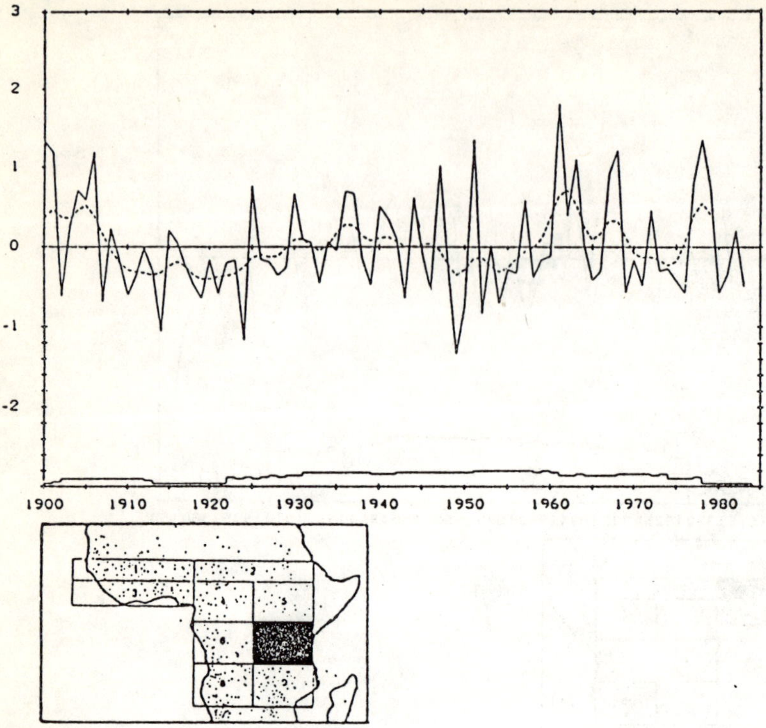
3
2
1
0
-1
-2
1900
1910
1920
1930
1940
1950
1960
1970
1980

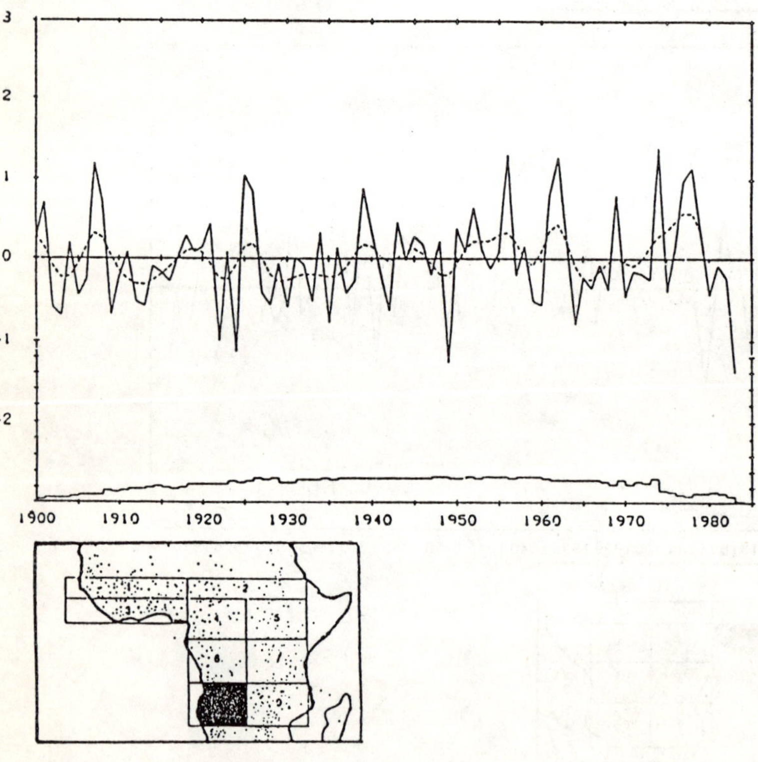
3
2
1
0
-1
-2
1900
1910
1920
1930
1940
1950
1960
1970
1980

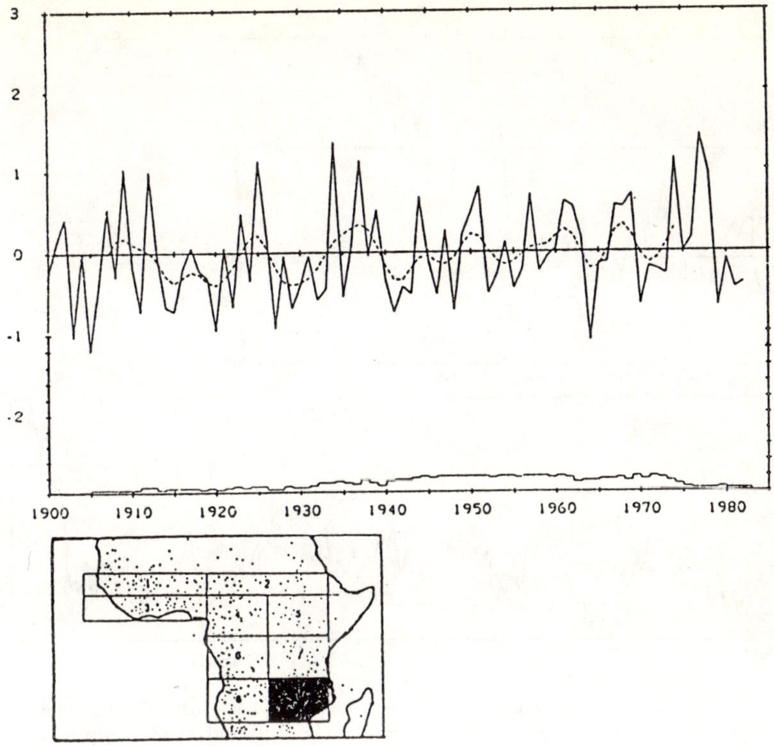

Fig. 6

has been a permanent decrease with seasonal fluctuations. The lowest level recently registered was in 1988. For the freshwater lakes, mainly for Naivasha, this low level probably reflects less a rainfall deficit than human overuse and deforestation.

"Another example comes from eastern Africa. Here 1984 drought conditions in the northern highlands of Ethiopia (WMO 1983b) and in Sudan were reflected in the lowest Nile flood level in 350 years. In the northern Sudan and much of Egypt, normal rainfall is insufficient to support agriculture, so the Nile waters are of crucial importance. It was reported that in 1984 only the discharge control capacity of the Aswan Dam saved Egypt from major water supply problems." (Farmer and Wigley 1985).

Lake Chad has fluctuations which demonstrate the climatic variability and the hydrological droughts at these latitudes.

Fig. 6. Normalized area average rainfall variations for three regions in southern tropical Africa. (Farmer and Wigley 1985)

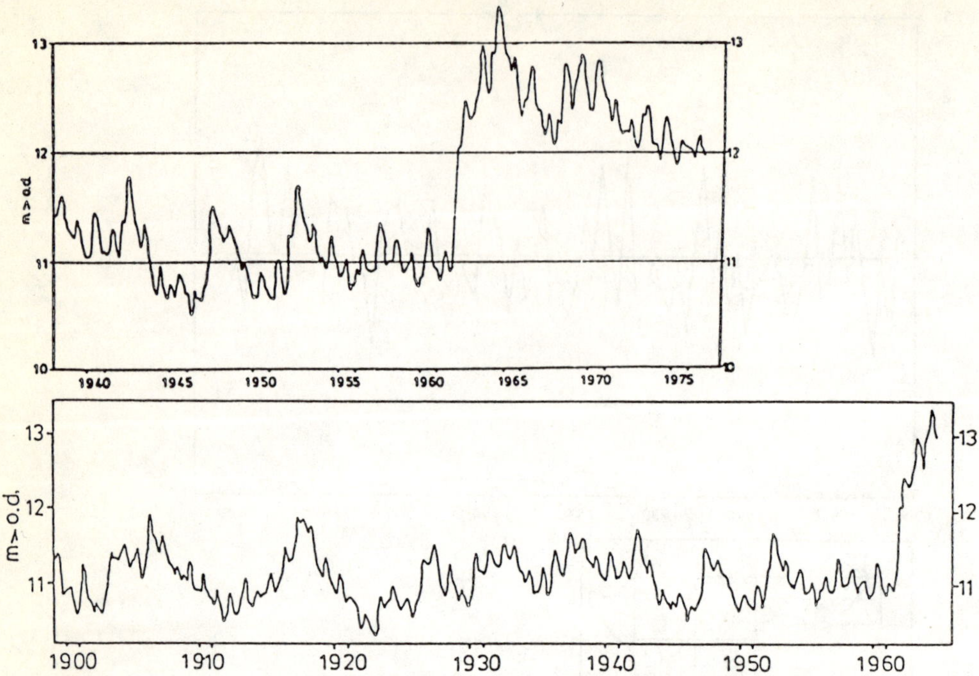

Fig. 7. Level of Lake Victoria at Jinja, Uganda. Data for 1900 to 1964 are from Thompson and Mörth (1965), with their units error corrected. The expanded later period, 1938 to September 1976, is from Farmer. (Farmer and Wigley 1985)

Lake Chad, one of the shallow tropical lakes, is located between 12° to 14°20 north and 13° to 15°30 east in an inundated sandsea. During dry periods the lake is divided into two parts separated by a dune ridge named "La grande barrière". Each basin contains water about 3 m deep and is surrounded by lines of islands which are the top of east-west transverse dunes, more or less submerged. In the south the submerged dunes are covered by aquatic plants: *Phragmites* and papyrus. When the surface reaches 20 500 km^2, its depth is 281.5 m. The water occupies the center of the basin but not its lowest point. This is located 500 km further northeast at the terminus of the Bahr-el-Ghazal valley in the Pays-Bas du Tchad near Largeau, at an altitude of 160 m. This situation explains why the wadi of Bahr-el Ghazal, during Quaternary times, had a flow in two directions.

The lake is located exactly at the threshold between arid and semi-arid ecozones. The Sahelian environment of the lake explains the fixed ergs of Kanem and Manga. The dry season is from November to June and the rainy season from May-June to September with about 30 rainy days. The temperatures are high with a potential evapotranspiration of 2200 mm at Bol.

Lake Chad is the residue of a vast paleo lake of 350 000 km^2. As early as 1909, Tilho (1910–1911) noted "the low mineralization of the lake in spite of a high evaporation." This is a result of the infiltration at the bottom of the lake (Engster and Hardie 1978).

The presence of water at a shallow depth, the climatic change, the topographical situation, and the hydrology combine to give to the lake its numerous fluctuations (Table 5).

When the level is low, the vegetated islands become active dunes and as early as 1972 a dune ridge system 5 to 20 km wide from the delta of Chari to Baga-Kawa emerged.

Table 5. Fluctuations of Lake Chad

Date	km^2	Level	Altitudes
ca. 5400 B.P.	350 000	High	320 m
2400–1900 B.P.		Low	
1800		Very high	286 ?
1823			282
1823–1850		Very low	
1850–1900		High	283–284
1908–1914		Very low	280
1950		Middle	281–282
1963	23 500	Short high level	283
1965–1971	19 000	Low	
1973		Very low	

Rivers also demonstrate the impact of drought. The annual discharge of the river Niger at Niamey has many variations and shows the last hydrological droughts (Sircoulon 1976). Between 1951 and 1970 the mean annual discharge of the river was 1100 $m^3\ s^{-1}$. The dry conditions in the 1940's and since the late 1960's stand out, and so also do the wet 1950's.

For 1973–74, after the first dry spell, on 4 July 1974 the flow dropped to a minimum of 0.6 $m^3\ s^{-1}$. This lower water level resulted from the lack of the usual January-February maximum flood (1700 $m^3\ s^{-1}$). The maximum flow 1560 $m^3\ s^{-1}$ occurred on 10 October 1973. This maximum flow of October occurred also in the hydrological year (July to June) 1984–1985. In 1985 the flow continually declined.

The river Senegal at Bakel for 1903/4 to 1983/4 (hydrological year: May to April) (Fig. 8), had low runoff periods below 700 $m^3\ s^{-1}$ in 1910–1917–1937–1949 and after 1968. Since 1968 the trend is towards a decreasing flow, which in 1985/86 did not exceed 215 $m^3\ s^{-1}$. The river flow data at Bakel reflect the Sahelian drought.

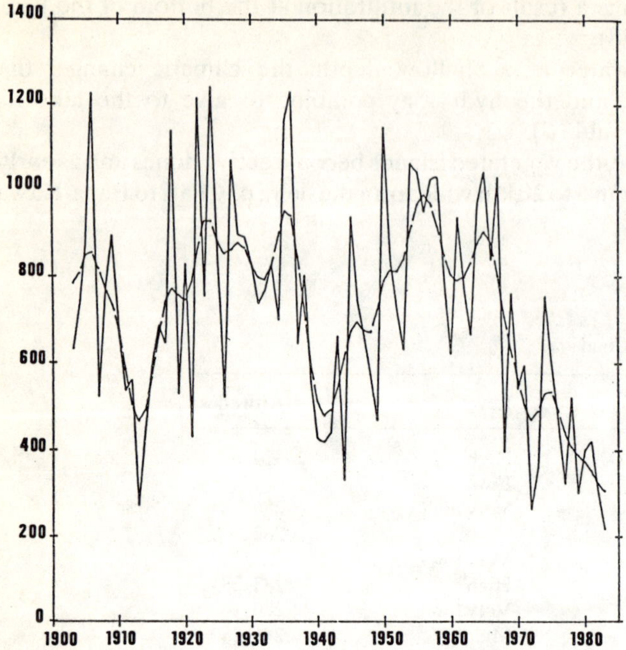

Fig. 8. Annual mean (May to April) flow for the Senegal River in m^3s^{-1} (Farmer and Wigley 1985). The *smooth curve* shows 9-year binomal filtered values. (Faure and Gac 1981; Gac, ORSTOM, Dakar, Senegal)

2.3.2.3 Agricultural Drought

Agricultural drought occurs when the water supply necessary for agriculture becomes scarce. An agricultural drought depends on the amount of rainfall expected and the use to which water is put, hence a shortage of water is felt because of human activity, whereas a meteorological drought creates stress on plant life unconnected with people and their needs. Wigley and Atkinson (1977) define agricultural drought in terms of soil moisture deficit and state that it occurs "when there is a consistently high soil moisture deficit over the growing season." In human terms, agricultural drought is a moisture deficit on a sufficient scale to cause disruption of the rural economy. In an extreme agricultural drought, crops fail, and animals, and perhaps people, die.

In their interesting concept of a changing drought threshold Palutikof et al. (1982) demonstrate how changes in population pressure and crop variety can alter the drought threshold. They analyzed the substitution of maize, a more water-demanding cash crop than millet, a traditional dry-area crop in Kenya, and observed that the drought susceptibility of three areas was seriously increased.

Two simple examples of agricultural drought may be given:

Example 1. Good rainfall in the 1950's and early 1960's in Africa led to an increase of the human and especially animal population with an extension of the grazing land toward the north and of the farming areas from the Sahelian to the Saharo-Sahelian ecozone. When at the end of the 1960's and the beginning of the 1970's there was a meteorological drought, pastoralists and farmers rushed toward the south, and created a dangerous concentration of livestock and population south of the 600-mm isohyet in Senegal, Mali, Niger, Chad, and Sudan. This southward drive had initially occurred in 1968 with the first harvest and grazing failure, but it increased in the following years with dramatic mortality of livestock and people.

Example 2. This example does not depend on rainfall deficit but on bad rainfall timing. Such an agricultural drought occurred in Kenya in the region of Kitui and is very well analyzed by Farmer and Wigley (1985).

The region of Kitui at equatorial latitudes has a bimodal seasonal rainfall distribution (Fig. 9) with a "long rainy season" (March to May) and short rains (October to December). In 1984 the annual total (1188 mm) was higher than the 1931–1960 mean (119%). The long rains from March to May in almost the whole of Kenya reached 208 mm only, compared with 400 mm mean between 1931 and 1960. The wettest month in 1984 was April with only 6 days of rains, but the extreme fall of 88.6 mm on 24 April (over one-third of the seasonal total) was

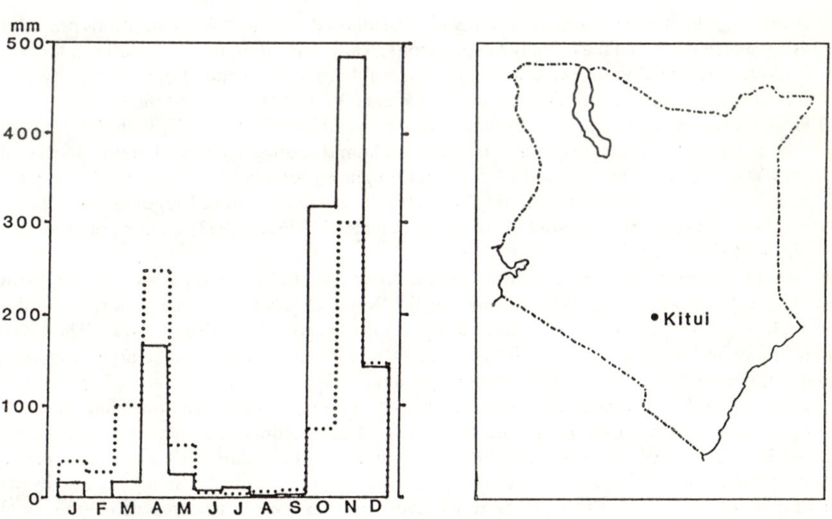

Fig. 9. Monthly rainfall values for Kitui Agricultural Office, Kenya, for 1984 (*solid line*) compared with the 1931–1960 mean (*dotted line*) (Farmer and Wigley 1985). 1984 had above-average annual rainfall, but the distribution within the year was such that agricultural drought conditions prevailed. Data supplied by the Kenya Meteorological Department

disastrous for crops. In addition, the short rains' distribution was also mediocre. The first month had an excess rainfall of 319 mm compared with a mean of 72 mm (1931–1960) with 153 mm in one rain on 21–22 October. To make it worse, the rain fell on mainly bare soil as it was the beginning of the wet season. So the coefficient of runoff was increased, with less infiltration and less water supply to the soils and crops. November had good regular rainfall, allowing good crop development; but December, with a nearly normal rainfall (142 mm), had prolonged dry spells (a 3-week period with only 3 rainy days) 2 days of which (11 and 20th) produced 123 mm, almost the monthly total (see Jackson 1989).

This distribution, resulting in poor maturation and low yields, shows that monthly and annual rainfall amounts must be correlated with distribution for the whole year. In spite of a rainfall above the annual average, agricultural drought prevailed in 1984 in Kenya because of an unfavorable distribution within the year. Kenya was not the only country affected by droughts in 1984. The FAO pointed out five African countries (Niger, Sudan, Ethiopia, Kenya, and Mozambique) suffering severe food shortage and 17 other countries with inadequate food supplies; a total of 22 countries with 250 million people.

2.4 The Droughts of the 20th Century Throughout the World

The report of the expert group meeting on the climatic situation and drought in Africa (WMO 1983b) gives an excellent summary of the climate variability in Africa throughout the last three decades:

> "The 1950's changes in Sahelian rainfall have been particularly dramatic. "Wet conditions prevailed during that decade, with decadal averages ranging from 15% above normal in the southern areas to 35 to 40% above normal along the Saharan fringe. Favorable conditions ended in 1960, and by 1968 the situation was very bad. Average deficits for the 6 years 1968 to 1973 matched the excesses of the 1950's, 15 to 40% below the mean. The extremes were reached in 1972 and 1973, with deficits of 50 to 60% in the northern Sahel and 30 to 35% in the more humid southern areas, where mean rainfall ranges from 400 to 1200 mm. In 1974 and 1975, when drought reportedly had ended, rainfall was still about 25% below the mean in the north and about 10% below in the south. Larger rainfall deficits occurred in subsequent years and continued until 1983, deficits of 1976 and 1977 matching or exceeding those of 1972 and 1973 in many areas.
>
> ". . . Sahel droughts sometimes coincided with abnormally dry conditions over much of the continent. Although the wet episode in the Sahel during the 1950's was paralleled by above-normal rainfall through much of southern Africa, it was a decade of generally dry conditions in east Africa. The 1960's were abnormally wet throughout east Africa and in many other equatorial regions, but drier conditions commenced then in the semi-arid subtropical regions of the continent. Drought conditions occurred early in that decade in southern Africa, a few years before they began in the Sahel. The early 1970's were a special case, however, in that they were years of nearly continent-wide drought, affecting northern Africa, east Africa, and especially northern margins of the Kalahari (but not those on the southern Kalahari). The last 8 years have variously seen drought not only in the Sahel, but also in east Africa and in southern Africa, although the most severe periods have not been synchronous in all three regions."

Short-term weather changes are common in all tropical latitudes.

Downing (1986) notes that in eastern Africa "drought in the 1970's was followed by severe drought in Ethiopia, Sudan, Somalia, and Kenya in 1983 and

1984, and mild to moderate drought in other countries and other years". In Ethiopia, 5 million people were estimated to have been affected by 1984 and this rose to 6.8 million by April 1986 (Warren and Agnew 1988a).

Reviewing the impact of drought in Africa, Glantz and Katz (1987) listed 13 countries affected in 1986. In 1983, Zimbabwe had a harvest of only one-third of 1980; in the 1980's Botswana and Mozambique had the lowest rainfall in recent history. Even South Africa had to import cereals in 1983 because of drought.

Warren and Agnew (1988a), quoting Milas (1983) and Hare (1983, 1984b), summarize the main droughts for Asia, Australia, and Latin America.

"In 1902, drought savaged herds and crops in Australia and there were dry spells during 1912–15, 1965–67, and 1972. The worst drought in two centuries ended in 1983, after crop production had fallen by 31% and farmers' incomes by 24%. Rainfall was half the normal amount and some areas experienced 4 years of drought" (Warren and Agnew 1988a).

"The drought had significant effects upon the Australian environment and its wildlife. The process of land degradation in Australia (Woods 1983) was considerably accelerated by wind erosion from drought-blasted grain fields and pastures. A dust storm on 8–9 February 1983 brought an estimated 150 000 metric tonnes of soil from farms in the northwest of Victoria to the capital of Melbourne and on out to sea (Anon 1983). A week later a series of massive bush fires covering some 500 000 ha in southeastern South Australia and Victoria, led to the death of 72 people and more than 300 000 animals, and caused approximately $400 million in property damage (Bardsley et al. 1983; Oliver et al. 1984). The causal link with the drought was quite clear. Less clear was the link between the drought and erosion losses (up to 48 metric tonnes of soil lost per hectare) from some of the burnt areas in the drought-breaking rains of March 1983" (Atkinson 1984).

"Wildlife losses from the drought must have been considerable, although documentation is sparse. Estimates of kangaroo populations within the commercial shooting zones over the period November 1982 to June 1983 suggested that the drought had caused the deaths of up to 70% of mobs (herds)." (Allan and Heathcote 1987).

"Dhar et al. (1979) and Krishnan (1979) noted the frequency with which droughts in India and Australia coincided (1965–67 and 1972–75). Vines (1986) found a number of cycles in Indian rainfall and believed that they could be explained by sunspot activity" (Warren and Agnew 1988a).

In India one of the worst droughts in recorded history occurred in 1899 when about three-fourths of the country experienced a rainfall deficit. Droughts in 1918 and 1972 affected 75 and 43% of the country (Sinha 1987). The 1979–80 drought was the most severe, with a water deficit of 20–54% in about 41% of the country. The 1982–83 drought, which affected 37% of India was one of the worst.

According to Warren and Agnew (1988a), in 1983 northeastern Brazil faced its worst drought since 1583. An estimated 10 million out of a population of 24 million people were destitute, and 90% of crops failed (Times 28/3/84). In Mexico in 1982, drought cut agricultural production by 12% and 8 million tonnes of staple foods had to be imported, costing $1.6 billion. Bolivia was also struck by its worst drought this century with drastic losses of livestock (Newsweek 20/6/83) and crops (Economist 6/4/83).

Causes and models proposed in the scientific literature to explain droughts are numerons but none is really satisfactory as a decisive explanation. Probably several causes are involved when a drought occurs.

2.5 The Vulnerability of Soils in Drylands

In arid and semi-arid areas the processes of soil formation are extremely slow: several hundred years may be necessary in the Sahel for genesis of the different soil horizons. Wind and water sorting of the weathered rock particles are the major factors of evolution. In higher rainfall areas with a complete vegetative cover, the biological process of soil formation becomes dominant. The vegetative cover is not seasonally destroyed and the soil is not exposed. As long as the crowns of the trees and the roots protect the soil, a place can look like a desert in the dry season but will regenerate in the wet season and is resistant to erosion. When the vegetative cover is destroyed, then wind and water erosion take place, unchecked, as in the desert.

The savanna soils are the result of a combination of biological and physical factors. The important difference between the soil of arid ecozones and that of humid ones is the following: in humid regions precipitation exceeds evaporation; there are soils which are subject to leaching with a permanent water table at which sesquioxide accumulation can form deep-seated horizons. Where evaporation exceeds precipitation in light soils, the plants intercept the soilwater and the nutrients, so that there is no accumulation of water in a water table, and all nutrients are available to the plants. The efficiency of the system is maximum and that is why the dry ecosystems are more productive than they would appear to be if judged by the standard of more humid climates. Furthermore the light sandy soils have a low soil moisture tension, so a very high proportion of precipitation is available to plants. However, any soil is ideal for the vegetation in the arid, the semi-arid, and the dry subhumid ecozones.

It is necessary to state that for the dry ecosystem no practical classification of soils exists, nevertheless, there are at least two unified soil classification systems for dry ecosystems, the FAO classification and Soil Taxonomy (US). We were unable to adopt one really consistent terminology throughout the book for all continents. Skoupy (1976) described soil classes 1,2,5,6,7,8, and 9. In the list below, I have added 3,4, and 10 to the classification of dry ecozone soils.

1. Desert gray soils. Gray soils characterized by humus deficiency in the surface horizon; the mother rock can be loess or loamy particles with $CaCO_3$ and $CaSO_4$. Heavy weathering results in the accumulation of clay minerals and the soluble salts accumulate at a depth of 100–200 cm below the surface.
2. Gray-cinnamonic and red-cinnamonic desert soils. Characterized by low humus content, carbonates are present in the whole profile and the surface horizon is sometimes decalcified. If irrigated, they are very fertile.
3. Desert sandy soils. These soils have different textures: sandy (erg), coarse-sandy up to stony (reg), and bouldery. If sand is blown away by wind, then mother rock, subsoil, or coarse angulous stones (hammada) appear on the denuded surface.

   Sandy soils are the most common in arid areas. They consist of more than 70% of sand particles and of less than 20% of clay particles. They have wide pores which allow rapid water infiltration. Only a small amount of water can be stored in these soils, but all of the water can be used by plants. Capillarity rise in sandy soil is very limited. Plant nutrients are stored only in small amounts.

4. Clay soils are formed with more than 40% of small particles separated by small-sized pores. In Sudan these soils are called dark cracking clay soils, in Somalia black cotton soils. They consist of heavy (often lacustrine or fluvial) clay deposits. During the dry season they shrink considerably while they dry out, and deep cracks develop. During the rainy season the soils swell; a relief of mounds and depressions is the result of these alternatives. Water infiltration and drainage are very limited, and after rainfall, water ponds the surface. The small-sized pores retain the water so strongly that plants cannot extract it easily. Capillary rise is pronounced in these soils. They also store a large amount of plant nutrients which are difficult to extract by plants.
5. Desert crust soils. These are most frequently formed by the compacting effect of $CaSO_4$ (gypsum) or $CaCO_3$ (calcrete or "caliche"). They are hard and thick, up to 2 m. When the calcareous crust is shallow, it cracks underfoot. The desert crust soils are characterized by their poor retention of water. In flood rains, surface runoff can reach extremely high values, and there is little or no infiltration. The reclamation of these soils requires expensive ameliorative measures.
6. Chestnut soils. These occur less frequently in the arid areas, e.g., on the border of the Sahara, in Iraq, Iran, Saudi Arabia, and elsewhere. They possess a relatively low humus content in their upper layer.
7. Chernozems. The chernozemic soils of these areas are sometimes classified as black soils, because unlike the genuine chernozems, they show some deviations. Thus, for instance, the tropical black soils have a low content of humus, $CaCO_3$ occurs in the whole profile, even in the upper horizon, and the carbonate horizon is lacking, especially in some margalitic soils. The thickness of the black-colored layer varies in the arid areas from 40 to 150 cm.
8. Terra rosa (red soils). These soils occur in the Mediterranean area. They are characteristic of the areas where the moisture period alternates with a drought period, whereby the coagulation of iron takes place. These soils occur on limestones and dolomites with higher $CaCO_3$ content.
9. Saline soils. The process of salinization is closely connected with irrigation management. These soils can have high potentials fertility but their amelioration is expensiv. The accumulation of salt occurs initially in one horizon under the surface. With the decrease in rainfall and the increase in evaporation, this horizon shifts to the surface, until a heavy salinization of the surface
horizon occurs with surface efflorescences of salt. These soils are represented by solonetzs with a higher Na content and solonchaks, the surface horizon of which contains the easily soluble chlorides, sulfates, and carbonates. With the solonetzization process is also connected the gilgai process (Australia). The intensive soil shrinkage in dry summer seasons results in pentangular cracks (e.g., in Iraq), that reach a depth of some centimeters. These cracks are gradually and spontaneously filled up with earthy materials which swell up in the winter rains. The surface becomes irregularly undulated (small loaf-shaped mounds) and the gilgai wedges become apparent in the soil

profile. Such soil can be improved by washing away gypsum and by the addition of organic matters through fertilizing.

10. Soils with a laterite or ferricrete formation. "The word laterite has been used for a variety of crustal materials. . . . often different from the ferriginous clayey material for which the term was originally coined by Buchanan (1807). Laterite has been defined by Prescott and Pendleton (1952) as a massive, vesicular or concretionary ironstone formation and it is in this sense that the word is now generally used" (Ollier and Gallaway 1990). Because of earlier inconsistent use, these authors propose to use the term ferricrete: "The term ferricrete was coined by Lamplugh (1902) for material cemented by iron oxides. The cement is ferriginous and the cemented material is usually but not necessarily iron-rich and can range from ferriginous concretions to non-ferriginous material." These formations in drylands can be considered to be formed under wetter conditions, mainly in areas where the annual rainfall exceeds 800 mm but with a well-defined rainy season.[6] They are characterized by leaching and migration of SiO_2 and the bases, whereas the sesquioxides remain on the spot in the hydrated forms. Al_2O_3 is partly bound with SiO_2. These soils have a low content of primary minerals and soluble substances, low humus content, low content of P_2O_5, a mutually unbalanced content of nutrients, and a relatively high degree of aggregate stability. They are characterized by the red color of the surface horizon.

    The profile consists in a shallow topsoil (0 to 15 cm) dark, red, crumbly, with loose, loamy texture. The topsoil in a normal soil profile is usually a well-defined layer which, thanks to its high content of organic matter, has a relatively abundant supply of plant nutrients. The topsoil may have an O horizon of undecomposed organic debris on top of the soil, beneath which is the A horizon in which organic matter is mixed with mineral matter. It is also characterized by a well-developed and stable structure which increases the water-holding capacity of the soil. This horizon overlies a subsoil with iron concretions or ironstone pan overlying mottled heavy clay. Water can infiltrate as long as the loamy top soil is not eroded. After erosion, of the topsoil, the ferricrete appears at the surface, blocks infiltration, and the soil becomes sterile.

The simplest definition of desertification is that of the transformation of a landscape which was not desert-like into one which is.

The three main aspects of an authentic desert landscape are: (1) Stony soils: reg stony plateaux called hamada when in limestone, and tassili when in sandstone, (2) active dune fields, (3) depletion and contraction of the vegetation, when the mineral content becomes the dominant feature in a landscape.

1. Reg, hamada, tassili are arid features. In semi-arid and dry-subhumid environments excessive erosion seldom leads to substrata exposure, bowalization is more common. The word bowalization in West Africa means the exposure of a ironstone. This occurs after exposure of the B horizon of a

[6] An exhaustive book on laterite is proposed by Tardy (1993).

lateritic or ferriginous soil. This B horizon, particularly rich in iron or aluminum when exposed hardens into a lateritic hardpan (or ferricrete or ironstone). In the case of bowalization, the soil is almost sterile, poor in nutrients; being very hard, it is also a physical barrier to root development and to water movements.

2. Active dune fields can appear in areas of degraded vegetated dunes and sandy alluvial deposits, or by extension of the margin of dune fields, but the phenomenon is very local: I have not seen a single example of spontaneous genesis of dunes, under human pressure, in continental subhumid and humid tropical areas.

   In the Sahel the reactivation of the vegetated dunes of the Old Red Dune System does not systematically lead to active dunes, but can have several other consequences: wind erosion and loss of the topsoil, exposure of the subsoil with formation at the surface of a hard-setting. "Soils with dense, sometimes weakly cemented surface horizons are identified as hard-setting" (Lal et al. 1989). The formation of hard-settings on the degraded dunes is a plague for agriculture. In the Sahel, hard-settings are associated with limited biological activity, soils low in organic matters, and with low fertility. When surface crusting occurs, establishment of plants is difficult because of low porosity, which results in a high runoff of precipitation. When cultivated, the hard-setting is destroyed, but reappears after the first rains. It removes one-third of the annual rainfall from plants (Thompson et al. 1982).

3. Depletion and contraction of the vegetation. The study of land occupation in western Mali shows that the percentage of denuded soils has increased very little during the last three decades and that, on the contrary, an increased density of vegetation in the depressions and valleys is the rule in the Sudanian ecosystem. Increase of the vegetation density in the lower part of the landscape can be considered as an indicator of land degradation. It means, in fact, in this case an increase of runoff on the slopes, an increase of the arrival of water and colluvial particles in the depressions and valleys, because of depletion of vegetation and water erosion on interfluves and slopes.

2.6 The Survival of Plants in Dry or Seasonally Dry Ecozones

All life depends on the presence of water. The availability of water is one of the overriding abiotic parameters. Plants consume water vapor by transpiration. In the dry ecosystems annual rainfall variability is high in quantity, and in time and space. The erratic characteristics of rainfall already mentioned for Kenya will be analyzed later in the framework of African climates north of the Equator.

Lack of water means that dry or seasonally dry ecozones pose special problems for plant life. In the Saharan and Sahelian ecozones sunlight and oxygenin the dry soil are abundant, but shortage of water and extremes of temperature are limiting factors. Xerophytic plants have adapted a wide range of solutions to these specific limiting factors. The most extreme adaptations are those of desert and semi-desert organisms, which have to survive extreme daytime heat and nocturnal cold, with little water available for transpiration.

Plants need a continuous inflow of carbon dioxide into their leaves and therefore need contact with the air outside. Through their stomata (pores in the epidermis of leaves) gas exchange between the plant and atmosphere is produced. Most plants close their stomata in the night, when there is no light for photosynthesis (production of organic compounds and oxygen with carbon dioxide, water, and solar energy). In dry environments they open the stomata in the morning and close them around midday when water loss through transpiration becomes too high. Possibilities for regulation by stomata are limited and xerophytic adaptations take over.

Xerophytes are plants adapted to dry areas. They can grow in moist habitat but their growth is slowed down. There is a gradual transition from true xerophytes of the arid environment to still xerophytic plants in the savannas of Sudanian ecozones. Most of the xerophytic structures lead to a reduction of photosynthesis.

Xerophytic adaptations defy the imagination; they consist of:

a) a modified anatomy of the leaves with thick wax layers covering all surfaces, hairs, or scales which reduce the airflow around the leaves, and lower density of stomata arranged on the underside of the leaf in deep troughs;
b) a reduction in size of the green part of the leaves, which can become composite with movable leaflets, which can close or open as in *Acacia* or can roll up as in grasses;
c) a very extensive root system;
d) storage of water in the roots or aerial part of the plant (succulents: plants with reduced leaf area and internal storage of water, e.g., *Euphorbia candelabrum*). Succulents are able to store carbon dioxide at night in their cells and use it during the day for photosynthesis with closed stomata. They can use water more efficiently by opening the stomata at night, as the transpiration at night is lower. Agave, Aloe, and Euphorbia have leaves oriented vertically so they are less exposed to sunlight. Papyrus, sugarcane, dura (sorghum), maize, and other tropical grasses belong to a group of plants which have found another type of adaptation to the loss of water vapor as a side effect of carbon dioxide influx into the leaves for photosynthesis. Instead of the classical C_3 pathway they are characterized by a C_4 pathway of carbon dioxide which allows them to continue to photosynthesize in warm and dry conditions with their stomata nearly closed. Succulents and some grasses grow better under the shade of trees (e.g., millet has a better yield when planted under scattered *Acacia albida*) and shade decreases 40% of the needs of irrigation. So these plants can be combined with a canopy of leguminous trees. This knowledge is empirical or based on scientific experimentation and the manager must learn to imitate the natural ecosystems to make them into productive agroforestry systems.

2.7 General Conclusion, Chapter 2

In drylands, droughts and climatic variability are the inherent parameters of the climate. Vegetative cover and soil are adapted to dryness; their prominent feature is not vulnerability. They may well not be as fragile as is often stated. Good resilience of plant communities has been observed over the last years in connection with good rainfall.

Chapter 3 of this book, which will analyze land degradation in drylands on different continents, will try to demonstrate how land degradation, related to dryness, is connected with human activities.

Chapter 3 Dimensions in Space of "Desertification" or Land Degradation: Their Degree and Specificity in Each Continent

"Man learned to cope with the environmental conditions almost all over the world by making mistakes. As long as man was active on a rather small scale, many of these mistakes were reversible. With the recent increase of scale of human activities the scale of the mistakes increased as well from 'trial and error' to 'trial and disaster'." (Van Noordwijk 1984)

3.1 Introduction: Global Dimension in Space of "Desertification"

The drylands of the world, i.e., the arid, the semi-arid and the dry subhumid ecozones, which according to our definition are threatened by land degradation, occupy 35% of the Earth's land surface and contain 20% of its population. Even though the only global assessment of the spatial extent of desertification, given by UNEP, has been critized, it deserves to be quoted (Ahmad and Kassas 1987).

1. Some three-quarters of 45 million km^2 of the world drylands, about 34 750 000 km^2 of rangeland, rainfed croplands, and irrigated land, an area approximatively the size of the American continents, are affected by desertification.
2. Between a quarter to a half of this area is severely affected, which means a loss of 25% or more of its potential productivity.
3. Every year, according to UNEP, 210 000 km^2 (21 million ha) of productive land are reduced to near or total uselessness while an additional 60 000 km^2 (6 million ha) are turned into barren desert.
4. Projections to the year 2000 indicate that such losses will continue or even increase unless remedial actions are taken. The rate of desertification could rise, as pressure on the land increases, particularly in developing countries, where population size is expected to double over the next two or three decades.

We know the difficulty and therefore the fragility of statistics for evaluation, and Forse (1989), speaking of UNEP, notes: "Some scientists question these statistics".

The question is: what is best? To have approximate statistics which have a value for political decisions but not for scientists. . . or to have nothing? At least they provide a start which can be confirmed or disproved.

Land degradation exists on all the continents but may be greater in Africa, where it results in famine. More than 80% of Africa's dry ecozones, rangelands, and rainfed croplands are said to be already affected by at least moderate degradation.

However, it must be pointed out that areas affected by land degradation are not all affected by the same processes. In Africa, according to the above UNEP statistics, only 16% of the continent's total dryland surface is subject to water action, while 45% of the total area is subject to wind action. Salinization affects 25% of the soil[7]. Animal pressure affects 61% of the continent and population pressure 45%.

To have a good qualitative approach to land degradation I shall try to analyze it continent by continent. I hesitated before deciding the order of presentation and did not find a really satisfactory criterion. I have finally decided to choose the degree of utilization of high technology, and propose the following succession:

1. Sub-Saharan Africa, where traditional extensive agriculture is dominant and where appropriate high technological methods have yet to be worked out except in some temperate areas, e.g., Highlands of Kenya, Zimbabwe, and South Africa.
2. Egypt will have a specific place because its agriculture is based on traditional intensive irrigation, where difficulties have been created by the misapplication of high technology.
3. China, with its sophisticated intensive, traditional agriculture, is in difficulties due to expansion into unsuitable areas. These are being rehabilitated by the application of modern methods of research and development based largely on irrigation.
4. The former USSR represents the transition from traditional low technological agriculture to highly technological industrialized rural activities.
5. Australia has the most highly technological capital-intensive pastoralism.
6. The western USA, the last case study, has an exclusively high technologically based and highly capital-intensive agriculture.

3.2 Sub-Saharan Africa: Traditional Extensive Agriculture, Unadapted High Technology, and Land Degradation

"Solitudinem faciunt, pacem appellant." Tacitus - *Agricola* (They made a wilderness and called it peace.)

The first section will deal with the bioclimatological ecozones of sub-Saharan Africa: Saharo-Sahelian, Sahelian, Sahelo-Sudanian. Then the question of paleodroughts, historical droughts, and the 20th century droughts will be analyzed. The third and following sections describe the ways in which man traditionally occupies the dry environment, what are its real potentialities and the way in which man and environment interact so as to highlight some basic recommendations for optimal use while still respecting the stringent climatic limitations. In the last section trials, errors, and disasters of present-day development will be shortly outlined.

[7]Dregne, in a personal correspondence (August 11, 1992), writes: "I believe the 25% number for salinization must refer to primary (natural-chotts, sebkhas) as well as human-induced. African drylands amount to 2138000000 ha. FAO says *all* irrigated land in Africa amounts to 10424000 ha (0.4%). Human-induced salinity in Africa occurs only on irrigated land. Even with chotts and sebkhas, 25% seems much too high.

3.2.1 Droughts in the Sahel

3.2.1.1 Paleo-Droughts and Historical Droughts

There is evidence from past times that rainfall variations and droughts are the rule in African climates, particularly at the latitude of the present arid, semi-arid, and subhumid ecosystems.

It is difficult to be in any way certain about climatic variation earlier than 20000 B.P. (B.P. means before present, which was conventionally decided to be the year 1950) because the indicators are rare and need interpretation. Scientists agree on three main arid phases at 70000 years B.P., 20000 to 15000 years B.P., and the last drought, which began around 5000 to 3500 years B.P. It is now generally accepted that the peak of the last glacial period 18000 B.P. corresponded to the drought in tropical Africa.

Probably during this dry phase (between 18 000 and 8000 B.P.) the southern limit of the Sahara (corresponding to the 150 mm isohyet) shifted about 1000 km towards the north and later between 6000 B.P. and the present about 600 km towards the south (Petit-Maire 1989). This assertion is justified by the presence of an aeolian sand-sheet at 5° south of the present active dunes. This sand-sheet can be followed from Senegal to Mali, Nigeria, Chad, and Sudan; its maximum depth of 60 m was observed in Niger (Mainguet 1982; Mainguet and Chemin 1983), its maximum extension was probably at Maiduguri in Nigeria (great erg of Hausaland).

Hurault (1972) demonstrates that this dry climatic phase expanded until the present Guinean ecozone (e.g., Banyo 6°03′N). Hurault mentions another drought in the same area at the beginning of the Christian era.

Nicholson and Flohn (1980) and Street-Perrott and Roberts (1983) show that after the glacial maximum there was increasing aridity at tropical latitudes culminating between 14 000 and 12 500 B.P. Later, between 12 500 and 5000 B.P., the climate became progressively more humid, with a belt of expanding lakes in Africa and Arabia. Schneider (1969) and Maley (1973, 1981) summarize the lake level variations. In the early Holocene (9000 B.P.) tropical Africa was humid and Lake Chad had a water surface of 320 000 km^2, more than 20 times the present area, and had a level more than 50 m above that of today.

During the wetter phase between 12 500 and 5000 B.P., two droughts were identified around 10 200 and 7400 B.P. After 5000–4000 B.P., the rainfall decreased with consequent lowering of the lake levels.

In a study of the middle and late Holocene, Pachur and Kröpelin (1987) on the basis of a north-south transect starting at the Gilf Kebir at 24 °N (Sudan) in the center of the presently hyper-arid eastern Sahara and ending in the Wadi Howar area at 17 °N, have shown that aridification affected the northern parts from 5000 years B.P. and the southern parts from about 3500 years B.P. The northern fringe of the paleo-Sahel shifted southwards approximately 500 km; neolithic man seems to have intensified the effects of aridification.

During historical times there is fuller information. Nicholson (1976, 1978, 1981) made an analysis that shows that there were wetter conditions than the present in the 16th, 17th, and 18th centuries, nevertheless with some major droughts around 1680, 1738–1756, and 1830. In the late 19th century, the average rainfall was probably 20–40% higher than today.

3.2.1.2 20th Century Droughts (Table 6)

In the ecozones in the southern Sahara, instrumental records are available from about 1900. Since the beginning of the century severe short-term climatic fluctuations have occurred:

1905-08 Low Lake Chad
1912–15 Severe dry period — drought
1916–24 Wet period
1925–28 Dry period
1929–39 Wet period
1940–44 Severe dry period — drought
1945–46 Wet period
1947–49 Dry period
1950–61 Wet period
1962–67 Intermediary period
1968–73 Drought
1974–76 Humid period
1977–85 Severe dry period — drought
1987–88 Humid period
After 1988 Intermediary period

Table 6. Highest rainfall deficit during droughts in the 20th century

Station	Year	Annual rainfall (mm)	Annual mean (mm)
Saint Louis	1914	144	330
Dakar	1972	117	500
Podor	1942	98	310
Tombouctou	1966	89	210
Zinder	1912	215	490
Niamey	1915	281	580
N'Djamena	1913	306	620

The Drought of 1910–16. The year 1913 is called in the Songhaï calendar the long Bere, which means the long famine, and in the Peul Calendar tasbane, which means calamitous. In fact, narrations of travelers and the military describe the rainfall deficit. Along the coast, in Saint Louis du Senegal, in 1914, the rainfall was only 144 mm instead of an annual average of 330 mm between 1964 and 1974. The rainfall of 1914 was the lowest of the series. Still higher deficits were measured: Niamey 1915, 281 mm, i.e., 48% of an annual average of 580 mm; N'Djamena 1913, 306 mm, i.e., 50% of an average of 620 mm.

The Drought of 1940–49. The Songhaï call the year 1942 Wande Wassu which means sack your wife (so as to relieve part of your load). The drought affected the whole Sahel in different degrees according to the areas.

The Drought of 1968–1986. The most impressive image of desertification is linked with the severe drought in the Sahel zone south of the Sahara, which started in 1968, and lasted until 1973 with 2 better years in 1974–1975; but the effects of drought were not overcome. There were again more or less continuous rainfall deficits until 1986, with a new peak of drought in 1982–83–84. Two good years in 1987–1988 followed the drought; they are the beginning of a better rainfall period but never as favourable as during the decade which preceded the drought.

The effect of drought is made worse by the fact that the 1960's were years of excess rainfall; for example, in 1967, Bamako was partly flooded. Nevertheless, as early as 1965 some parts of Mali and Chad had only very low rainfall, while in 1968 the deficit was general from Mauritania to Senegal and Chad but with considerable heterogeneity in its incidence. In Niger, the area between Maradi and N'Guigmi had good rainfall. In 1969, Niger and Chad had normal rains in the Saharo-Sahelian ecozone (100- to 300-mm isohyet).

In 1970 all the countries of the Sahel had low rainfall except Chad. In 1971 there was an intensification of the drought extending to the north of Benin, Cameroon, the Central African Republic, and Zaire. 1972 was the culminating year for the drought. The situation in the Sahel was the more dramatic the nearer a place was to the true desert.

1. For the ecozone between the 100–300 isohyets the annual rainfall was between 45 and 80 mm with a deficit of 75%.
2. For the ecozone between the 500–600 isohyets the deficit was from 25 to 40%.

In 1973 the rainfall was still short in the whole Sahel, except between Tema and Kaya (Burkina Faso) and south of Chad. Globally, the drought persistently increased.

In 1974 the situation seemed to recover, except in Senegal, where the deficit was over 30% and in Mauritania from Nouakchott to Atar, where it exceeded 50%.

In the whole Sahel the (1974) drought was more difficult to endure as the 1950's had had a rainfall 20% higher than the average. The result was that the nomad Peuls (Fufulbe or Fulani) from northern Nigeria and southern Niger pushed further and further northwards to the Air mountains. The grazing and agricultural areas of the Sahel received decreased rainfall: 20% less in 1968 but 33% less than in the 1950's and in 1972–73, 50% less than the average. When the drought continued in 1974 and then, in discontinuous spells, until 1986, the same temporal and spatial heterogeneities were observed throughout the semi-arid and subhumid ecozones south of the Sahara.

The drought was mainly the result of the shortening of the rainy season. Later, the drought resulted not only from a shorter rainy season but also from a decrease in the number of heavy rains of more than 40 mm. The quantity of rains of 20 mm or less did not diminish.

These details are worth study, because the most recent drought is in fact the first about which precise data exist (Table 7). Several explanations were given for the drought which began in 1969. The natural causes of the deficit which were suggested were:

1. Scarcity of the atmospheric depressions over the Weddell sea (southern part of the Atlantic Ocean); the pressures over the south Atlantic were higher.
2. Weakness of the subtropical high pressure areas.
3. Lower equatorial low pressure areas on the Gulf of Guinea.
4. Reinforced high subtropical pressures over the Sahara.

The human-induced causes of the deficit were alleged to be:

1. Degradation of the vegetative cover and higher albedo (reflective capacity) over the Sahara and its semi-arid and subhumid boundaries which upset the rainfall mechanisms. Bare soils are better reflectors and reflect the incident sun radiation.
2. Changes in the surface moisture.
3. Deforestation in the southern, wetter parts of West Africa.
4. Increasing atmospheric carbon dioxide concentration.

These anthropogenetic factors have all been suggested as possible causes of desertification, but with equivocal results. The causes of drought are complex and not attributable to one single factor. None of the models suggested gives a satisfactory explanation.

3.2.1.3 Drought and Shifting Isohyets in Eastern Mali. Where Are the Most Severe Effects of Drought?

The last drought had consequences reaching from the Sahelian steppes, to the Sudanian savannas, and to the Guinean forests, but the effects are different from one ecozone to the other. Is indeed the Sahel, the most affected ecozone as stated in the literature and assessed by governments and international bodies?

To try to obtain a scientific answer to such a problem some questions must be answered: do all the ecozones react to drought at the same speed and in the same way? Do the different combinations of climatic parameters and socio-economical one have the same indicators of degradation?

Controlling desertification or land degradation cannot be brought to a successful outcome without a thorough knowledge of the movements of the isohyets. We have to know the location of the average isohyets, the location of the isohyets during rainfall deficit periods, and also the location during good rainfall periods.

The example chosen of climatic variations is in Mali. The analysis of the rainfall variations and of the modifications of the landscape in the last three decades in eastern Mali between 15°30 and 11°45 N with air photographs and satellite images have tried to provide an answer to this question of the relation between climatic variations and meteorological drought.

Table 7. Total annual rainfalls and minimum by stations in the western Sahel

Station	Average 1931–1960	Minimum rainfall 1968–1974		Absolute minimum	
	(Asecna[a])	Year	Mean (mm a^{-1})	Year	Median (mm a^{-1})
MAURITANIA					
Nouakchott	138	1974	36		
Akjoujt	106	1974	20		
Aleg	265	1973	162		
Atar	104	1972	36		
Boghe	332	1972	116	1972	112
Boutilimit	203	1973	42		
Kaedi	401	1972	130	1920	120
Kiffa	351	1972	119	1972	119
Mederdra	251	1972	86		
Moudjeria	236	1972	65		
Néma	315	1974	182	1928	167
Rosso	290	1972	53		
Tidjikja	142	1974	62		
Timbedra	388	1974	159		
SENEGAL					
Coki	533	1972	188		
Dagana	330	1972	81		
Dahra	511	1973	238		
Dakar Yoff	578	1972	117	1972	117
Linguere	535	1972	245		
Louga	473	1972	156	1972	156
Matam	537	1972	176		
M'Bao Thiaroye	569	1972	181		
Podor	336	1972	110		
Saint Louis	347	1972	152	1914	144
Yang Yang	527	1972	157	1972	156
MALI					
Ansongo	302				
Bandiagara	592	1972	442	1938	315
Bourem	167	1973	87		
Dire	243	1972	157		
Douentza	510	1972	245		
Gao	263	1973	144	1931	60
Goundam	252	1973	83		
Gourma-Rharous	176	1973	134		
Hombori	414	1972	300		
Kabara	220	1973	104	1966	89
Ke-Macina	572	1973	385		
Kidal	138	1972	92		
Menaka	262	1972	162		
Mopti	546	1973	326	1973	326
Mourdiah	565	1972	385		
Niafunké	328	1972	64		
BURKINA FASO					
Dori	555	1968	368	1926	244
NIGER					
Agadès	168	1972	74	1970	40
Birni N'Konni	591	1973	289		

Table 7. (*Continued*)

Station	Average 1931–1960	Minimum rainfall 1968–1974		Absolute minimum	
	(Asecna[a])	Year	Mean (mm a^{-1})	Year	Median (mm a^{-1})
Filingué	514	1973	216		
Gouré	410	1973	130		
Madaoua	513	1973	156		
Mainé Soroa	423	1973	262		
N'Guigmi	225	1972	69	1928	41
Niamey ville	573	1973	371	1915	281
Tahoua	403	1973	245	1942	209
Tanout	290	1973	201		
Tera	514	1974	330		
Tessaoua	552	1973	251		
Tillabery	506	1973	337		
Zinder	508	1973	298	1912	215
CHAD					
Abéché	495	1973	186	1973	188
Ati	425	1973	163		
Bol	341	1972	62	1913	46
Fada	97	1972	0		

[a]ASECNA, Agence pour la Sécurité de la Navigation Aérienne.

Figure 10 shows the location of the isohyets at three different periods:

1. An average of almost six decades, 1922–1980 with wet and dry spells,
2. An average of a more humid decade, 1951–1960 with a shift of the isohyets towards the north.
3. An average of a drier decade, 1971–1980 with a shift of the isohyets towards the south.

These isohyets were established with a very scattered network of synoptic stations, but NOAA imagery confirms the results. The mean shift of the isohyets is 100 km. The shift reaches its maximum 150 to 160 km at the latitude of the 1000 to 1200 mm per year isohyets and its minimum of 40 to 60 km at the latitude of the 800 mm isohyet.

These values are of great importance. They establish that **the region of maximum stability of the climate is in the dry subhumid zone, and the minimum in the Sahelian and the Sudano-Guinean zones.**

The comparison of these three series of isohyets shows the high variations in the rainfall averages at the same point during a dry or a rainy decade. The most southerly location of the 400-mm isohyet of a dry period, is almost the same as that of the 600-mm isohyet of a rainy period. The same occurs with the dry years' 600 mm isohyet relative to the 800-mm isohyet of a rainy period and also for the dry years' isohyets of 600 mm and of 1000 mm. On the contrary, for the 1000- and 1200-mm isohyets, the shift of the isohyets of dry years crosses those of rainy years.

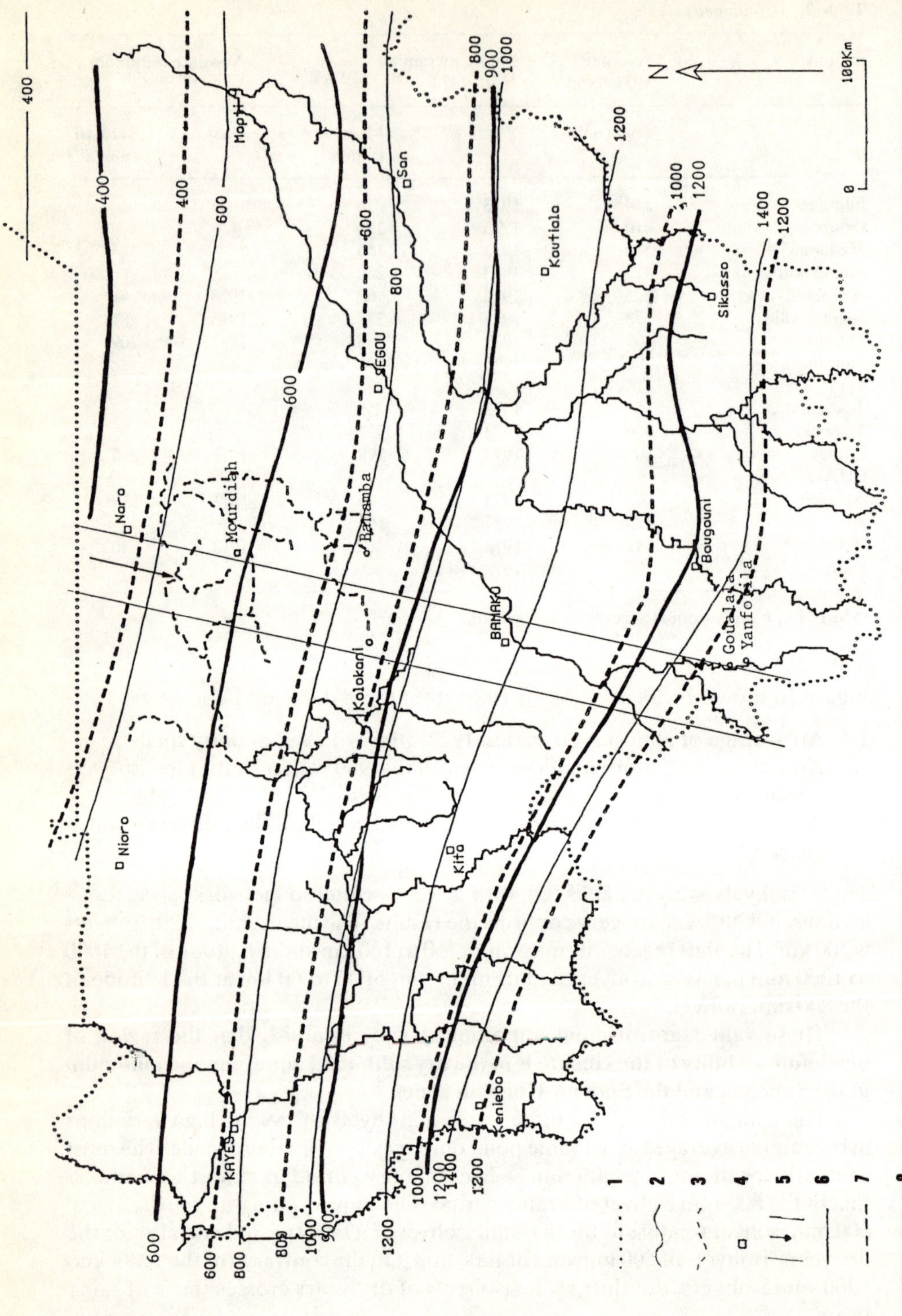
Nara
Mourdiah
Nioro
Kayes
Mopti
San
Segou
Baramba
Kolokani
Bamako
Kita
Keniebo
Koutiala
Sikasso
Bougouni
Goualala
Yanfolila
N
0
100Km
1
2
3
4
5
6
7
8

As seen in Kitui (see Sect. 2.3.2.3, example 2), even the total annual rainfall is not the only climatic parameter to take into account; the consequence of rainfall shifts for agricultural activities is certainly less detrimental in the south, in the Guinean ecozone, where the annual rainfall varies between 1200 to 1400 mm than in the Sahelian ecozone, where according to the decade, the average annual rainfall may be less than 400 or 600 mm.

On a regional or national scale, environmental changes even during several decades are very often so slight that they are difficult to perceive on satellite images. In the absence of records of different dates the task is more complicated.

Multitemporal and multiscale projects are recommended to show these transformations clearly. A project of this type was launched in 1987 by UNEP under the leadership of the Institut Géographique National (IGN), Paris, with ORSTOM and Laboratoire de Géographie Physique Zonale (LGPZ), University of Rheims (France).

1. The transect defined in Mali follows the Spot path no. 40 from row 319 (15°40 N) border of Mauritanie, to the row 328 (10°50 N) border of Guinea (Fig. 10).
2. The three types of documents used were air photographs of the 1950's, air photographs of November 1987 (infrared to the scale of 1:36 000, panchromatic to the scale of 1:50 000) and, from the same date, Spot Satellite Images which were enhanced.

Some important results were obtained (Table 8):

1. The limits of the areas over which wind and water erosion operate was ascertained with considerable accuracy:

a) 14°30 N, southern limit of exclusive wind erosion. It corresponds to the mean 500 mm isohyet per year in the Sahelian zone. Slightly more south very local wind erosion remains visible in bare fields at the latitude of the mean of the 700 mm isohyet.

According to Spot image analysis, in this transect, 14°10 N is also the southern limit of the Sahelian, paleoaeolian sand-sheet. It corresponds to the actual mean 600-mm isohyet. Accumulated during a drier phase of the climate with a maximum rainfall of 150 mm, the fixed sand-sheet indicates a paleodrought responsible for the deposit and a meridional limit of the Sahara farther south than at present.

b) Around 14 °N between the mean annual (wet year) 650 and 750-mm isohyets the Sudanian zone begins. This is also the zone, according to our documents, where water erosion becomes dangerous. The zone of maximum degradation

◄

Fig. 10. The average isohyets west of Mali from 1922 to 1980, for the dry spell of 1971–1980 and for the decade 1951–60. (After documents of CTFT/CIRAD 1987). *1* International border; *2* water course; *3* dry valley or intermittent water course; *4* synoptic weather station; *5* SPOT path no. 40; *6* isohyet for the period 1951–60; *7* isohyet for the period 1971–80; *8* isohyet for the period 1922–80

Table 8. Boundary of dynamic processes and potentially dangerous mechanisms according to the latitude of the isohyet in western Mali

Latitude		Potentially dangerous mechanisms	Boundaries of dynamic processes	Isohyet in mm Average 1922–80	Surplus 1957–60	Deficit 1971–80
14° 30′ N	1	Exclusive aeolian erosion	1. Northern boundary of traumatizing surface water degradation mechanisms discernible on satellite and aerial photograph images	500	600	450
14° 10′	2	Interaction of aeolian and surface water mechanisms	2. Southern boundary of the inherited continuous sand sheet	600	700	500
14° 00′	3		3. Northern boundary of the maximum surface water degradation area	650	750	550
13° 50′	4		4. Southern boundary of perceptible indices of aeolian re-shaping, by deflation, of cultivated land	a 700	800	600 a
13° 35′	5	Surface water erosion	5. Southern boundary of the maximum surface water degradation area	800	850	700
13° 20′	6		6. Southern boundary of surface water degradation mechanisms dicernible on satellite images and aerial photographs	900	1 000	800
13° 00′ N	7	Diffuse surface water erosion Colluvial deposition	7. Northern boundary of discernible hard, lateritic crust on satellite images and aerial photographs	950	1 100	850

a: maximum degree of degradation.

by water erosion reaches the latitude of 13°35 N, which corresponds to the south Sahelian zone with the mean 800 mm isohyet (850 mm during rainy spells).

Still further south, water erosion is still visible on satellite images and air photographs, but it is a less potentially dangerous process.

c) In the north Guinean zone, south of a mean annual rainfall of 900–1000 mm, the degradation is less. Nevertheless, colluvial erosion of material on slopes is detectable through the increasing surface of iron crust outcrops in the last three decades.

The observation of the extent of the ecozones dominated by wind erosion, a combination of wind and water erosion, and water erosion reveals that the zone between 650 and 800 mm per year, which is north Sudanian, is the most affected by degradation.

In this ecozone the soil and vegetation deterioration has a triple aspect:

a) The tops of the slopes, where rill wash or sheet-wash are at a maximum, are the area of maximum degradation.
b) The interfluves are often bare.
c) A striped texture called "brousse tigrée" on the slopes is also an indicator of deterioration.

Paradoxically, in the network of gullies, already vegetated in the 1950's, the vegetative cover is denser in 1987. But even this observation means degradation, because with an increasing runoff more water arrives in the valleys, the unfavorable infiltration/runoff ratio being the main cause. Therefore the regrowth of vegetation is improved. The improvement is not over the whole landscape but very precisely localized in the depressions (valleys included).

It is also at these latitudes that the first high population densities appear: more than 10 inhabitants per km^2 and even locally more than 20. We have observed earlier that the zone around the annual 800-mm isohyet has a less variable rainfall, which makes it an attractive zone for the people when drought occurs further north.

The rate of change in the area and the status of cultivated land was analyzed (Fig. 11) by assessing the cultivated land and the fields cultivated annually for the last three decades between 1950 and 1987. A double series of changes appear:

1. In the 1950's the percentage of cultivated land increased, from northern to southern latitudes: from 28% in the north at the latitude of Nara (15°40 to 15°10 N), which is in the south Sahelian zone, to 65% in the area of Mourdiah (14°20 to 13°40 N), which is in the north Sudanian zone, to 74% in the south of the transect, the area of Yanfolila (11°50 to 11°20 N) in the north Sudano-Guinean zone.
2. By 1987 the situation has changed: since the 1950's the area of cultivated land has doubled to 55% in the south Sahelian zone (area of Nara). In the north Sudanian zone (area of Mourdiah) the cultivated land occupies 71% of the landscape which represents a negligible increase compared with the 1950's (65%). There is almost no change for Yanfolila in the north Sudano-Guinean zone.

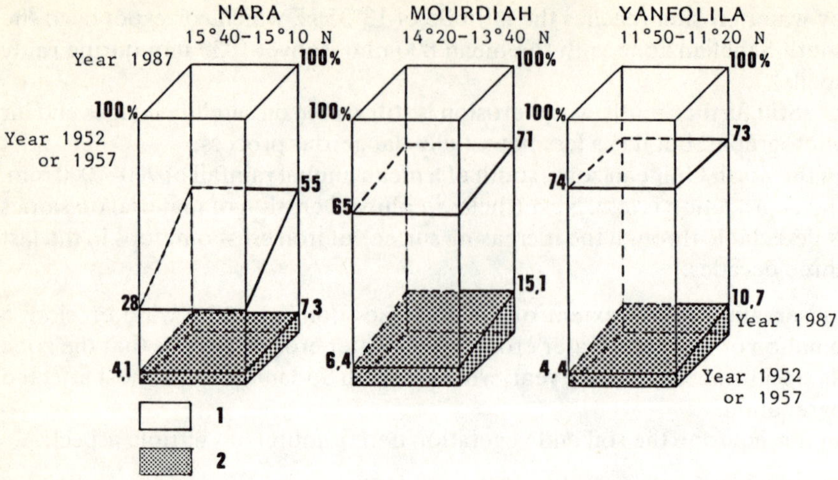

Fig. 11. Land occupation in western Mali. Comparison between 1952 or 1957 and 1987. *1* Total cultivated land; *2* land cultivated in 1 year (1952 or 1957) and 1987

Briefly, the influence of latitude disappears south of the 800-mm isohyet (around 14 °N); and the cultivated land has not increased. This observation can be explained by pressure on the land or even land shortage and also because above 800 mm annual rainfall, deficit of precipitation has less influence. In the 1950's all the available cultivable land was already used.

Mourdiah (north Sudanian zone) has the highest percentage of fields cultivated annually at 6.4% of the space, as it had since the 1950's, compared to around 4% at other latitudes. The latitudes of Mourdiah are therefore the most exploited. They are also the latitudes where fallow periods have become shorter and most reduced.

To sum up, **in the driest ecozone (400-mm isohyet) north of the transect, the result of the increase of population was an increase of the cultivated surfaces. At the latitude of the 800-mm isohyet (north Sudanian), which is the zone of maximum human pressure, there was no increase of cultivated land because all the available land has been cultivated for several decades. Here the answer to population increase was the decrease of the fallow period.**

3.2.2 The Bioclimatological Ecozones

1. The Saharan hyperarid ecozone at its northern boundary (because the rainy season is an autumn and in winter) can be delimited by the 100-mm annual rainfall isohyet and 100- to 150-mm isohyet at its southern limit (because the rainy season is in summer). In these areas the rainfall is very unreliable, and completely dry years can occur. The rainy season lasts annually for a maximum of 30 days. The coefficient of variation of rainfall is more than 80%. The temperature fluctuates over a wide range with hot days and cool nights, which results in

scarcity of the vegetation. Natural vegetation is of the contracted type, limited to wadis and oases. Some shrubs can be found, but most plants here are annuals (grasses which flower in their first year and subsequently die) and are visible only in the rainy season, surviving the dry season as seeds. After occasional rains, short-living grass is visible. In winter when the air is cool at night, dew is formed and grasses can grow for a short time after rainfall.

In the Saharo-Sahelian subzone between the 150- and 250-mm isohyets, the vegetation is irregularly distributed in patches in the valley floors, in depressions or small holes. This distribution in pockets of more abundant growth which plays a crucial role in normal times could be mapped. The pockets are the last grazing points in drought, supplying the greater proportion of the fodder in the Saharo-Sahelian ecosystem.

The most efficient use of this concentrated vegetation is nomadism. The maintenance of mobility is the best answer to this natural grazing. Ranching systems are not advised. They can only be developed when rainfall is higher and allows irrigation.

The probability of success of rainfed agriculture is almost nil. Highly developed systems of irrigation were used in the oases to allow agriculture.

The soil, generally poorly developed, due to little rainfall and vegetation, varies from pure sand to stony surfaces: regs (surfaces of rounded pebbles or angulous blocks), hamadas (limestone plateaux), and tassilis (sandstone plateaux). Brown soil of crumbly structure, usually rich in salt crust, can also occur.

Wind erosion is dominant. Sand and dust storms are problems for all life forms.

2. The semi-arid (Saharo-Sahelian) ecozone has annual rainfalls of 100 to 400 mm with a coefficient of variation between 30 to 80%. The growing season is short but numerous short-life plants grow well, especially in wet years. Some woody plants: *Acacia tortilis, Acacia mellifera,* and *Commiphora africana* are found. Large parts of the semi-arid ecozone are treeless grasslands: grass steppe, probably due to human activities, and partly composed of perennials (plants which live more than 1 year and which have a stock of reserves in the underground parts and survive the dry season resting, to resume growth at the start of the new rains.)

Pastoralism is here the least problematic type of land use; rainfed agriculture is more uncertain, and commercial agriculture almost impossible.

In the subzone north of the 400-mm isohyet, regular irrigation requires 1700 to 2200 mm of water because evapotranspiration is high and rainfall low (Davy et al. 1976). When the grass steppe in this subzone is destroyed without soil deterioration, 2 or 3 good rainy years or 3 to 5 years of exclusion from use are sufficient to rehabilitate the vegetative cover in sandy soils. The status becomes irreversible when soil resilience is destroyed.

3. The Sahelo-Sudanian ecozone, also called low rainfall woodland, sahel-savanna, or thorn scrub according to the type of vegetation, has an annual rainfall between 400 to 600 mm with a coefficient of variation from 25 to 35%. The

growing season has 90 to 120 days. Water erosion is here more severe than wind erosion. The woodland savanna north of the Sahara is composed of Aleppo pine and thuya (*Callitris quadrivalvis*). South of the Sahara, the Sahel savanna is composed of short annual grasses up to 1 m tall, perennial grasses in the depressions (*Andropogon gayanus*) and trees up to 10 m high: combretaceas, baobab trees (*Adansonia*), *Acacia mellifera* on clay soil and *Acacia senegal* on sandy soil, *Hyphaene thebiaca* (Dom palm) in flooded areas. Rainfed agriculture and sedentary livestock are possible. The southern limit of camel-raising corrresponds to 600 mm annual rainfall and is generally the northern limit of *Glossina*, which causes trypanosomiasis.

4. The dry subhumid ecozone or north Sudan savanna or intermediate rainfall woodland has an annual rainfall of 600 to 1000 mm. The grasses, much taller (2 m), are mostly perennials instead of annuals. The trees are several acacia species (*Acacia seyal*), with *Balanites aegyptiaca* and *Borassus aethiopium* along the streams and rivers. Apart from rainfall, the main factor determining the savanna vegetation is fire.

Usually in the ecozones where annual rainfalls reach 800 mm, and where drainage of excess water is possible, the soil has a profile composed of a shallow red sandy-loam topsoil (0–15 cm) overlying a layer of pea-iron nodules or more consolidated ironstone subsoil. On these soils the plant cover is not very dense. Water can infiltrate without much difficulty and its availability to the vegetation is good when the topsoil is intact. In contrast, when degradation occurs with the disappearance of the topsoil, the subsoil outcrops and becomes a hard iron pan into which rainfall cannot infiltrate.

In this ecozone pastures become denser, but of lower quality. Penning et al. (1982), analyzing the productivity of Sahelian pastures, demonstrate their better nutritional quality than the savanna further south. In the Sahelo-Sudanian ecozone, the most widespread evidence of degradation on pastures is not denudation but plant substitution. In East Africa, where overgrazing is the rule, may be replaced palatable grasses by shrubs (Pratt and Gwynne 1977). In the rift Valley in East Africa near Lake Naivasha, the slopes appear nicely vegetated on satellite images, with a signature indicating dense biomass disguising plant substitution: the vegetative cover has kept its density but lost is productivity.

In the north of Kenya, in the semi-arid area of Marsabit, an investigation of the pastures in May 1988, after a good rainy season, showed that the pasture had a good rate of recovery even after the drought years of 1984–1985.

In the Sahelo-Sudanian subzone, irrigation is efficient and can provide a supplement for a longer season of fresh grazing and fodder production.

3.2.3 Traditional Land Occupation. Potentialities and Agricultural Limits in High Climatic Risk Drylands

This section is an analysis of the traditional uses of dry environments, ecozone by ecozone, and of their potentialities and limitations for agricultural development.

Man, more than any other living organism, has an overwhelming influence on the ecosystems, and the scale of human activity is constantly increasing from:

1. the hunter-gatherer stage with fishing, hunting, gathering, which have little effect on the biotic factors (living organisms influencing the ecosystem),
2. to the animal domestication stage with a more pronounced effect on the biotic part of the ecosystem,
3. up to the crop domestication stage, which can have a disastrous effect on biotic factors but also on abiotic factors (nonliving factors influencing the ecosystem) because of the influence on the soil.

Resources for human use have been divided into those which are renewable and those which are nonrenewable. This division is relative: trees, edible plants, and domesticated animals are renewable resources as long as they are not overexploited. Water can theoretically be included, i.e., the water cycle of rainfall, evaporation, infiltration, runoff, flow to the sea, would be a renewable resource. However, when underground water is used, the classification is more doubtful; is this water a fossil supply or is it, after use, replenished? Many renewable resources are becoming nonrenewable because the environment is spoiled by human activity, producing pollution of water and air, and causing soil erosion.

The four traditional types of land use in sub-Saharan Africa nowadays, hunting, fishing, domesticated stock-keeping, and rainfed agriculture, will be described successively.

3.2.3.1 Hunting

As long as the population remained low, traditional hunting did not affect game animals; but the introduction of guns and the commercial trade in ivory and rhino horns has changed the population of these wild animals. In the early century, in the Sahel, their weight was 400 to 500 kg km^{-2} (this weight still exists in the Kalahari). Presently, the numbers of large mammals have been reduced to 60 kg km^{-2} (Le Houérou 1989, pers. commun.). Hunters are also under pressure because hunting grounds are being taken for other purposes.

3.2.3.2 Fishing

Fish is an important source of food and protein. Unfortunately, a number of nomads dislike fish intensely and do not eat it. The Toubous, living near the lakes of Ounianga Serir and Ounianga Kebir (Chad), cannot even stand the smell of fish, which is abundant in their lakes.

However, many people do eat fish there and they have also developed many methods of catching it by using poisonous plants, nets, dykes, and baskets (some of which are real works of art) and of preserving it by smoking, salting, and drying. Nevertheless, fishing is affected by changes in water quantity and quality in natural lakes and rivers and by impoundment in reservoirs.

3.2.3.3 Domesticated Stock-Keeping

There is a wide range of domesticated stock in the drylands of Africa: camel, donkey, horse, cow, sheep, goat, and chicken, which must all more or less adapt to the limiting factors: scarcity of water, fodder quality and supply, and diseases.

The density of plants increases with rainfall, but the quality of these plants may diminish. Fodder reaches the highest quality in the areas of low rainfall (Saharo-Sahelian zone) and deteriorates towards the south. The low biomass of the northern Sahel zone is of high protein content. In the savanna, where the biomass is high, the quality is good only at the start of the rainy season. Later on, the hard leaves and stems have a low protein content and offer poor grazing qualities.

Camels and sheep are most able to withstand the heat and sandstorms, need less water, and can walk longer distances. In the cool season camels can stay without drinking for 15 to 20 days, in the hot season about 3 days. Horses and donkeys are not resistant to the pests and diseases of the southern Sahelian and Sudanian ecozones; they are restricted to drier zones. Cows need drinking water every other day and therefore cannot graze more than 5 to 8 km from a waterpoint. Camels and goats, feeding on shrubs, bushes, and trees, are browsers, while cows and sheep, mainly feeding on grasses and herbs, are grazers.

Describing the domestic stock of Sudan, Van Noordwijk (1984), in a short summary, gives the relations between rainfall and pastoral activities:

> "In the desert with an average rainfall of less than 50 mm, only occasional nomadic grazing of camels is possible. The herds have to search for the places where rain fell that particular year, as rainfall is very much localized. The herds travel several hundreds of kilometers per year. In areas with 50–100 mm of rain on average, nomadic camel-keeping with some sheep is still the only possible use of the land. Fifty ha or more is needed for one livestock unit (the equivalent of one cow or seven sheep).
> In regions with 200–400 mm of rain, the requirement for one livestock unit is still in the order of 10–15 ha and where there is 400–600 mm of rain, 6–12 ha is necessary. When the rainfall exceeds 200 mm, cattle keeping becomes possible, provided there are water points.
> In the region with over 600 mm of rain, diseases and forage quality become the main limiting factor, rather than shortage of food or drinking water. Biting flies are the main reason for cattle migrations here. The main disease is 'nagana' or trypanosomiasis, which is transferred by the tse-tse fly *Glossina morsitans*. Other biting flies, mosquitoes, and ticks are additional factors limiting the use of the wetter grazing land. In the floodplain the limiting factor usually is the availability of nonflooded grazing land in the wet season, although dry season grazing land may be in short supply as well." (Van Noordwijk 1986).

With annual rainfall lower than 300 mm, pastoralism must be considered as the best answer for human survival: nomadism and transhumance are the two different types; the nomads are camel people.

Dryness increases the limiting factors of food and water supply, and movements on various scales become necessary. At the fringe of the desert the nomads have no permanent place of residence; they move according to regular routes modified by local rainfall patterns. Semi-nomadism is the type of land use which corresponds to the Saharo-Sahelian zones. The animal-owner has a permanent place of settlement where he grows crops during the rainy season, but travels with his herds during the longest part of the year.

The cattle people are the people of transhumance: in dry seasons the pastoralists send their herds, tended by herdsmen, towards the south in more humid zones where the tse-tse fly becomes a limitation in the rainy season. In contrast, in the rainy season they move to the north, where there is good pasture and water free of this nuisance.

Pastoral people have a high level of knowledge of their environment, of such things as plants and plant communities, palatable or poisonous plants, and medicinal plants for animal diseases. For the pastoralists cattle are not only a source of food. They are also highly important in social life for status, prestige, and as a capital reserve.

3.2.3.4 Agriculture in Sub-Saharan Drylands

When annual rainfall exceeds 250 to 300 mm, rainfed farming can share space with pastoralism and becomes exclusive when parasites prevent cattle raising.

Several domesticated crops are native to the Sahelian and Sudanian zones of Africa: sorghum, pennisetum, and eleusine originated from wild grasses of east Africa. Their wild relatives are still growing.

Sesame, lentil, and yam are also native to Africa. Groundnut, maize, sweet potato, and cassava, domesticated in the Americas, spread into Africa after the 15th century. Other useful wild plants can still be domesticated by future generations.

Domesticated crops are completely dependent on human activities for seed dispersal and for protection from other than human consumers, from pests and diseases, and from competition with other plants (weeding). After 2 years of cropping, the system deteriorates because of nutrient shortage, loss of soil structure, weeds, parasites, pests, and diseases.

The traditional answer to these difficulties is shifting cultivation: after a few years land is depleted in nutrients and organic matter and people move on to new land. Fallow alternates with cropping. If the land is left fallow for a long time [at least 20 to 30 years in the sandy soils north of Zinder (Niger)] the soil fertility can be fully restored. When the pressure on the land is increased, the duration of the fallow is reduced because people need to come back to the plot sooner.

Presently, around the villages of the Sahel and the north Sudanian zone, the duration of the fallow is not sufficient to restore the fertility of the soil, and rings of deteriorated soil surround the villages. When the fields lie directly around the homestead, the human wastes are used as a manure, making vegetable cultivation possible.

Other solutions for a better conservation of fertility exist. Keeping the cattle on the crop fields at night after the harvest is a classic method in the Erg Hausa in Niger and in Bahr-el-Ghazal province (Sudan). Sometimes animal manure and wood or grass are collected, brought in, and burnt in the field, providing additional sources of nutrients. Flooding along the main rivers, Nile and Niger, brings water and nutrients to areas which can be permanently cultivated. Green manure and mainly leguminous crops improve the nitrogen status of the soil and can at the same time be good fodder for cattle.

The root system of the crops usually develops too late in the rainy season to prevent leaching of nutrients. Trees, on the contrary, can pump up nutrients from deep down: *Acacia albida* with its leaves in the dry season, its valuable pods, and its nodules on roots, is a good soil protector, nourisher, and fodder producer.

Shifting cultivation with shortening of the fallow under the population pressure on the land is nowadays not an efficient system and needs to be replaced by permanent cultivation based on crop rotation. Permanent cultivation means land occupation for more than two-thirds of the time.

Mixed cropping is another African way in which man makes use of the natural resources. Monoculture is not traditional in Africa; the fields are used for several crops at the same time. The mixed cropping has several advantages: spreading the risk — if one crop fails, others may compensate. Crops are hidden visually from pests or by smell confusion by being so mixed. Crops with complementary root systems can use the soil more completely, and with complementary growth seasons can use light and water more continuously. The disadvantages of mixed cropping are negligible. It requires more labor, and mechanization is not easy. Mixed cropping is certainly one of the most efficient and sophisticated systems of using the land.

The "gardens" of gum arabic: *Acacia senegal* (Van Noordwijk 1984) are a good example of mixed farming in the Sudan savanna zone (400–600 mm rain per year, 8–11 dry months). Naturally, this savanna has short annual grasses under 1 m and scattered trees up to 10 m high. *Acacia senegal* can be tapped for gum over a period of 6–10 years. Then the deteriorated trees are removed. *Balanites aegyptiaca* are kept. On the cleared land sorghum and bulrush millet are grown mixed with groundnuts, simsim, watermelon, and lubia beans. The cropping can last for 4–10 years before the fields are abandoned to fallow of *Cenchrus biflorus*, *Acacia seyal*, and *Acacia senegal*. The trees must be protected in the first 3 years against burning and grazing. When they are 8 years old, tapping starts again and the whole cycle can be repeated. *Acacia seyal*, which restores soil fertility, is also a good fuelwood and produces the precious gum.

In a dryer ecozone, the Saharo-Sahelian ecosystem of Niger north of Zinder, the simultaneous growing of crops and some of the animal husbandry occur around the villages so that milk and manure to fertilize the crop are to hand. The majority of the cattle graze at a considerable distance from the village, where they are tended by the youth. Van Noordwijk (1984) describes the complexity of the farming systems of the Twic-Dinka in the northwestern side of a flood plain in the Gogrial district (Sudan). The calendar (Table 9) shows a total occupation of time and space. The quality of rotation results from its flexibility. There is no rigid order, but by observation of variations in the weather, and of the density and types of weeds, an "opportunist" strategy is used.

This section has shown, for the Sahelo-Sudanian ecozone the relationship between high population pressure, possibly due to the stability of the 800-mm isohyet, and the most severe land degradation.

Throughout the world, environmental problems of development arise merely from the increase in the number of people. Some authorities reject this statement, saying that the average population density in sub-Saharan Africa is low. This may well be true, but many parts of the continent are forbidding and cannot be used for living at all (e.g., rocky plateaux, swamps). In other inhabited parts the carrying capacity of the environment already seems to have been reached, at least with the agricultural techniques available. In the flood plain, conflict between pastoralists and farmers is frequent, showing the saturation of all suitable land (north Central African Republic, north Cameroon). In the steppe zone the pressure on the land has already led to degradation (south Tunisia).

3.2.4 What Are the Risks and How Can They Be Avoided? (Table 10)

Traditional land use in the drylands of Africa has been described in the preceding section. It is clear from that description that there are many positive aspects which must be kept in these traditional methods. Nevertheless, extra pressures are arising from population increase and from modern-style economies, which increase the risks. The risks and ways of avoiding them will be briefly outlined and illustrated before degradation itself is discussed.

3.2.4.1 In the Sahel

In the Sahel between the annual 150- and 600-mm isohyets, the basic objective for future environmental management must be conservation of vegetation and soil.

In the northern part of the Sahel between 150 and 250–300 mm annual rainfall:

1. A widely scattered vegetative cover on regosols and sand dunes, which varies in amount from year to year due to the erratic nature of the rainfall, offers sufficient resources for pastoralism, herding of camels and sheep.
2. Cultivation must be eliminated except where irrigation is possible: along major river valleys and flood plains, e.g., the rivers Senegal, Gambia, Niger, Volta, Logone, and Chari, some swamp lands of lake Chad, some dry valleys with a notable inferoflux as in the Dallols in Niger and in scattered depressions receiving runoff water or with a high water table, where special ecological conditions, which differ from those of the ecozone, exist.
3. Because potential wind erosion is high, integration of windbreaks and small scale reafforestation projects should be widely adopted, with agriculture in the form of shelterbelts. This type of association can be successfully achieved by the association of irrigated agriculture and shelterbelts. The southern oases of the Taklamakan desert, western China, are good examples.
4. Use of solar energy as an alternative to fuel wood might be systematically considered.

Table 9. Agricultural calendar of Twic-Dinka (Gogrial district, Sudan). (Van Noordwijk 1984)

	January	February	March	April	May	June	July	August	September	October	November	December
Rains												
Floods												
Cattle	majority of cattle in cattlecamps		returning to village		young men stay out for fattening competition				majority of cattle to cattlecamps to the N and E.			
					majority of cattle around homesteads							
	few cows in village for milk and manure								few cows in village for milk and manure			
Chicken	eggs eaten by children						breeding season (kites are away) →					
Crops	irrigated tobacco		clearing land		cultivation:		weeding		harvest:			irrigated tobacco
					← cam-dura →					storing		
	—	manuring →			← rabdit-dura →					the ears		
	threshing grains				← maize, beans →							threshing grains
					gourds →							
Wild plants	cum, lang, cuei, thau	storage	waterlily rhizomes						akuoth, grass seeds	storage		wild rice seeds
Fuel	dried durasticks →		← firewood →								← dried durasticks —	
Fishing	lungfish caught		in rivers and big pools (war, D.)						building dykes (thiik, D.)		fishing in weirs (rok, D.)	
Hunting		hunting antelopes in groups										
Building activities	grass cut; mud walls-roofpoles-thatching roofs											grass cut; poles collected
Social	family visits						competition					dances

Table 10. Meteorological data and optimum agricultural potential in Sahelian and Sudanian ecozones

Zone	Annual rainfall (mm)	Radiation ratio Cal cm^{-2} day^{-1}	Temperature (°C) Dry season	Rainy season	Length of rainy season	Soils	Potential evapotranspiration	Agriculture — optimum potential
Saharo-Sahelian	100– 250	507	30–35	20		Regosols	Very high	Grazing — nomadism
Sahelian	250– 400	500	30	20–25	62 days max: July August	Regosols	Always > to Rainfall	Grazing pastoralism. Grazing recommended with management of irrigation along the migration pathways of herds Cowpea — millet
	400– 600	490	30	25	72 days July to Sept.	Arenosols Yermosols Solonchak Fluvisols	Limited excess but possible	Half sedentary pastoralism — rainfed agriculture possible — irrigation possible with excess rainfall — cowpeas — millet — groundnut
Sahelo-Sudanian	600– 900	485	25–30	25	110 days July to Sept.	Fluvisols Vertisols Luvisols Gleysols	High	Sedentary pastoralism — rainfed diversified agriculture — northern limit of cotton — short growing period crops (sorghum) — long growing period plants (cotton) — in the higher rainfall areas
	900–1100	470	25–30	25	120 days June to beginning Oct.		Monthly mean 160 mm Excess moisture between July and Sept.	
Sudano-Guinean	1100–1500	< 470	25	25	150 days May to Oct.	Nitrosol Ferralitic-fersialitic soil — brown soil	High the maxima are more spaced time	Sedentary pastoralism — possibility of double crop with rainfed agriculture — excess rainfall of 200–600 mm 1 out of 2 years. Growing cycle exceeds a mean of 122 days, but high soil erosion potential. Rotation of legumes and fallow is recommended — rice and sugar cane — high irrigation potential can allow one crop during the dry season

Southwards the density of the vegetative cover increases until it becomes an open grass steppe, and then a wooded steppe near the 400-mm isohyet with the main soil types becoming: arenosols, yermosols, solonchaks, and fluvisols.

1. In the northern half small patches of agriculture can be envisaged in the depressions, benefiting from a local accumulation of runoff moisture. Only short-cycle crops can be recommended such as cowpeas (niébé) and millet. Substantial reduction of yields occurs in about 50% of years due to inadequate rainfall and invasive pests, e.g., locusts.
2. Degradation of soil surface by wind and water erosion, where land is overexploited by frequent cropping and denuded by animals consuming harvest residues, should be prevented.
3. In view of climatic risks, the 400-mm isohyet might be considered as the reasonable northern limit for rainfed agriculture. South of this limit, traditional agriculture is possible only during the short rainy season with its intrinsic late starts and early conclusions resulting in harvest failure because of shortening of the growing season due to failure of the rains.

   Thanks to medium sandy soils in the department of Zinder (Niger) cultivation of local short-cycle varieties of millet under rainfed conditions are possible with very low yields in dry years. When mean annual rainfall in the departments of Maradi, Niamey, and Tahoua (Niger) reaches 400–500 mm, a rapid rise of yield with rainfall of 2.1 kg ha^{-1} mm^{-1} is obtained to an optimum of 475 kg ha^{-1} at 225 mm of April to July rainfall (Davy et al. 1976).
4. Between the 400- and 600-mm isohyets cultivation might become more widespread. The growing season lasts 60 to 80–100 days, but short-cycle crops should still constitute the main rainfed crop: cowpeas, millet, groundnuts. Sahel temperatures, radiation amounts and day-lengths favor all-year growth of irrigated crops.

   "Characteristics of recommended varieties are firstly, fast growth after germination so that full ground cover is obtained to utilize solar radiation fully in the short period in which water limitations are minimal. Secondly, early flowering in late August (early September in Senegal) is necessary to permit grain development before end of season dessication sets in. Nonphotosensitive varieties are to be preferred but photosensitive varieties should be used to hasten flovering if late sowings are made even if diminished plant development results in lower yields." WMO (1978).

   However, the temperatures are often too high for wheat, except in the middle of winter. Since bread is becoming the staple cereal in the towns, the development of a heat-tolerant wheat is highly desirable particularly for sub-Saharan Africa.
5. Average rainfall over ten stations in Senegal and 34 stations in Niger and annual national yields of bulrush millet and sorghum for 23 harvests have shown an association of increasing yield of 2 kg ha^{-1} mm^{-1} rain before 31 July and 0.6 kg ha^{-1} mm^{-1} annual rainfall. It is in successful germination and establishment of the young crop that foundations for a good harvest are laid (Davy et al. 1976).

Such simple results are a particularly useful method to predict yields by the end of July. As soon as records of rainfall are obtained, an approximate estimate of the coming harvest can be made and crop failure predicted, even if rainfed agriculture is considered as having unpredictable yields.

6. Because the limiting factor is water supply, runoff retention measures and increasing absorption of the rain into the soil would be useful. Crop programming must take both temperature and water limitations into account.

   The unreliability of rainfall at the start and at the end of the rainy season can be compensated for by a beneficial supplementary irrigation.

3.2.4.2 In the Sudanian Zone

In the Sudanian zone between the 600- and 1100-mm isohyets the objective would be full utilization of water in agriculture.

1. In the northern half of the Sudanian zone, between 600 and 900 mm annual rainfall, the growing season can exceed 100 days and wooded savanna becomes dense forest on fluvisols, vertisols, luvisols, and gleysols. Longer-cycle crop varieties and wider diversification are possible: short-period sorghum (up to the northern limit of the Sudanian zone) and early varieties of maize, rice, sunflower, and in higher rainfall, cotton.
2. The potential for the increase of agricultural production of annual crops is greater here, better than further north or further south. The limitation in this ecosystem is the intensity of land use. Since most of the cultivable land in this zone is already being farmed, it is emphasized that:
   a) use of chemical fertilizers is economically productive because rainfall produces neither excessive leaching, nor limitation of the growing season;
   b) irrigation schemes are efficient because they are in a zone of moderate solar radiation and evapotranspiration;
   c) in drier years when long-period grain crops would fail, short-season crops or fodder crops can be envisaged. This suggests the provision of seeds of different varieties adapted to rainfall fluctuations, the local varieties being adapted to a mean rainfall.

3.2.4.3 The Sudano-Guinean Subzone

The Sudano-Guinean subzone of 900 to 1100 mm or more has a rainy season sufficiently long and regular to satisfy the requirements of all common tropical long-cycle crops (rice, cotton, and sugarcane). The shorter-cycle ones (sorghum, maize, groundnut, cowpea, and millet) are eliminated by the excessive rains. The disadvantages of the subzone are: (1) the heavy cloud cover of July-August-September which reduces solar radiation and photosynthesis; (2) leaching of nutrients in July and August precisely at the optimum time for plant growth. Soil fertility maintenance is a high priority. Rotations, including leguminous crops and ploughing in of crop residues, are the best solution. Chemical fertilizers must be used with parsimony; (3) excess rainfall can result in water erosion especially when heavy storms fall on saturated soil. One of the solutions is to maintain the crop cover as long as possible. Photosensitive varieties are generally to be preferred; if they are planted at the first rain of 20 mm, they will benefit from a long growing season and complete the vegetative cycle at the best time for harvesting at the beginning of the dry season. This limit of 20 mm is empirical. According to field observation, "re-sowing of seeds was rarely necessary when the first sowing coincided with rain of about 20 mm in 1 or 2 days." (Davy et al. 1976); (4) waterlogging occurs in heavy soil and requires drainage; (5) diseases and pests during the maturation phase of a crop require a careful choice of sowing dates and of resistant varieties. Double cropping can be possible with irrigation in the dry season.

3.2.5 Environmental Difficulties of Development. Disasters: What Disasters?

Large areas of the dry savanna zone have lost a part of their nutritive vegetative cover and are locally denuded or have lower yields because land is over-used. Degradation appears on patches of land which gradually unite to form a vulnerable zone.

Factors which in drylands and on the wetter side of the drylands zone are thought to be responsible for this land degradation include climatic changes, overgrazing, overcropping, deforestation, over-exploitation of water. All these factors are interacting.

According to the observations of the effect of the three rainy seasons (1986–88), the vegetation seems to recover quantitatively in the Sahel, but quantitatively does not mean qualitatively. The range specialists claim that some very nutritive plants have disappeared. The question as to the role of climatic fluctuations in desertification and of the irreversibility of degradation are still being discussed.

The deterioration of the steppe of *Rhanterium suaveolens* south of Tunisia in a semi-arid ecozone with 150 mm annual rainfall for example, is considered by Le Floch and Floret (1972) to be irreversible degradation due to overcultivation and overgrazing followed by erosion of the top soil by wind and dune reactivation (Rapp 1974). Le Floch and Floret consider the degradation irreversible because the rehabilitation of the pristine steppe seems illusory and expensive. Nevertheless, different steps of rehabilitation were successfully experimented with for the choice of species, criteria of productivity, durability of plants, resistance to drought and cattle, and efficiency for optimum water use. In addition, local varieties (*Artemisia, Plantago,* or *Argyrolobium*, and among the grasses *Aristida ciliata, Cenchrus ciliaris,* and *Digitaria nodosa*) were tested. Several other experiments were completed: resistance of the plants toward hydric stress at different steps of development, cultivation in pots, in nurseries, with or without irrigation, and studies of the plasticity of the ecotypes from different more or less continental or more or less arid origins. With the good results obtained can we still speak of irreversibility, knowing that rehabilitation is a gradual process, with different stages which can take a very long time? Is it not satisfactory to stop soil degradation and to produce a sustainable vegetative cover?

3.2.5.1 Overgrazing

A rough estimation shows that from the beginning of the century, the number of sheep and goats has increased by a factor of 5, the number of cattle and camel by a factor of 10. In the meantime the extent of the grazing areas available has not increased. With the exception of regions around the towns or around permanent water sources, there is the impression that overgrazing in environmental degradation has nevertheless been overestimated in south sub-Saharan Africa. Moreover, overgrazing leads to reduction of the carrying capacity, reduction of palatable plants, with the nutritive ones decreasing and preventing rehabilitation, as does the trampling, which reduces infiltration of rainwater. A further concentration of animals on the remaining overgrazed land inhibits the chance of vegetative recovery.

Pastoral people try to maximize the number of livestock, irrespective of quality, because they believe that disaster strikes in percentages. In the Steppe of Algeria since the beginning of the century the number of sheep has reached 10 million head. This number fell to 4 million during the droughts. Thanks to the possibility of importating concentrated fodder, 50% of which is paid for by the government, no sheep deaths are registered, and natural regulation does not occur any more. These results can look positive, but in fact they are negative, because of their drastic effect on the vegetative cover made fragile by drought. Other causes of degradation of the Steppe of Algeria are the clearing of land and plowing. Thirty to 40% of the Steppe is now cultivated, but the proper use of this land must remain pastoral (Lahouati 1989, pers. commun.).

It is certain that disaster is at a maximum in the pastoral communities. The ability of dryland environment to sustain pastoral populations has drastically decreased.

Excessive concentration of pastoralists and their herds is a constant difficulty in the rangeland around boreholes and settlements. When the traditional tribal leaders who had been strong enough to control the land and water use disappeared, the sensible balance between the size of the herd and the quality of grass and water available also disappeared. Government introduction of boreholes, available to all users regardless of pasture supply, resulted in local overgrazing because when they tried to prevent access to the boreholes in the dry season the herders refused to leave and were strong enough to prevent the government administrators from turning off the tap (e.g., south of Aïr, Niger). Degradation around boreholes resulted in loss of herds when drought occurred. The Sahel provides numerous examples of animals starving in the presence of what is still an adequate water resource, because there was, at the same time, insufficient fodder.

The settlement of nomads is another negative proposal. It does not mean abandonment of the attachment to animals but, on the contrary, the keeping of small stock in the vicinity of the settlement. These animals can cause a degraded area around the agglomerations. To settle nomads is, in fact, a mistake. Traditionally, pastoral herds grazed one area totally, then moved to another. Now they are confined by conversion of rangeland to mechanized dry farming, by the conversion of flood plains to irrigated agriculture, by the increase of wildlife reserves for purposes of tourism. The introduction of heavy vehicles into pastoral activities also has negative consequences: increasing pressure on the rangeland by moving herds quickly, and increasing herd survival under drought conditions by bringing water and fodder to them. The pastoralists are the group which has suffered most under the forces promoting change in the rural world.

3.2.5.2 Overcropping and Deforestation

In time of drought, rainfed farmers cultivate a larger area to try and maintain their total yield. They can do this because the weed growth is also correspondingly less in time of drought. This leads to the exposure of larger areas of soil to erosion. With developed farming, the temptation is not to use a greater area but to increase the use of fertilizers and pesticides.

In the semi-arid and dry subhumid ecosystems of Africa deforestation and overcropping are probably another main factor in land degradation. Too much land is cleared of the natural vegetation mainly in areas where a good crop is only possible in an exceptionally good rainy year; in a dry or normal year the fields are open to wind or water erosion. Large-scale mechanized farming schemes are far worse than small traditional fields amidst bush because all the vegetation is cut, leaving no obstruction to wind erosion. Overcropping is mainly the result of mechanized farming methods not adapted to the environment and of using land which is not really suitable for growing crops because the rainfall is unreliable. To lack of water, overcropping adds lack of suitable soil. When degraded, the soils have lost their nutrient supply.

All types of vegetative cover can be removed to create cropfields but also to supply fuel wood and charcoal for cooking. Around the urban areas rings of fuel wood and charcoal collection can exceed 100 km (120 km around Bamako, 200 km around Khartoum) (see also p. 111).

The traditional way of life has been upset by (1) modern medicine, reducing the death rates of both people and their animals and so increasing the population density of both, and (2) replacement of the subsistence farming system by modern monoculture. In Senegal, the groundnuts basin demonstrated the damaging effects of such specialization on one crop as did cotton in the Sudan.

Traditional farming systems were used to overcome the problems of fluctuating rainfall, fragile soil, pests, and diseases. Abandonment of subsistence farming for a concentration on one crop brings multiple dangers: the loss of a healthy diet and local fulfillment of basic needs, with the increase of risks, with unforeseeable climatic fluctuations, with a vulnerable strategy in face of fluctuating international prices, and environmental degradation because of mediocre and discontinuous soil protection.

3.2.6 Conclusion

All these areas are prone to considerable fluctuations of rainfall. Problems arise, because in a period of good rainfall people flood in to occupy the land. When the rainfall deteriorates there are no mechanisms to painlessly reduce the population to a level which the land can then support. The advice from the experts to the government must be: "**You must create a system of reserves which can alleviate the situation when the drought occurs.**" If the fluctuation persists over a long period (e.g., 15 years) neither people nor government seem to be able to work with that kind of long-term objective. But when the situation becomes better again, the capacity for recuperation of the environment in fact exists.

We have seen that Africa has experienced many droughts throughout its history; but drought does not automatically lead to land degradation and famine. The role of drought continues to be questioned.

From the different definitions of drought I retain two meanings: (1) a period of abnormally low rainfall for a particular agricultural purpose and (2) its practical meaning as an accelerator of land degradation and famine-inducer for a population in the main involved in subsistence agriculture. Drought is not a mechanism of desertification, it is a cause of reduction of water supplies or of the decline of its effectiveness in an agricultural economy which has already lost its balance because of over-exploitation. Harrison (1987) considers prolonged drought as a factor of intensification in a general trend of economic mismanagement. In this way drought affects society's development.

Climatic variability (such as drought) in time and space, e.g., rainfall deficit or excessive precipitation, has devastating impacts and more dramatic effects on the developing countries and their development. Weather changes, especially drought, as well as having environmental impacts (land quality, dry watering points, widespread soil erosion, soil compaction, salinization, reduced forage for livestock, reduced firewood, the withering and devastion of crops) also influence, in various degrees, all socio-economic parameters (acreage grazed and planted, food export and import, food balance, farm grain storage, consumption, balance of payment, debt repayment, increased prices, degradation of urban areas, communication axes, migration rates, labor supply, etc.).

Climatic variations result in wide annual crop yield variations. From the beginning of the drought to 1978 in Senegal:

"yields of millet and sorghum, major components of local diet, have varied from a low of 170 kg ha^{-1} in the Diourbel and Theis Departments to a high of 1000 kg ha^{-1} in the Casamance. In Niger, national average yields of millet have ranged from 297 kg ha^{-1} with 286 mm average total rainfall in 1973 to 523 kg ha^{-1} with 520 mm rainfall in 1967. Such wide variations in national cereal production are quite unknown in more climatically favored countries". (WMO 1978).

Another example can be described in sub-Saharan Africa. During the 1950's and early 1960's more favorable (wetter) conditions led governments and farmers to expand northwards and to cultivate seldom-used rangelands and to compete with the pastoralists for their wet-season pastures. The former rangelands were in the long run unfavorable for sustainable agriculture. In the late 1960's, with the onset of the multiyear drought, the impacts of inappropriate land use practices became obvious. Accelerated degradation of vegetative cover and competition for space between herders and farmers occurred. This competition still remains active: at the end of April 1989 disputes broke out in Mauritania between Mauritanian and Senegalese, 200 people were killed and 27 000 asked for repatriation in both countries.

Drought is a recurrent feature of African climates. Similar droughts have occurred in the past, but no available analysis suggests that the phenomenon reappears within a well-defined period. So droughts must be considered as a part of life in the drylands and future planning should assume that similar low rainfall levels will periodically occur.

3.3 Egypt: An Example of the Misapplication of High Technology Leading to Man-Made "Desertification"

Egypt is the country of dry Africa where land degradation has reached the highest degree. Analysis of causes, processes, and effects will reveal its magnitude.

3.3.1 Causes of Land Degradation

1. Small surface of cultivable land. Egypt has 100 million ha of which the only 3 million which are cultivable are virtually all concentrated in the Nile Valley and Delta.
2. Arid climate, which forces Egyptian agriculture to be dependent on irrigation. The average rainfall ranges from practically zero in Upper Egypt to around 25 mm at Cairo and 190 mm in Alexandria.
3. Rapid population growth. Around 1 million people every 10 months. At 1900 the land/man ratio was 0.20 ha/man; by 1983 it had decreased to about 0.06 ha/man (Kishk 1984).
4. Small farms. More than 80% of the farms have less than 1 ha (Kishk 1984). Such a small surface leads to overuse with permanent irrigation, especially since the building of the Aswan High Dam.
5. Agriculture is the main source of livelihood for more than 50% of the population and the main source of foreign exchange necessary to finance development.
6. Use of marginal land. Expansion of agriculture on new poor and vulnerable land, mainly sandy or calcareous substrata.
7. Uncontrolled expansion of urban and rural settlements at the cost of cultivable land. A comparison of Landsat images of 1972 and 1978 indicates an increase of 12 140 ha per year (Kishk 1984).
8. Exploitation of clayey soils for industrial production of bricks and building material. These soils are mainly fertile alluvial soils of the Nile plain. The layer of 1 m thick is extracted from the top soil. Under this layer is often a

salty sterile horizon which is then exposed as a desert-like bare surface. The use of alluvium for brick-making is now forbidden.

9. Non-existence of conservation measures of the land. For the poor Egyptian farmers, land conservation measures are a luxury they cannot afford.

3.3.2 Processes of Land Degradation

3.3.2.1 Water Table Rising

Because of their location in the Nile plain, the good alluvial soils can be easily irrigated. Irrigation in Egypt began several thousand years before the present day. The water table has risen mainly since the building of the Aswan dam to within a few feet of the soil surface. Over-irrigation, perennial irrigation, without rest for the soil, abandonment of the fallow system, and inadequate drainage have brought three mechanisms of degradation well known in many irrigation schemes: waterlogging, salinization, and alkalinization. Kishk (1984) considers that almost all irrigated land in Egypt is salt-affected and at least 1 350 000 ha (one-half) is dangerously affected with a loss of production of 20%.

3.3.2.2 Absence of Nile Alluvium Resulting in Wind Erosion

In the original Aswan dam, the water flows out through sluices at the bottom of the dam. This scoured out the silt and caused it to pass down the Nile to re-vitalize the irrigated lands of the Delta and its associated fisheries. In the High Dam at Aswan, the Nile flows over the top, dropping its load of silt from the eroded mountains of Ethiopia behind the dam (which will in time silt up) instead of being, every year, during the natural flood, spread out on sandy alluvium. This annual protection has disappeared and the fertile cultivated land is now eroded by deflation and aeolian winnowing, which export the organic matter and the fine particles of the top soil. Wind erosion is the process which affects the largest surface in Egypt.

3.3.2.3 Loss of Nutrients

Increase of leaching results in the removal of a great amount of nutrients from the soil, without any systematic replacement. The absence of Nile alluvium brought from upstream since the completion of the dam is another cause of decrease of nutrients.

3.3.3 Effects of Land Degradation

Soil and water salinization, soil pollution, and loss of soil fertility are the main consequences of land degradation.

Soil pollution. Pesticides have been intensively used on a high percentage of fields, mainly cotton, about 536 000 ha (Kishk 1984).

Loss of soil fertility. Kishk (1984) made the assumption that 10% of the production in Egypt was lost every year because of loss of soil fertility.

In the delta, as a result of water overuse, marine water incursion is very threatening. The consequences are not quantitatively known to date.

3.3.4 Conclusion

The case study of Egypt can be considered as the opposite of the case study of Australia which will be analyzed later. In Australia there is a large land surface with little human population but overstocked with cattle and sheep; in the case of Egypt there is a small cultivable land surface and consequent overuse of the soil by a population engaged in an endemic demographic explosion, which triggers off land degradation and "desertification". This has not been helped by the cutting off of the Gift of The Nile, i.e., alluvium, by the high dam.

"Desertification is also seen at the Siwa oasis in Egypt. Many of the peripheral villages of the oasis have been abandoned and the inhabitants have moved to the main centers or have left the area completely. The orchards and palm groves were left unattended and eventually the trees died. The main reason for the desertification has been the high salinity of the groundwater, which has caused a rapid rise in soil salinity." (Beaumont 1986).

At the beginning of the 1980's Ghonaim and Gabriel (1980) suggested two main causes of salinization and degradation:

1. The departure of many of the able-bodied men to the Libyan oil fields; thus labor became scarce, the drainage networks were not maintained, and salinity built up.
2. Increase in animal husbandry to supply the meat needs of Libya with the result: overgrazing and depletion of pastures and fodder resources.

3.4 China: A Sophisticated Traditional Intensive Agriculture Beginning to Be Modified by High Technological Methods and "Desertification" or Land Degradation

3.4.1 Introduction: The Agricultural Origin of Land Degradation in China

After the former USSR and Canada, China is the third largest country in the world with about 9.6 million km^2 (6.5% of the land area of the world).

"China has now about 100 million ha of cropland which has been developed for 7000 years" . . ."most of the arable land in China has now been cultivated" . . ."practically no more virgin land and vegetation still exist and mankind has made its imprint nearly everywhere." (Zhao Songqiao 1986).

The Chinese agricultural history of more than 70 centuries and the engagement in agriculture of more than four-fifths of the population explain how agriculture, the major human activity, has exerted the greatest impact, both positive and negative, on the geographical and physical environment.

Zhu Zhenda and Liu Shu (1983), summarizing the distribution of degraded land (which they call desertified, meaning wind-eroded land only), give the most comprehensive assessment.

"Most of the regions in Northwest China, North of Northern China, and West of Northeast China belong to desertified lands. . . . It is estimated that the desertified lands formed in the historical periods occupy about 120 000 km^2. . . . The increase in desertified lands in approximately the last 50 years covers about 50 000 km^2, the areas of the two items amounts to 170 000 km^2, plus 158 000 km^2 of latent desertified land, the total areas of desertified lands in Northern China is increased to 328 000 km^2 (32.8 million ha), which occupies 3.4% of the total land area in China (i.e., 24.7% of the total area of sand desert, gravel desert, and desertified land in China, and which covers 10.3% of the land area in northern China). It involves 35 million people in 207 cities, provinces, and districts in the following provinces and autonomous regions: Xinjiang Uighur Autonomous Region, Inner Mongolia, Ningxia Hui Autonomous Region, and Qinghai, Gansu, Shaanxi Shanxi, Heilongjiang, Liaoning, Jilin, Hebei, and Henan Provinces." (Table 11).

Later, in 1988, the same authors note that

"the desertified land in north China is mainly distributed over the western part of the Northeastern Plain, and the vast area to the east of the Tarim Basin, to the north of the Great Wall and the Kunlung Mountain ranges and to the south of the Sino-Mongolian and Sino-Soviet borderline. It is also scattered over eastern Henan and the middle and southern part of the Hebei Plain, the area being relatively small." (Table 12).

There is a good correlation between arid, semi-arid, and dry subhumid ecozones and land degradation. The Chinese link deserts and land degradation as shown by their statistics when they speak of "prevention and control of deserts and desertification in China".

"The total area of deserts and desertified lands amounts to 1 095 000 km^2, about 11.4% of the total area of China, of which 49.8% are sandy deserts (including sand dunes and wind erosion lands), 41% gravel and rock deserts, and 9.2% desertified lands." (Fig. 12).

Chinese farmers, like all the other land users of the planet, have made mistakes in using natural resources and land. The mistakes usually come from the absence of alternatives, the necessity for the prolonged utilization of the soil, food shortage, and general socio-economical difficulties.

Drought and land degradation have been recognized by the Chinese as an old problem for at least the last two millenia (Fig. 13). The destructive process of environmental degradation is mainly contemporaneous with disturbances and upheavals aggravated by droughts (Zhu Zhenda and Liu Shu 1983).

Table 11. Surface (km^2) of sandy deserts and gobis in northern China. (After Zhu Zhenda et al. 1986)

Province (region)	Total area (km^2)	Sand deserts including dunes and eroded land	Gobi area
Xinjiang	713 000	420 000	293 000
Gansu	68 000	19 000	49 000
Qinghai	75 000	38 000	37 000
Inner Mongolia	401 000	213 000	188 000
Ningxia	6500	4000	2500
Jilin	3600	3600	0
Liaoning	1700	1700	0
Shaanxi	11 000	11 000	0
Heilongjian	2600	2600	0
Total	1 282 400[a]	712 900	569 500

[a]The area of newly desertified lands is not included in the above figure.

Table 12. Surface of the major deserts in China. (Zhu Zhenda et al. 1986)

Desert	Area	Desert	Area
Taklimakan Desert	337 600	Horqin Sandy Land	42 300
Gurban Tunggut Desert	48 800	Otindag Sandy Land	21 400
Badain Jaran Desert	44 300	Kumtag Desert	19 500
Tengger Desert	42 700	Ulan Buh Desert	9900
Mu Us Sandy Land	32 100	Sand Desert and Yardangs Qaidam Basi	34 900

Figure 13 reveals the same variability and absence of cycles on a bi-millenial scale as does the shorter African record.

The periods of drought are also those during which the winds are more frequent and stronger. This point is confirmed by our observations in Niger. For the 12 synoptic stations of Niger, the wind roses drawn for a humid decade, compared with the wind roses of the drought years 1968–1973, show a systematic increase in frequency and speed of wind during the dry years.

The deserts of China belong to the temperate zone desert type. They are distributed in continental intermountain basins and in continental plateauz. A distinction between the western areas west of 106 °E, which is the arid desert zone, and the eastern semi-arid and subhumid steppes is made systematically in the Chinese geographical literature.

West of 106 °E the deserts are about 90% of the total area of Chinese deserts (Meckelein 1986), with sandy desert Shamo and gravel and rock desert Gobi. The aridity index is around 0.03–0.20 and falls below 0.02 west of the Tarim Basin. Vegetative cover is sparse, shifting dunes predominate of 75% of the total area of Shamo (Fig. 14).

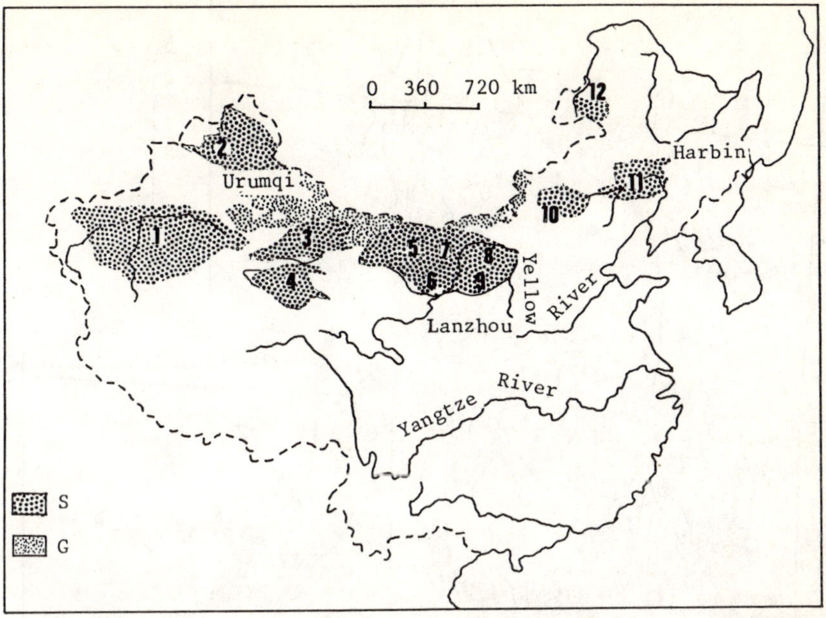

Fig. 12. Distribution of the drylands in China. (Zhu Zhenda et al. 1986). *1* Taklimakan Desert; *2* Gurban Tonggut Desert; *3* Kumtag Desert; *4* the deserts in Qaidam Basin; *5* Badain Jaran Desert; *6* Tengger Desert; *7* Ulan Buh Desert; *8* Qubqi Desert; *9* Mu Us Sandy Land; *10* Otindag Sandy Land; *11* Horqin Sandy Land; *12* Hulun Buir Sandy Land. *S* Shamo (sandy desert); *G* Gobi (Stony desert)

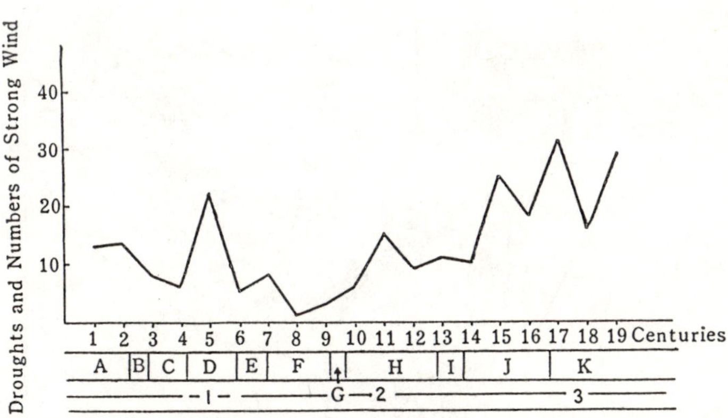

Fig. 13. Diagram of the periods of drought and stronger winds and their historical occurrence. (Zhu Zhenda and Liu Shu 1983). *A* The Eastern Han Dynasty; *B* the Three Kingdoms; *C* the Jin Dynasty; *D* the Northern and Southern Dynasties; *E* the Sui Dynasty; *F* the Tang Dynasty; *G* the Five Dynasties; *H* the Song Dynasty; *I* the Yuan Dynasty; *J* the Ming Dynasty; *K* the Qing Dynasty. *1,2,3* periods of desertification

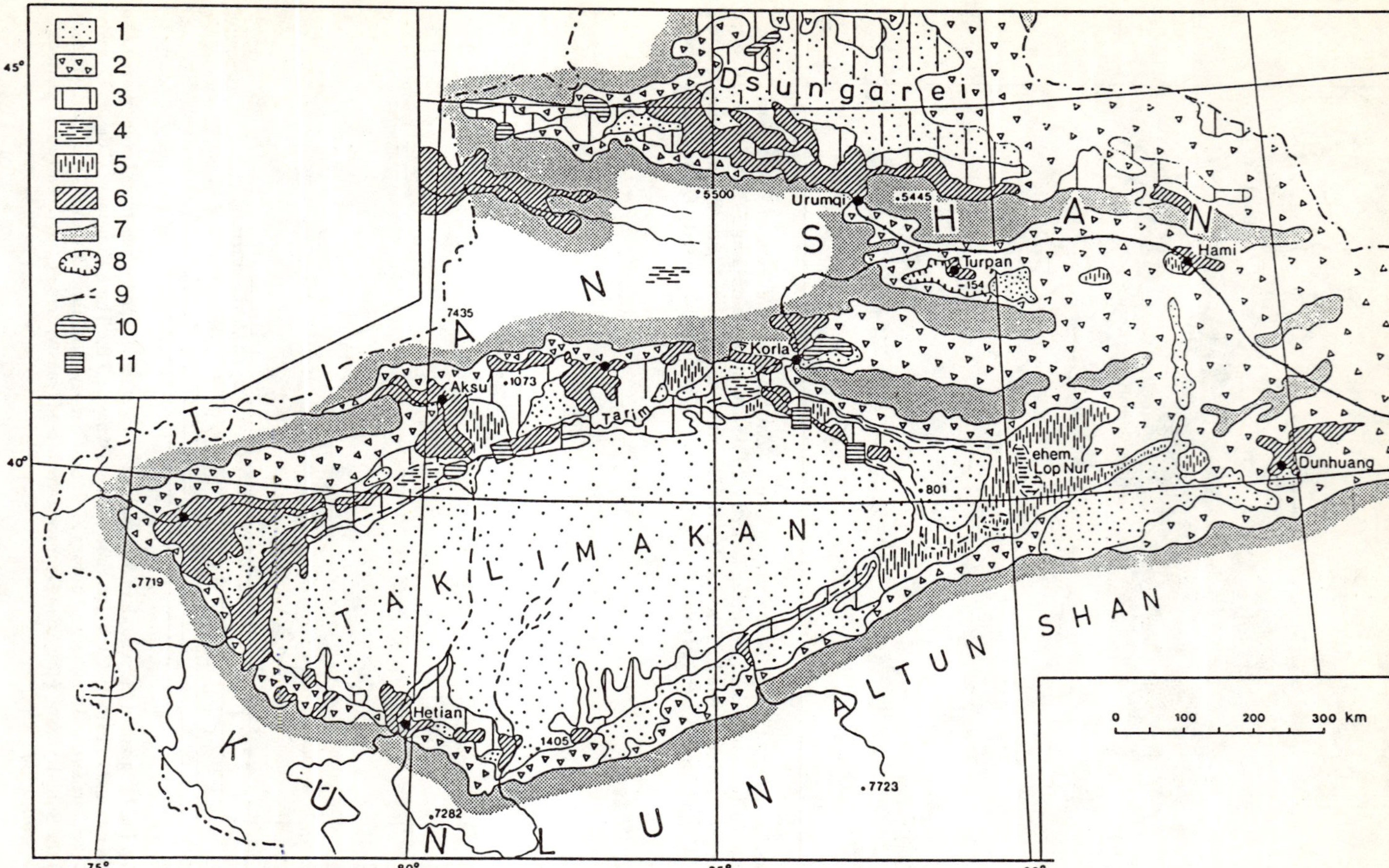

Fig. 14. Sinkiang Desert. (After Mecklein 1986). *1* Shamo (Sandy Desert); *2* Gobi (Stony Desert); *3* sand sheet with steppic vegetation; *4* flooded area; *5* saline area; *6* oasis; *7* mountain foot; *8* depression; *9* river; *10* lake; *11* water reservoir

The active dunes of the Taklamakan (or Taklimakan) reach a height of 100 to 150 m, and in the Badain Jaran desert 200 to 300 m. Fixed and semi-fixed dunes are characteristic of the Junggar Basin and around other lake basins in arid zones and along the margins of alluvial cones. Some rivers which flow deep in the desert are fringed by populars (*Populus euphratica*) and tamarisks.

To the east of 106 °E, the semi-arid and subhumid steppes receive an annual rainfall of about 200 to 250 mm. The aridity index is about 0.5–0.75. Grasses, herbs, shrubs, and trees grow there. Fixed and semi-fixed sand dune areas and sand plains are also the predominant degraded areas.

3.4.2 Classification of the Desertified Areas in China

From west to east the desertified or degraded areas extend for 50° in latitude. This dispersion explains why they differ so obviously and why it is so difficult to summarize their differences.

Zhu Zhenda and Liu Shu (1988) have proposed a classification of desertified land according to its geographical location and origin.

Three types of desertified lands are distinguished:

— Degradation which appears in deserts and semi-arid steppes (transition between true semi-arid and desert). The steppes which are distributed in the eastern part of Wushao and of the Helan Mountain Ranges mainly in Inner Mongolia, in the oases and the alluvial plains along the southern part of the Sino-Mongolian border, in the northern part of the Great Wall and in the eastern part of the Sino-Soviet border belong to this type.

The degradation of these vegetated areas has human causes: over-cultivation, overgrazing and excessive firewood collection. Human-influenced factors play the decisive role through the destruction of the balance of the ecosystem.

The processes of degradation are physical, due mainly to wind erosion, which damages the soils and is responsible for shifting sand and deterioration of the sandy topography with the genesis of active dunes.

1. The degraded land appears in the oasis areas around and in the arid ecosystems and along the desert alluvial plains. The oases and the alluvial plains of the southern Taklamakan desert at the foot of the Kunlung Mountains Range are examples.
2. Degradation appears in subhumid zones. Sandy soils are the most affected. The degraded land has generally a strip-like distribution because it follows the ancient courses of dry rivers. (Fig. 15). The surface soil is composed of sandy deposits either from floods or from ancient river courses, as in the eastern part of the Henan Province and the southern part of the Central Hebei plain. When, during the dry season, this sandy material is denuded and wind force is at a maximum, wind erosion can exert very severe degradation.

Ninety percent of the degraded land in China is distributed in the semi-arid and subhumid steppe areas.

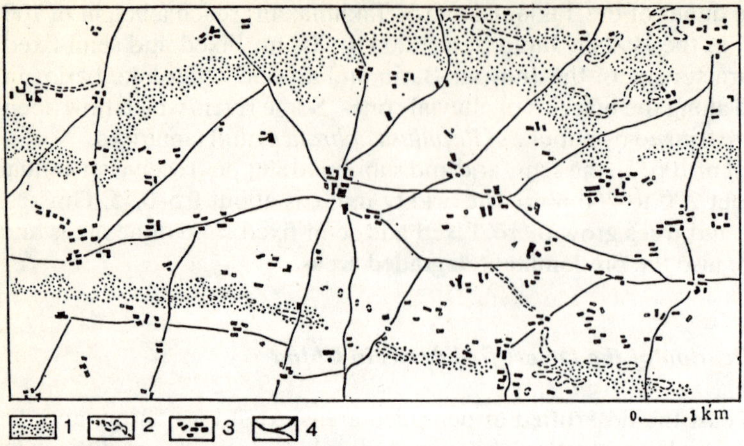

Fig. 15. Lace-like desertified lands along the ancient river courses in the Plain of Eastern Henan Province. (Zhu Zhenda and Liu Shu 1983). *1* Degraded land; *2* ruins along ancient river beds; *3* actual settlements; *4* road or path

3.4.2.1 Deserted Land and Desertified Land in the Arid Ecosystems

1. *The Oases of Taklamakan.* In arid ecosystems, sedentary settlement and permanent presence are only possible when based on irrigation. The exploitation of these areas for agricultural development is divided into two stages:

a) management of natural oases during the historical stages by use of available surface water and a shallow underground water level,
b) the modern stage based on boreholes from deep aquifers, unfortunately often without the guarantee of water renewal.

In the title of this section the word "deserted" has been used with the meaning of abandoned by people. A five-step scheme describes the evolution from development to desertion of the oases when the balance between human activities and natural resources is disrupted:

1. Existence in arid ecosystem of natural oases along a valley seasonally flooded by allochtonous rivers.
2. Installation of human settlements based on agricultural development with the help of irrigation from shallow underground water level.
3. Increasing scarcity of water because of precipitation changes or overuse.
4. People can no longer survive and abandon the site. Cultivated vegetative cover disappears.
5. On the exposed, loose land surface, aeolian mechanisms become predominant. A desert-like landscape replaces the natural oasis.

In this case we prefer to suggest the word deserted, which we consider better than desertified. Management of these oases is indeed a human gain in a vulnerable environment. The least lack of attention leads to reconquest by the desert.

The oases, probably the major human imprint on deserts, reached their highest development in China mainly along the margins of the Taklamakan desert, along the ancient silk roads. The density of the ancient, old, and new oases in the Tarim Basin, represented on the map (Fig. 16) by Zhao Songqiao (1986) demonstrates that the phenomenon oasis is a deep-rooted oecumene in Chinese history. Modern oases are confined to the fringe of the desert, whereas numerous ancient oases, almost 150 km from the fringe of the desert, are now buried under the sand.

Modern oases, developed before 1949, continue to exist as agricultural areas mainly located along the Tarim River and its tributaries north of the Taklamakan. In the Tarim plain, since 1955, more than 140 000 ha of farmland of oasis type have been reclaimed (Zhao Songqiao 1986), thanks to sophisticated irrigation systems but with insufficient control of irrigation water, salinization, and shifting sand.

In this oasis type of desertified land, the areas of deterioration are patchy, along ancient water systems, centered around the oasis. They are independent units which are not linked together in the landscape.

Zhu Zhenda and Liu Shu (1983) enumerate the different oases which have disappeared in the southern Tarim Basin:

"Loulan at the lower reaches of the Kongque River, Jingjue at the lower reaches of the Niya River, Andir at lower reach of the Andir River, Kelatun at lower reach of the Keriya River, Wusu Taji at the lower reach of the Qira river and Damu Canal, Akesibili, Rewak, and Tetelkram at the lower reaches of Lao Damu Canal and Yulong Kashi River, ancient Pishan at the lower reach of the Hueschi Tag River. . . . . the desertified lands in the northern Tarim Basin, cultivated land areas in the Han (Han and Easter Han Dynasty, 206 B.C.-A.D. 220) and the Tang Dynasties (Tang Dynasty A.D. 618–907) also centered respectively on the areas at the lower reaches of the Dina River in Luntai and of the Weigan River in Kuqa, Xinhe, and Xayar".

The oases located at the southern piedmont of the Tienshan Mt. Range in the northern Tarim Basin (Kuga, Xinhe and Xayar for example) were also created during the Han and Tang dynasties at the fan-fringe plains where the underground water level was at its highest. Under the influence of long-term irrational irrigation, salinization of soils and waterlogging appeared. The cultivated land was abandoned and aeolian movement of the sand in the alluvial plains was initiated. Desert-like landscapes appeared, extended by sand encroachment.

These oases are all located in arid and hyper-arid desert zones. Their disappearance is linked with river course changes and scarce water supply.

When the oases are located in deep sand deposits, and when water becomes scarcer and degradation of the irrigation system occurs, the density of the vegetative cover decreases. Aeolian action amplifies the processes of degradation, creating in a monodirectional wind regime of moving dunes: barchans and barchanic dunes, transverse ridges and along the river banks and on the dry delta

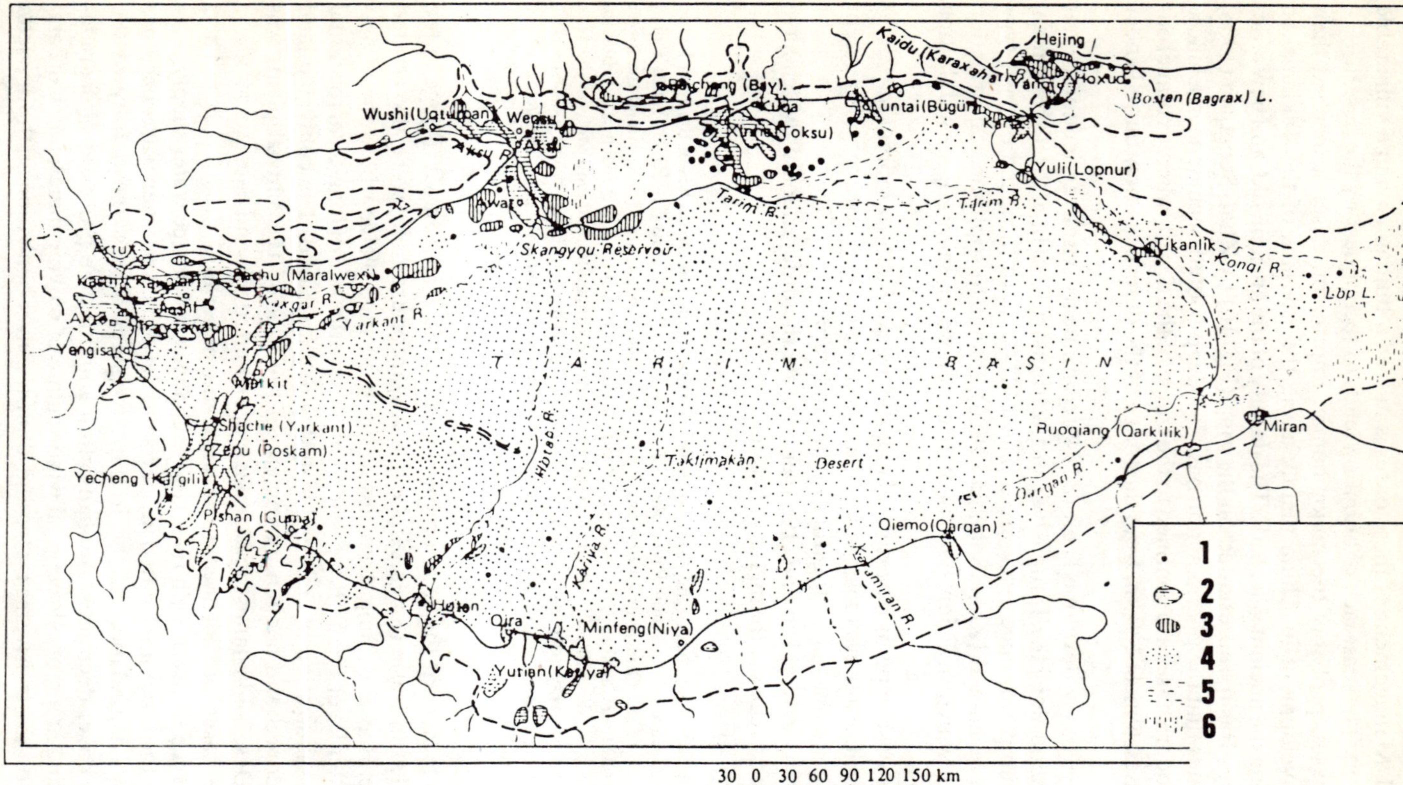

Fig. 16. Ancient, disappeared, old, and new oases in Taklimakan. (Zhao Songqiao 1986). **1** Ancient, disappeared oases; **2** old oases; **3** new oases; **4** sandy desert; **5** marshland; **6** salty marsh

abraded shrub-sand mounds (Oases of Loulan, Kelatun, Jingjue, created during the Han dynasty, are examples).

When the soils are more clayey, lower dunes appear, the same sand mounds but also corasion (corasion results from the scouring effect of saltating sand) landscapes with aeolian mounds, yardang-like or rebdou[8]-like (Table 13).

Table 13 shows how in only two decades the sandy landscapes of the lower reach of the Tarim river suffered deteriorating changes from wind erosion.

On a sketch map of the distribution in southern Taklamakan of ancient towns (Shanshan, Qiemo, and Jingjue), prosperous in the Han dynasty and now covered by sand dunes (Zhu Zhenda and Liu Shu 1983), it can be seen how the contemporary cities have developed in a fringe which is 30 km further south and nearer to the foot of the Kunlung Mountain than the ancient ones (Fig. 17). This new fringe of oases leads to the interpretation of desert encroachment: 30 km in 2000 years, which is an average of 15 m per year. Can this sand progression and oasis migration be classified as desertification? I am inclined to answer "no".

In fact, the oases are created on marginal land where silty clayey sandy soils and water are available, but where the permanent deposit of sandy material and

Table 13. Deterioration of the surface morphology in the plains of the lower reach of the Tarim river. (After Zhu Zhenda and Liu Shu 1983)

Year	Percentage of the areas occupied by each surface morphology (%)		
	Shifting dunes	Shrub sand mounds with lower shifting dunes	Sand-blown ridges shrub sand mounds
1958	12.5	40.4	47.1
1978	20.2	43.4	36.4

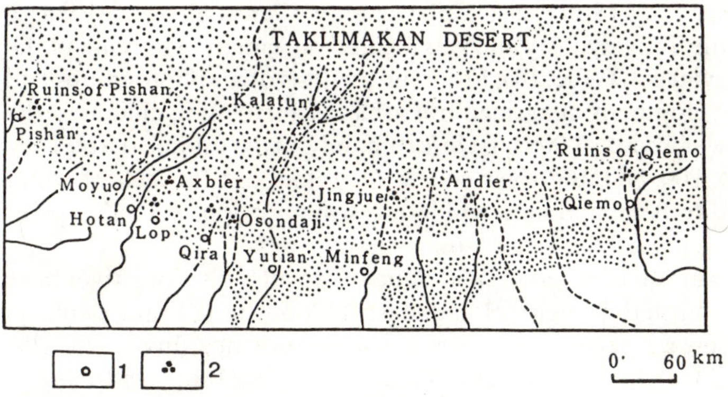

Fig. 17. Distribution of ancient towns in southern Taklimakan desert. (After Zhu Zenda and Liu Shu 1983). *1* Contemporary towns; *2* ruins of ancient towns

[8]Rebdou is a sand edifice fixed by a vegetal.

the increase of sand-sheet can only be avoided with difficulty by human settlements. The sand is alluvial material constantly brought by the rivers originating from the glaciers of surrounding mountains or from periglacial stages, for example the Yarkang River and the Hotan River from the Kunlung Mountains. Even the Tarim River is mainly fed by the Aksu River, which originates in the Tianshan Mountains. The rivers have a high discharge with a high solid load of coarse material produced mainly at periglacial altitudes. In the case of all these inland basins encircled by lofty mountains the supply of sand is permanent and inexorable.

During dry periods the water supply from the rivers decreases, the sources dry up, and the lower reaches of the rivers migrate upstream. The rivers, instead of flowing far into the desert, stop in their course nearer and nearer the piedmont, obliging the oases to move upstream. During these dry spells the vegetation is looser, the winds are exacerbated, and wind erosion easily reworks the alluvial deposits.

Oases are similar to obstacles placed by man on wind passes. They are obstacles to the free circulation of sand and responsible for sand accumulation against tree plantations, buildings, and walls. The human settlements require permanent management and more and more attention as the oasian space increases with the population increase. This conflict between land use and natural conditions (latent natural factors resulting in desertification) can be called desertification when natural conditions triumph over human ability to protect their achievements.

2. *The Oases of the Hexi Corridor and Other Cities in Subhumid Steppes.* A second set of oases are those of the Hexi Corridor (Gansu Province) and Western Alxan League (Inner Mongolia). The Hexi Corridor runs along the Northern foot of the Qilian Mountains rising to 5000 m. This location means permanent snow and water supply for the foot plains. The corridor has a hyperarid to arid climate with a rainfall of 50 to 100 mm. Thank to the water coming from the mountains during the Han dynasty were oases created. During this dynasty, the Huns, who were grazing there, were pushed back towards the north of the desert.

The town of Sueyang (Southeast An'xi county, Gansu Province) was a flourishing oasis in the Tang dynasty and was located on a main river bank of the Changma River alluvial cone (Fig. 18). From the analysis of air photographs and according to the development of agricultural areas and irrigation channels, the surface of the oasis was about 33 000 ha. Unfortunately, in the middle of the Tang Dynasty, about A.D. 700–800, the river course migrated and for lack of water supply the oasis had to be abandoned and left to wind erosion. A landscape of shrub-sand mounds replaced the rich oasis.

Other examples of ruins buried in sand dunes were all built along river beds which dried up or migrated: Longle (Han dynasty), Shouchang (Tang dynasty) both in Gansu Province, Luetue Town from the same province, Juyan (Fig. 19) and Heicheng (Fig. 20) at the lower reach of the Ruoshui River (Inner Mongolia) were cultivated between the Xia and the Eastern Han dynasty (2205 B.C. to A.D. 220).

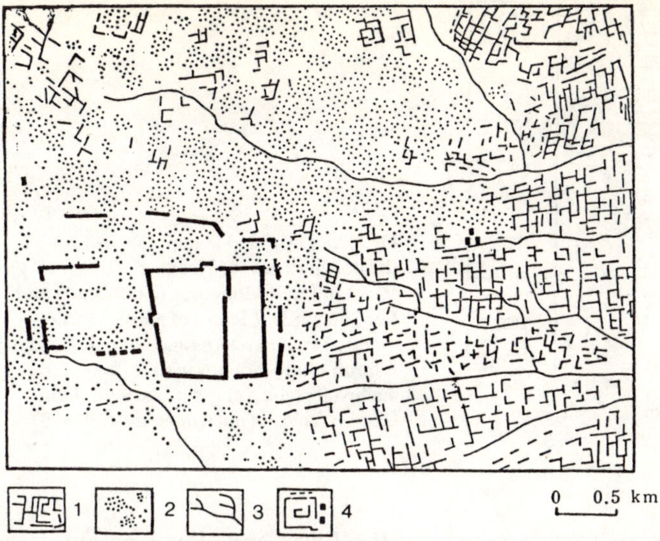

Fig. 18. The ruins of Sueyang. Tillage areas and irrigation canals. (Zhu Zhenda and Liu Shu 1983). *1* Ruins of eroded fields; *2* shrub and shrub sandy mounds; *3* traces of ancient canals; *4* abandoned city walls

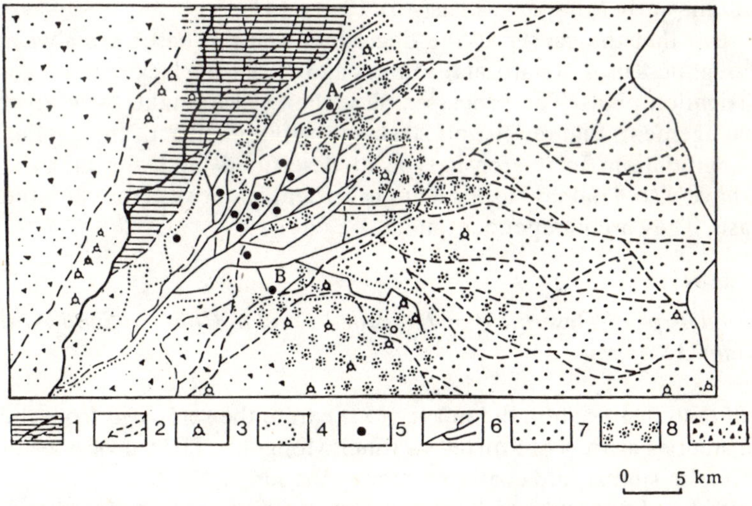

Fig. 19. Degradation of cultivated land around urban centers of Juyan and Heicheng (Black City) (Inner Mongolia). The subhumid steppe is scattered with shifting dunes and shrub sandy mounds resulting from deteriorating wind action. (Zhu Zenda and Liu Shu 1983). *1* Water systems and delta in the eastern tributary of contemporary Ejin river; *2* dried riverbed of the eastern tributary in the lower reach of Ruoshui river; *3* ruins of Beacon fire tower; *4* ancient cultivated land; *5* ruins of towns and settlements; *6* ancient canals; *7* shifting sand dunes; *8* shrub sandy mounds; *9* alluvial plain. *A* Ancient Juyan; *B* ancient Heicheng (Black City)

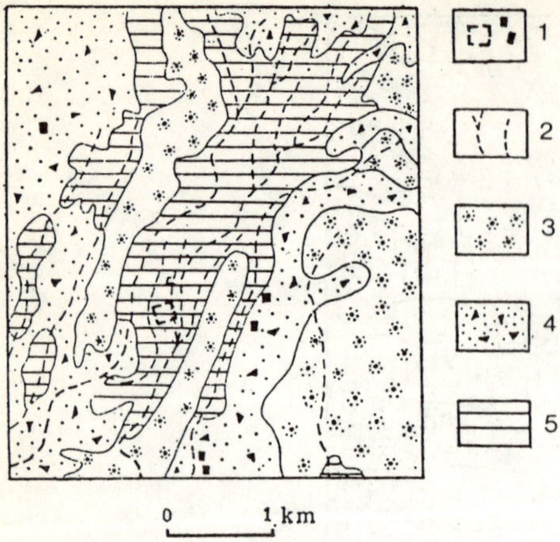

Fig. 20. Degraded land in the arid zone in the northern part of Heicheng (Black City), Western Inner Mongolia. (Zhu Zenda and Liu Shu 1983). *1* Ruins of ancient settlements; *2* water systems and irrigation canals; *3* shrub sandy mounds with shifting dunes; *4* alluvial plain; *5* clay surface with slight erosion (ruins of ancient cultivated land)

The analysis of aerial photographies of the Juyan and Heicheng old oases reveals that the human settlements were located on an alluvial cone depending on the Ejin River, previously called Jo-Shui. The assessment of the old agricultural system shows that the occupied surface was about 30 000 ha.

Heicheng was the capital of the western Xia and Yuan dynasties. The abandonment of Juyan and Heicheng was caused by the degradation of the irrigation system because of war and turmoils in the end of the Yuan (A.D. 1279–1368) and the early Ming dynasties (A.D. 1368–1644).

We have seen that the causes of the degradation of the oases are always linked with thoughtless uses of natural resources and increasing lack of management of the irrigation plans. The processes of aeolian erosion take over when there is mismanagement. Figures 19 and 20 show the dominance of the aeolian processes of degradation. In this oasis type of desertified land, the areas of deterioration have also a patchy disposal along ancient water systems centered around the oasis. They are independent units.

3.4.2.2 A Second Type of Desertified or Degraded Land in China in Semi-Arid and Dry Subhumid Steppes

The analysis of land degradation did not try to separate the semi-arid from the dry subhumid steppes as they are mixed in Inner Mongolia. Their processes of degradation are also similar and erosion is mainly the same.

The semi-arid and dry subhumid steppes, east of 106 °E, are dispersed over the eastern area of Wushasling and Helan Mountain Ranges and to the north of the Great Wall as far as the Sino-Mongolian border. The deteriorated land in the subhumid steppes corresponds mainly to sandy deposits from river floods, along the ancient river course. The old course of the Huanghe River in the eastern Henan plain is the most famous example.

Inner Mongolia, called by the Chinese Nei Mongol, has a surface area of 710 000 km^2 and is the northernmost part of China. It is mainly composed of two plateaux of 1000 and 1500 m: the Nei Mongol plateau and the Ordos plateau, with large river valleys, mainly the Huanghe and Xi Liaohe. Nei Mongol has been the habitat of pastoral people since neolithic times. Along the plateaux borders and in river plains, agriculture has developed in the past 100 to 200 years. A broad mixed dry agro-pastoral belt and an intensive irrigated agricultural belt have been created. Recently, numerous industrial and mining plants have been established.

From north to south the climate changes from cool-temperate to temperate and warm-temperate, and from east to west shifts from subhumid to semi-arid and arid. The climate shows a great contrast between long and severe winters from November to March (in January the mean temperature is –28 °C in the north, –09 °C in the south and minimum –50 °C) and mild summers from middle June to middle or late August (with a mean July temperate of 18 °C in the north and 23 °C in the south). The annual rainfall varies from 150 to 200 mm in the west to 400-450 mm in the east. Spring and winter are dry with strong winds.

The area is naturally a grassland. The temperate grassy steppe passes into a temperate desert steppe with drought-tolerant vegetation on the western Nei Mongol and Ordos plateaux. Frequent droughts do not favor dry farming, and pastoralism stays dominant. As there are sometimes 12 or more months without precipitation in the west, farming without irrigation is not possible.

The degraded areas are mainly fixed and semi-fixed dunes and sandy plains: for example, the shifting sand area in the southern part of the Mu Us sandy land along the Great Wall was denuded in the last three centuries, and the shifting sand in the Khorcin sandy area in the last 100–150 years (Fig. 21).

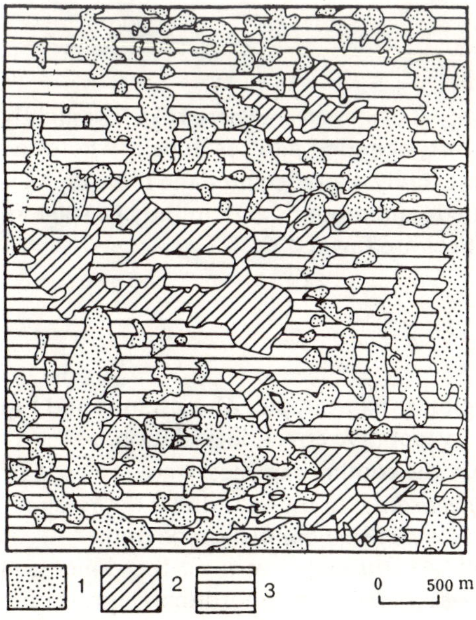

Fig. 21. Severe degraded land in the steppe zone. Rangeland of Naiman Banner, middle Korqin sandy land, Inner Mongolia. (Zhu Zhenda and Liu Shu 1983). *1* Shifting sand dunes; *2* flat sandy land; *3* semi-fixed sand dunes

In the 1950's the landscape was still an undulating sandy steppe without shifting sands. After cultivation during the last three decades, the surface changed and 54.8% of the land is now occupied by shifting sand (Zhu Zhenda and Liu Shu 1983).

Table 14 shows how in two decades fixed sand can be changed into an active aeolian landscape.

Table 14. Deterioration of the surface morphology on both sides of the Great Wall in northwestern Yanchi County. (After Zhu Zhenda and Liu Shu 1983)

Year	Percentage of the areas occupied by each surface morphology (%)			
	Shifting dunes	Semi-fixed dunes	Fixed dunes	Sandy flat lands
1956	16.1	3.7	45.5	34.7
1976	45.3	13.9	31.2	9.6

3.4.3 The Major Indicators of Degradation

These are of three types:

1. Changes in the topography of the landscapes.
2. Changes in the composition of the vegetative cover.
3. Changes in the composition of the top soil.

3.4.3.1 Changes in Topography

The sandy fixed semi-arid, subhumid steppes originally had a more or less slightly undulating surface. When over-use occurs, denuded patches appear where the wind-blown sand becomes mobile. In a severely degraded area the fixed undulations are replaced by fixed, than denuded sand mounds and, in the last stage of active degradation, barchanic sand edifices are formed.

3.4.3.2 Changes in Composition of the Vegetative Cover

Taking the Yanchi County Ningxia Region as an example, Zhu Zhenda and Liu Shu (1983) describe the initial phases of vegetative cover degradation.

Stipa breviflora, Lespedeza dahurica, the common species, were replaced by *Glycyrrhiza uralensis* and *Sophora alepecuroides* in the interdunal depressions. The good edible plants *Stipa gobica* and *Lespedeza potanii* disappeared.
by dunes, the evolution continued with *Artemisia ordosica* on the dunes and *Glycyrrhiza uralensis* and *Sophora alepecuroides* in the interdunal depressions. The good edible plants *Stipa gobica* and *Lespedeza potanii* disappeared.

The net vegetative cover is about 35 to 45%. With increasing density of the dunes only a few psammophytes persist: *Artemisia ordosica* and *Salix psammophila* with a cover of about 10%. On the shrub sand mounds *Caragana* spp. prevails. On the gravel plains in the Eastern Alxan League in Inner Mongolia the plant cover decreased from 45–50% to 5–10% in four decades. Overgrazing results in a cover of predominantly *Artemisia* spp. and the disappearance of the edible *Reaumuria soongolica* and *Salsola passerina* (Tables 15 and 16).

Table 15. Impoverishment of organic matter in the topsoil (Ordos steppe). (After Zhu Zhenda and Liu Shu 1983)

Site	Surface features	Depth of sampling (cm)	Percentage of organic matter
Gongkahan	Undesertified undulant highland	0 – 10	1.408
	Ridge where desertification is developed	0 – 8	0.745
Sumitu	Fixed dune on ridge	0 – 1	1.403
	Sandy land initiates to cause desertification	0 – 10	0.236
Zhuhe	Residual ridge	0 – 1	0.712
	Ridge initiated to cause desertification	0 – 6	0.114
Qarhan Taolaohai	Undesertified land	0 – 20	0.770
	The land has been desertified	0 – 20	0.157
Qarhan Nor	Undesertified ridge	0 – 17	1.485
	The ridge has been desertified	0 – 16	0.953

Table 16. Mechanical changes in surface composition in the Ordos steppe. (Zhu Zhenda and Liu Shu 1983)

Region	Surface features	Depth of sampling (cm)	Particle size range (Percentage of different grain sizes) (mm) (%)			
			1–0.25	0.25–0.05	0.05–0.01	<0.01
Wengongliang' Ordos Steppe	Uncultivated grassland	0–1	/	74.47	22.70	2.83
	Arable land influenced by desertification after cultivation	0–1	/	90.85	7.73	1.42
Mangha tu' Ordos Steppe	Uncultivated grassland	0–20	1.23	68.56	19.56	10.65
	Arable land influenced by desertification after cultivation	0–20	1.50	81.52	11.64	5.34
A state-owned farm in eastern Xilin Guole Steppe	Uncultivated landgrass	0–5	7.2	43.4	28.2	21.2
	Arable land influenced by desertification after cultivation	0–5	23.8	66.0	3.9	6.3

3.4.3.3 Changes in Texture of the Top Soil

The effects of wind erosion on the top soil are of three types:

1. Export of organic matter. Measurements in the Ordos Steppe (Zhu Zhenda and Liu Shu, 1983) (Fig. 22) quantify the considerable loss of organic matter on all the sites where vegetation has disappeared and the surface has been reworked into active even mobile sand forms.
2. Changes in the texture, with increase of the coarse sandy material of over 50 μm and in contrast, a decrease in the percentage of fine material particles smaller than 50 μm (Fig. 23).
3. Changes in the porosity of the top soil from 7 to 1%.

3.4.4 Human Factors Leading to Environmental Change in Semi-Arid and Dry Subhumid Zones in China

Chinese scientists distinguish between historical and contemporary desertification. Historical degradation occurred in arid and semi-arid ecozones, where people lived at that time. Contemporary land degradation occurs in subhumid ecozones because these are now inhabited.

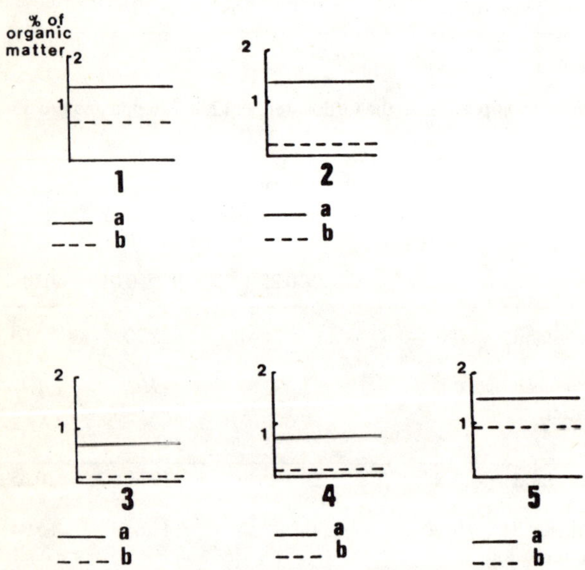

Fig. 22. Loss of organic matter in the topsoil of cultivated soils in Ordos Steppe (China). *1* Gongkahan, *a* undesertified highland, *b* ridge where degradation is developed; *2* Sumitu, *a* fixed dune on ridge, *b* sandy land initiates to cause degradation; *3* Zhuhe, *a* residual ridge, *b* starting of degradation; *4* Qarhan Taolaohai, *a* undesertified land, *b* the land has been degraded; *5* Qarhan Nor, *a* undesertified ridge, *b* the ridge has been degraded

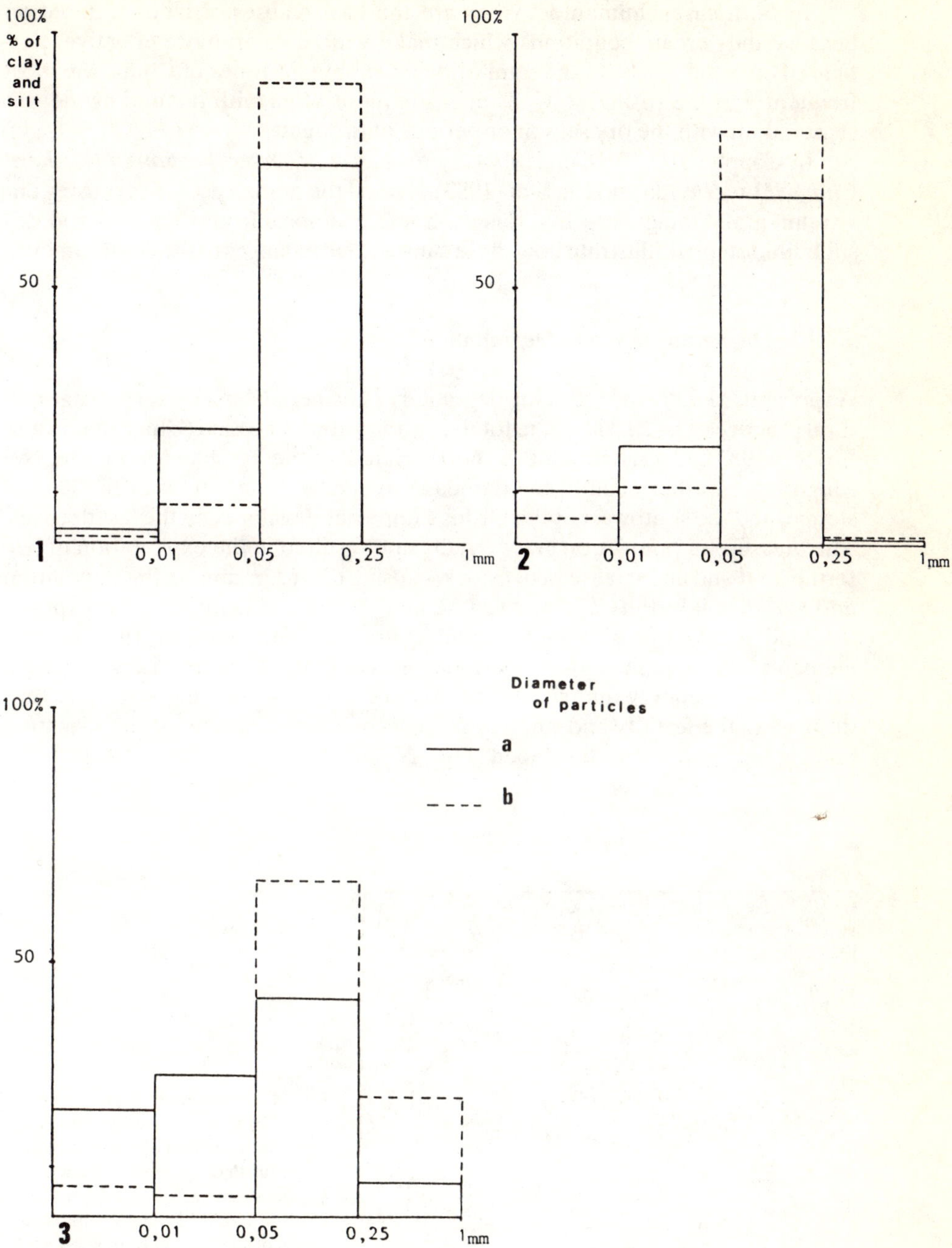

Fig. 23. Loss of clay and silt in cultivated soils. Increase of the percentage of sand in the topsoil in Ordos Steppe (China). *1* Wengongliang Ordos Steppe; **2** Mangha tu Ordos Steppe; *3* Eastern Xilin Guole Steppe. *a* Uncultivated grassland; *b* arable land influenced by degradation after cultivation

In both cases, human activities are the basic causes of land degradation because they create conditions which make wind erosion more effective, particularly in sandy soils. In the semi-arid and subhumid zones of China, the most frequent and the strongest winds coincide more often with the end of the wet season than with the dry season or periods of drought.

In Chapter III, *The Contemporary Processes of Desertification in Northern China*, Zhu Zhenda and Liu Shu (1983) divided the human activities causing environmental damage into five classes. Some examples from semi-arid and dry subhumid steppes illustrate how these causes affect changes in the environment.

3.4.4.1 Overcultivation and Degradation

At present, 11 735 km^2 are being degraded in northern China by current agricultural practices, i.e., 23.3% of the total degraded areas in China (Zhu Zhenda and Liu Shu 1983). Degradation of the northern half of the Yinshan Mountains, Nei Mogol, the southern half of the Otindag sandy desert, the center of the Korgin steppe, and the southwest of the Ordos steppe has developed in the last decades. The pressure of population grew rapidly and resulted in the exploitation of less fertile land and an increase in livestock, causing disorders among the population and loss of soil fertility and yields. Use and plowing of sandy surfaces exposed the land to wind erosion and, according to the classic pattern, the nutritive elements, the organic matter, the moisture content, all the particles playing a major role in energy and substance exchanges in dry farming, were lost thus destroying the fertility and stability of the ecosystem. The soil lost its structure, and aeolian land forms developed (Figs. 24 and 25).

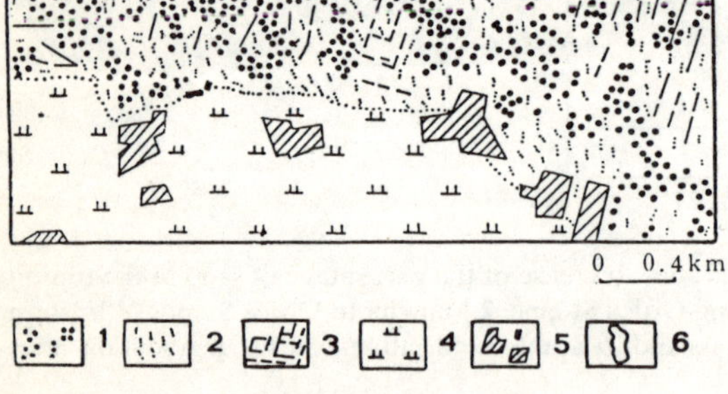

Fig. 24. Extension of degradation caused by overcultivation in flat sandy areas at the margin of degraded areas in the north of Zhangwu County, Liaoning province. (Zhu Zhenda and Liu Shu 1983). *1* Degraded land composed of shifting sands in patches; *2* slight undulating sand flat land; *3* abandoned land; *4* farmland; *5* actual settlements; *6* water accumulating on low land

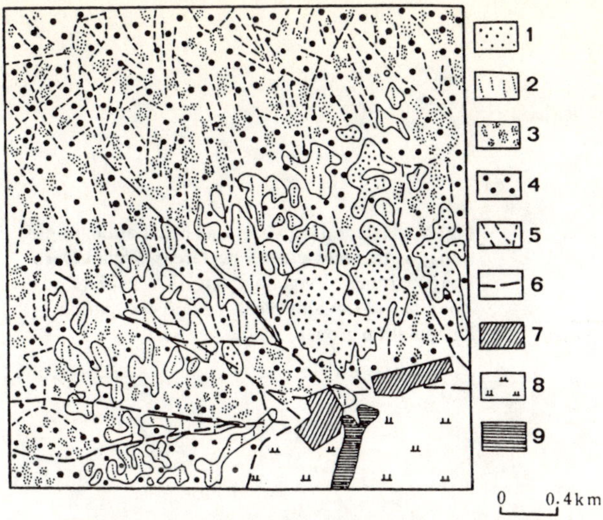

Fig. 25. Aggravation of degradation degree near actual settlements in the northwest of Zhangwu County, Liaoning province. (Zhu Zhenda and Liu Shu 1983). *1* Shifting sand dunes; *2* undulating sandy lands (shifting sands); *3* shifting sands in spots; *4* semi-fixed sand dunes; *5* traces of aeolian erosion after cultivation and abandoned lands; *6* roads; *7* actual settlements; *8* farmlands; *9* water

Figure 26 shows the links between (1) demographic explosion, (2) enlarged cultivated land, (3) decreased yield, (4) increased wind erosion. The negative results of such a scheme were measured in Erfonzi Commune, Wuchuan County, Nei Mongol: the annual average depth of soil removed by wind erosion is 3 cm, the loss of organic matter 2626.5 kg, of nitrogen contents 169.05 kg, and of clay 25.95 tons ha^{-1}. The relationship between depth of erosion, loss of organic matter, and decrease of yield is shown in Fig. 27.

This local effect of wind is not the only one. The particles eroded by wind can be transported and accumulated far down-wind, forming there deposits which can be reworked as sand patches or active dunes.

We have seen in Niger (Africa) that the deposits can occur several tens, even hundreds of kilometers leeward of the erosion area. Beijing nowadays has frequent dust storms, the origin of which is mainly land degradation in Inner Mongolia (Liu Tungsheng 1988).

The Mu-Us sandy land, located at the southern margin of the Ordos plateau, is an example of wind erosion in a semi-arid ecozone. Zhao Songqiao (1986) notes: "Since the Tang Dynasty, for more than 1000 years, it has been subjected to the southward moving of a sandy desert, with shifting sands drifting more than 100 km southward". The 60-km-wide belt of shifting sands which has developed along the Great Wall can be attributed to the accelerated removal of natural vegetation in the last three centuries.

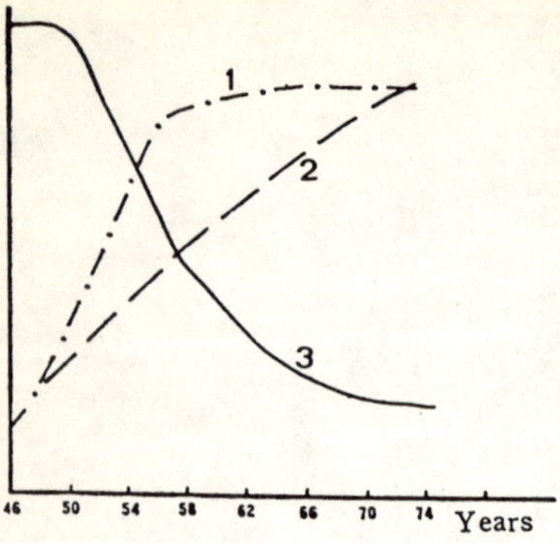

Fig. 26. Relation between land occupation, population growth, and yield per hectare in Wulan Qarb League, Inner Mongolia, from 1946–1974. (Zhu Zhenda and Liu Shu 1983). *1* Land occupation; *2* population growth; *3* yield per hectare

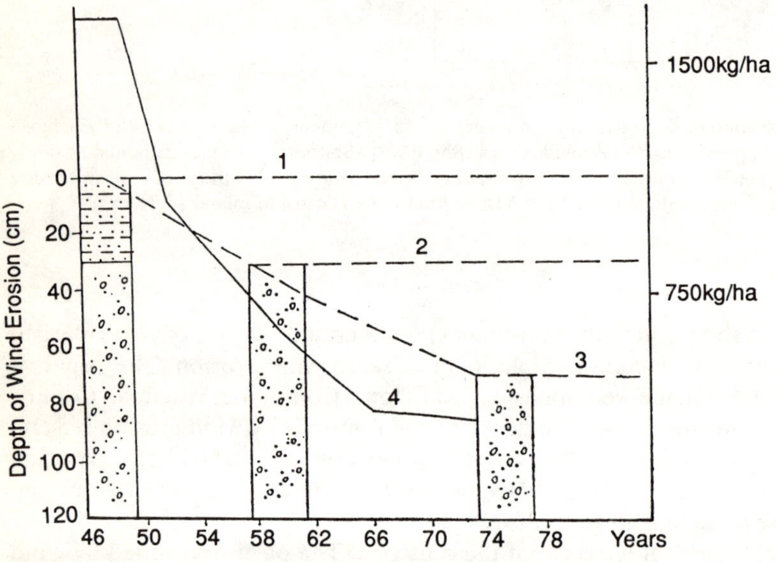

Fig. 27. Depth of wind erosion and declining yield in Gongji Tang Commune, Siziwang Banner, Inner Mongolia. (Zhu Zhenda and Liu Shu 1983). *1* Level of the topsoil before cultivation; *2* level of the surface from 1958–62; *3* level of the surface from 1974–78; *4* yield per hectare

3.4.4.2 Overgrazing and Degradation

Overgrazing has the same effects in temperate, dry and subhumid steppes as overcultivation; it initiates wind attack, particularly on sandy soils. Some statistics are given by Zhu Zhenda and Liu Shu (1983): the degraded steppe or grasslands cover an area of about 14 821 km^2, i.e., 29.4% of northern China is degraded land. Whereas in semi-arid areas with a humidity index of 0.20–0.50, 4 ha are necessary for grazing one animal, in the early 1950's in China each animal shared 3.33 ha of grassland, which decreased in the late 1970's to 1.11 ha. Such animal pressure resulted in quantitative and qualitative steppe degradation. According to Zhu Zhenda and Liu Shu, the original *Stipa* spp. has been replaced by *Artemisia frigida* and weeds. From 1957 to 1973 shrubs decreased by 76.2%. The vegetation became more sparse and shorter because of livestock multiplication and competition between cultivation and grazing. Animal trampling and consequent destruction of the top soil structure initiated wind erosion. A series of degradation rings surrounds waterwells (Fig. 28). Field observations show a central area with a radius of 300 to 500 m with little vegetation, often only bare ground (sand or gravel); a second ring with a radius of 500 to 1000 m, where some more resistant grass or shrubs are scattered, and then at a distance of 1000 to 3000 m from the well, vegetation gradually returns to normal. By comparison, I observed in Botswana (Kalahari desert), in the Bookspik area, the rings of degradation around a well reaching a radius of 10 km before the steppe had totally recovered its original vegetation.

However, it is extremely uncertain whether we can assume that overgrazing is the cause of land degradation or desertification. In China, as well as in Africa, overgrazing is strongly related to rainfall. North of Mali in the Saharo-Sahelian ecosystem, in sandy soil, Guyot (pers. commun.) observed a vast sandy surface completely denuded and degraded. Suddenly he arrived in an area where the grass had recovered and was dense. According to the nomads, a local heavy rainfall occurred some weeks before. This grassy area was detectable on the Spot satellite image.

In the case of overgrazing it is also difficult to estimate the threshold of irreversibility. Nowhere in the literature is there any assessment of the number of years of fallow necessary for recovery before the onset of irreversible degradation.

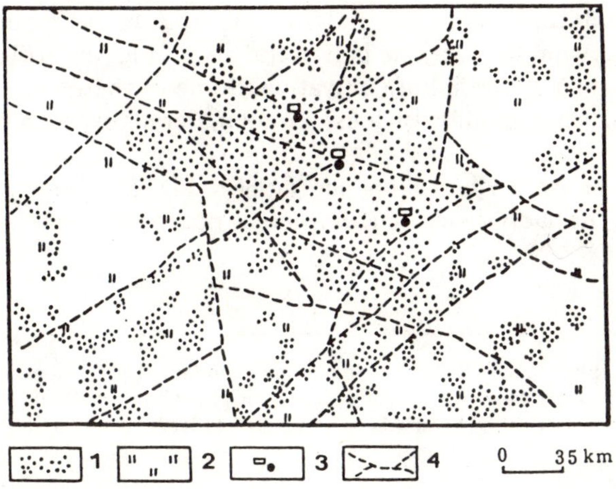

Fig. 28. Degradation caused by overgrazing. *1* Barren alluvial surface; *2* grazing land; *3* livestock sheds and wells; *4* road and path

3.4.4.3 Firewood Collection and Degradation

The Chinese consider firewood collection to be the most important human factor of degradation. "In Shengmu county, Shaanxi Province, there were 22 667 ha of fixed dunes growing natural *Cupressus nephrolepis* in 1949, but now the survival is only 22%, i.e., 73% has been lost by firewood collection. In Alxan League, Inner Mongolia, 60% of the woodland of *Haloxylon ammodendron* was reduced as a result of firewood collection. It is estimated that 10 kg of *Cupressus nephrolepis* is needed per day for consumption by a family of five people, so that each family would expose 0.2–27 ha of fixed sand dunes growing *Cupressus nephrolepis* per year." (Zhu Zhenda and Liu Shu 1983). To firewood collection these authors add wild plant collecting for medical purposes (*Glycyrrhiza uralensis, Cistanche deserticola,* and Ephedra sinica) and of lichens and fungi for food.

3.4.4.4 Land Degradation by Misuse of Water Resources

Misuse of water resources has different aspects:

1. Reduction of vegetative cover, disappearance of plants, increasing the runoff/infiltration ratio.
2. With 350 to 650 mm of annual rainfall, the loess plateau in a semi-arid to dry subhumid ecozone is certainly the most devastating example of water erosion in the world. Loess is the wind deposition of particles of silt and clay, and loess soils are considered among the most productive soils. They cover a large area in northern China in and around the bend of the Yellow River.

   Loess soils are highly susceptible to water erosion. Water erosion (gullies and sheet erosion) carries 1.6 billion tons of silt per year to the Huanghe River, which means 400 kg m^{-3} during the most severe period of annual erosion (pers. commun. from Zhu Zhenda). Even though the Chinese are masters in building terraces, these conservation measures are not widespread, and owing to the position of most cropland on gentle slopes, "about one half of the total cropland has suffered from soil erosion, and about a quarter of the total cropland has lost one half of its top stratum of fertile black earth after fewer than 100 years of cultivation. Some of these areas have lost all their top soil, leaving yellowish loess parent materials exposed". (Zhao Songqiao 1986).

   The impressive figure from Zhao Songqiao (1986) (Fig. 29) demonstrates better than words the badly dissected loess plateau near Suid, Shaanxi, 110–111 °E, 37–38 –N.

Fig. 29. Water erosion in the loess plateau near Suide, Shaanxi. (Zhao Songqiao 1986). *1* loessic Yuan; *2* loessic Liang; *3* loessic Liang-mao; *4* loessic mao; *5* loess terrace; *6* gully; *7* dry small valley; *8* small valley; *9* broad valley; *10* gorge; *11* top of liang; *12* top of mao; *13* gully; *14* narrow valley; *15* fault line

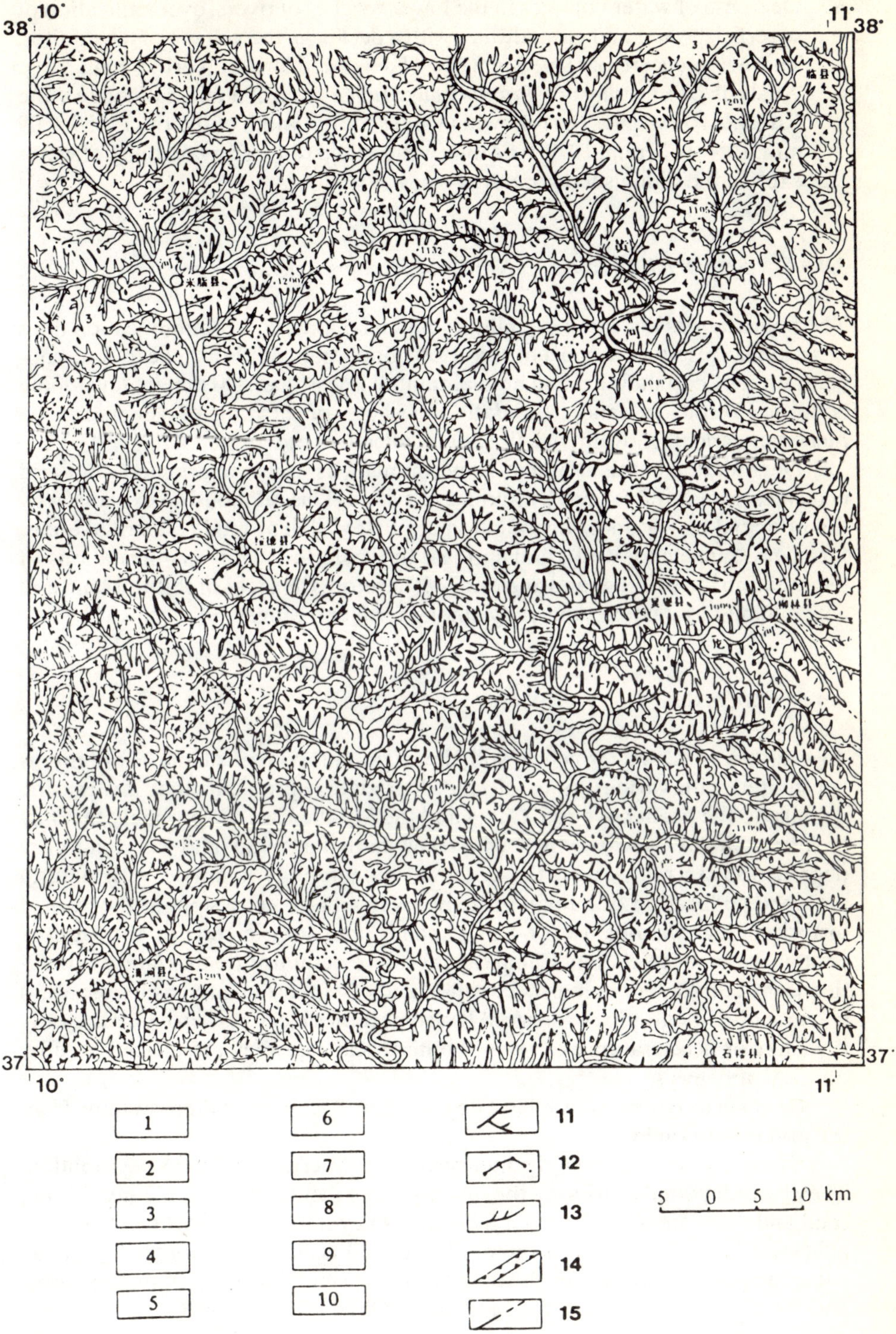

10°
38°
11°
38°
37
37°
10°
11°
1
2
3
4
5
6
7
8
9
10
11
12
13
14
15
5 0 5 10 km

3. Decrease of water volumes in the lower reaches of rivers, over-utilization of superficial water and groundwater with a decline in the volume of the aquifer. These aspects are observed in 8.6% of degraded land in North China.
4. Analysis of tree rings of *Populus euphratica* from the alluvial cone of the Ruoshui River gives the following values for the width of the ring: 1959: 1.09 mm, 1966: 0.77 mm, 1979: 0.62 mm. Misuse and reduced availability of water explain the diminished growth of trees.

3.4.4.5 Land Deterioration and Nonrural Human Activities

The nonrural activities: mines, mills, industrial plants, and communication lines near towns result in a concentration of population. Without exception, concentration of population when annual rainfall is less than 800 mm in developing countries results in land degradation because of clearing of vegetation by machinery and its movements, firewood collection, etc.

Degradation in this subhumid environment, as well as in arid and semi-arid ecozones, is also influenced by human activities. This relationship explains why the principal indicators of degradation are first visible around residential areas, farmland, rangeland, wells, and springs. In China the major manifestations of degradation are shifting sand and dunes. Therefore the areas of shifting sand and active dunes are scattered patchily among farm and rangeland.

3.4.5 Conclusion

Agriculture in China, which is still the main component of the economy, is confronted with various kinds of land degradation problems that are causing damage to the productive soil. China and the Middle East are probably the only countries in the world, which can show both historical and contemporary degradation. In both cases, the potential vulnerability is linked with natural parameters: location of human settlements in an arid ecosystem in the case of historical degradation; use of fragile sandy and loessic soils in the case of contemporary degradation. China can therefore demonstrate that land degradation is the corollary of human occupation and of expansion into unsuitable areas because of demographic explosion. It also becomes clear that droughts are only revealers or accelerators even if in the historical case they are exacerbated by political upheavals.

China is a country where land degradation affects several ecosystems from the arid to subhumid.

Chinese scientists have distinguished several criteria of land degradation. They consider wind erosion as the major process, and its effects, mainly shifting sand and dune areas, as the main indicator. They have attacked the problem by analysis of surface wind erosion, depth of sand accumulation, intensity of the wind currents, conditions for the appearance of sand dunes (Di Xingmin 1988), and the degree of activity of the sand dunes (Wang Kangfu 1988).

It is noteworthy that Chinese scientists are active at all stages in the battle against land degradation: analysis of the problem, research for solutions, and implementation of practical planning. Soil protection and the development of intensive irrigation systems are the solutions often chosen.

3.5 Transition from Traditional Low Technological to High Industrial Agriculture and Land Degradation in the Former USSR

Measures to combat desertification in the former USSR aim at protection, rehabilitation, and development of arid and semi-arid lands. This is the main reason why, in order to understand land degradation in the former USSR, I shall deal with those Soviet dry ecozones which are on the Asian arid diagonal, and draw up an inventory of the natural productive forces before investigating the causes, mechanisms, and effects of land degradation.

3.5.1 Dry Ecozones in the Former USSR

In the former Soviet Union the drylands occupy mainly the Turanian plain. They extend from the east Caspian shores to the Tien-Shan and Altai foothills (Fig. 30).

"In the USSR deserts and semi-deserts occupy an area of around 300 million ha or 14% of its entire territory (Babaev and Freikin 1977). Nearly all of these are situated in the Kazakh SSR, the Uzbek SSR, and the Turkmen SSR. Within the desert belt (between 51° and 80 °E), there are the following well-known desert areas, featuring different physiographic peculiarities: Kamenistry Ustjurt, the Circum-Caspian sand clay plain, the sandy Trans-Caspian Karakum, Kyzylkum, Mouyounkum, Circum-Aral Karakum, and Sary-Ishiko-Tau, the stony Betpak-Dala, the clay Golodnaya Steppe, the Karakalpak Steppe, and the Soundukli sands, besides other smaller areas." (Fig. 31).

The northern boundary of the drylands coincides with the 180-mm mean annual isohyet (Fedorovitch et al. 1963).

The large extent of this arid zone in latitude between 36° and 48° and even up to 50 °N (about 1500 km from north to south) and in longitude between 53° and 83 °E (about 2500 km from west to east) explains the wide range of climates, soils, and vegetation found in the area. The common factor is the low amount of precipitation, mainly in spring, late autumn, and winter, with a dry season from mid-May to mid-October and a rainy season for the rest of the year. The annual rainfall is around 100 mm in the central regions (Karakum and Kyzylkum) and between 80 to 250 mm in piedmont areas (Fig. 32).

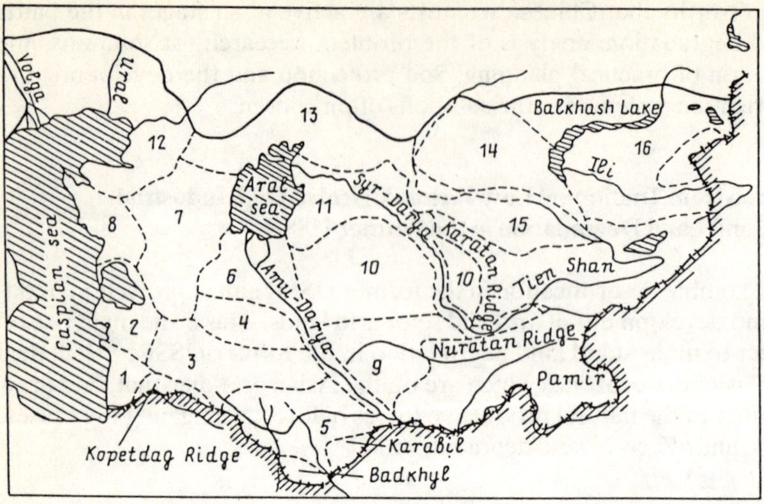

Fig. 30. Nomenclature of the natural regions in the drylands of Central Asia and Kazakhstan. (UNEP-UNEPCOM 1980). *1* Southeast Near-Caspian; *2* Karabogaz Gulf Region; *3* Central Kara-Kum; *4* Zaunguz Kara-Kum; *5* southeast Kara-Kum; *6* Amu-Darya delta and valley; *7* Ustiurt; *8* Mangyshlak; *9* South Kyzyl-Kum; *10* North Kyzyl-Kum; *11* Syr-Darya delta and valley; *12* North Near-Caspian; *13* Near-Aralian Kara-Kum; *14* Betpak-Dala; *15* Muiunkum sands; *16* Lake Balkhash region

3.5.2 Aridity and Other Physical Factors Leading to Land Degradation

3.5.2.1 The Climatic Factors (Table 17)

The continental location, the extension in latitude and longitude, coupled with mountain barriers in the south, southeast, and east, but opening towards the Arctic in the north, are the main factors which contribute to increased aridity. The dry areas of the former USSR, central Asia, and Kazakhstan, can be subdivided into two subzones: the northern and the southern (Table 17, Fig. 32).

Figure 32 quantifies the aridity, the mountain area in the southeast being less dry than the rest. The climatic difference between the northern and the southern subzones is mainly in the length of the growing period, which is long in the north, varying from 5 to 8 months with the absolute minimum temperature reaching –35 °C in the north to –17 °C in the south in January. Absolute maximum temperatures in July are not so different between north and south and differ less, viz. 42 to 49 °C. The mean annual temperature increases southwards from 5 ° to 16.6 °C. Dryness is especially marked in summer when the air humidity can vary between 30% and 2–3% (on some days) and in early autumn also, because of the high temperature and high evaporation rates, from 1400 to 2500 mm per year (Babaev 1986a) when a seasonal soil drought desiccates the top soil and increases plant transpiration. In Karakum the soil surface becomes extremely hot during

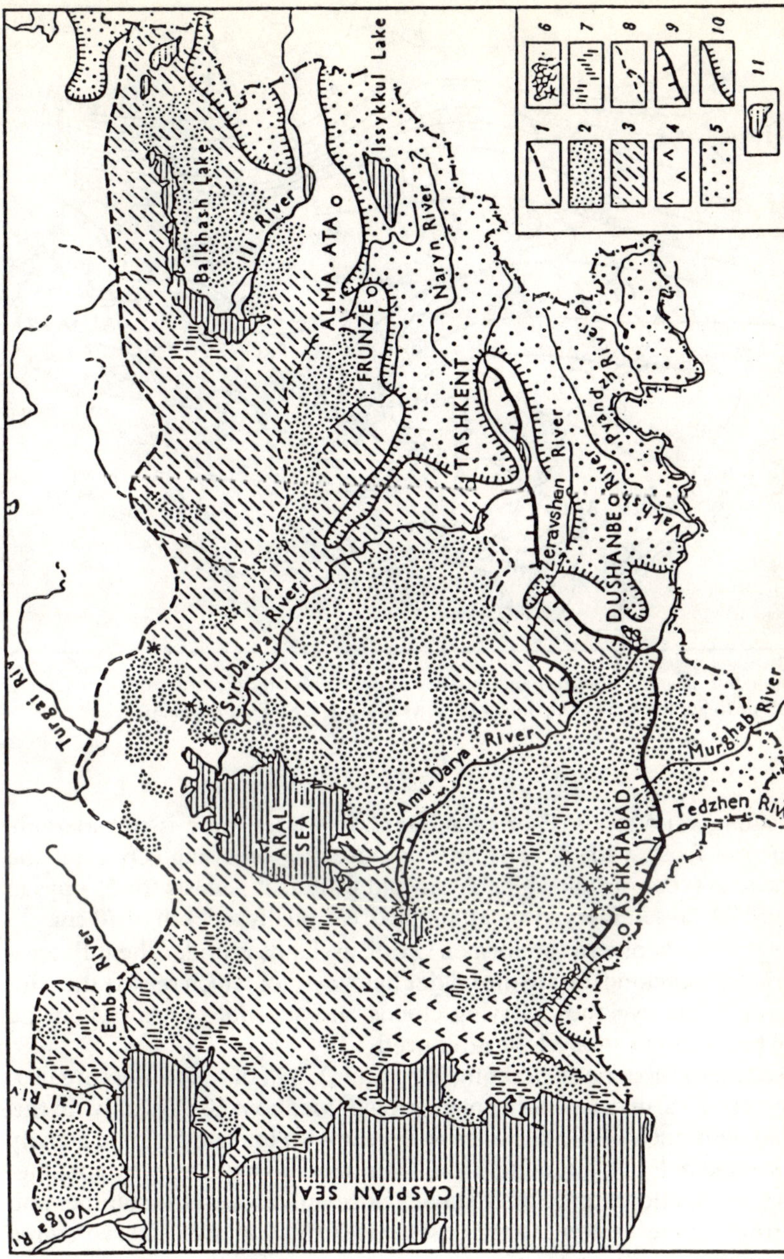

Fig. 31. Natural regions in the drylands of Central Asia. (Orlovsky, UNEPCOM 1982b). *1* Northern boundary of deserts; *2* sand; *3* clay; *4* gypseous; *5* loess; *6* takyr; *7* solonchak; *8* intermittent and dry channels; *9* main canals; *10* mountain regions; *11* seas and lakes

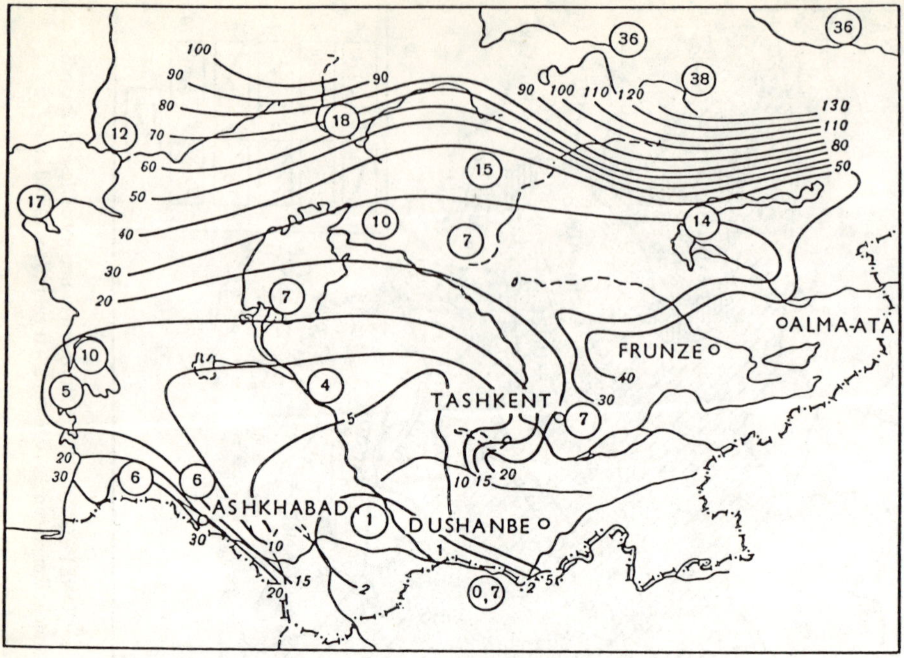

Fig. 32. Precipitation in mm (isohyets) and number of days with precipiration >0.1 mm (*circled figures*) from June to September. (Chelpanova 1963 in Orlovsky, UNEPCOM 1982b)

the day in summer, up to 70 °C or even 80 °C. In winter there is considerable snow cover and frost, with adverse effects on cattle breeding. Over all the drylands the snow cover is unstable, but in cold winters it persists for 35 days in Karakum and 70 days in Kyzylkum, attaining occasionally a depth of 70 cm.

The northern subzone of the former USSR drylands is under the influence of the Siberian anticyclone during the colder half year, for which reason the winters are severe. The cyclones carrying clouds and snowfalls occur when the influence of the Siberian anticyclone is reduced.

In the southern subzone (Karakum, Kyzylkum), half of the winters are usually mild and even warm, suspending plant growth for about 10 days only. In this area the Iranian branch of the polar front develops an intense cyclonic activity during the second half of winter and spring months. This explains why ephemerals go into a second florescence after staying dormant through the summer and start new growth in the autumn. In the oases, with irrigation, two and even three harvests of quick-growing crops are possible on the same field.

Grigoriev and Budyko (1959), in their classification of climates, summarize this drylands climate perfectly. The northern subzone is characterized by a dry climate with a warm summer and a winter with little snow, while the southern subzone has a dry climate with a very warm summer and a moderately dry winter.

Table 17. Major climatic indices for deserts of Central Asia and Kazakhstan. (Orlovsky, UNEPCOM 1982b)

Station	Air temperature, °C			Average absolute minimum	Absolute maximum	Period with average daily air temperature	Vegetation period	Sum of air temperatures	Precipitation, mm		
	Annual	January	July			<0 °C	(t ≥ 10 °C)	≥ 10 °C	Annual	Warm season	Cold season
						Northern subzone					
Teren-Kuduk	4.8	−15.4	24.5	−35	43	151	157	1973	171	102	69
Ak-Tumsuk	8.9	−7.9	25.2	−26	42	107	173	3466	129	77	52
Sam	8.9	−9.1	27.0	−28	46	126	179	3773	136	82	54
Kungrad	9.9	−6.7	25.3	−26	44	106	186	3751	95	46	49
Kunya-Urgench	109	−6.0	26.6	−24	45	102	193	4042	76	31	45
						Southern subzone					
Ekedje	13.2	−3.9	29.7	−22	46	81	205	4697	100	40	60
Darvaza	14.8	−2.0	31.2	−19	46	56	217	3123	97	39	58
Akmolla	15.8	−1.0	32.2	−20	47	44	224	5393	91	34	57
Cheshme	15.7	−0.2	31.5	−20	49	29	225	5292	97	38	59
Bairam-Ali	16.0	1.5	30.2	−16	48	0	234	5268	135	40	95
Serakhs	16.6	3.0	30.2	−15	48	0	242	5402	178	51	127
Takhta-Bazar	16.5	2.7	31.2	−17	48	0	236	5370	241	53	188

The arid zone has an annual total rainfall between 80 to 100 and 200 to 250 mm in the piedmont loess. Its aridity index ranges from 3 to 20 (from Saharan to Sahelian.

The annual regime of precipitation varies widely: in the northern subzone spring precipitations are higher than those of the cold season. In the southern subzone rainfall of the cold period prevails. The year is divided into two seasons: a dry one from mid-May to mid-October and a humid one for the rest of the year.

The sparse vegetation of perennials and annuals allows pastoral nomadism and irrigation farming.

Aridity, scanty vegetation, the high susceptibility of the sandy, gypsum- and loess-rich desert to erosion, poor natural drainage, poor development of physical and biological processes of soil formation, low humus content, and high salinity contribute to marked wind erosion, causing sand and dust storms. The winds are stronger in summer:

"Strong winds always cause dust storms. These occur in summertime, with velocities of 7–10 m s^{-1} and higher, and last 2 to 6 h. In the Kyzylkum and northern Karakum, dust storms occur on 20–30 days a year, in the Southeastern Karakum 40–50 (in the Repetek area, on as many as 60 days), and in the western part of Central Asia on 50–60 days a year". (Babaev 1986a)

According to Sapozhnikova (1970), the maximum average annual number of days with dust storms in Central Asia is recorded in central and southeastern Karakum (Cheshme 1948: 113 days; Repetek 1939: 106 days), in western Turkmenistan in the Kopet-Dag piedmont areas (Molla-Kara 1939: 146 days) and on the shores of the Aral Sea (Muinak 1958: 121 days). According to Romanov (1960), the average figure for Central Asia can reach 250 days per year. In the Karakum, dust storms occur mostly in spring, in Kyzylkum in summer, in the piedmont areas in autumn. In most cases the duration of a storm does not exceed 3 h. Nevertheless, in the west and extreme southeast of Turkmenistan, the longest sand storms (5%) last more than 24 h and can exceed 70 h (Fig. 33).

3.5.2.2 Soils

Shallow thickness (50 to 200 mm), negligible humus quantity (0.3–0.5%), lack of noticeable leaching capability, absence of structure (sandy texture and 4 to 6% of silty fraction) potential high salinity, and low fertility are the characteristics of dryland sandy soils.

The latitudinal differences in rainfall and in temperature can be seen as the main causes of soil differences between the northern and the southern subzones of the dry ecozones. The northern subzone has mainly gray-brown desert soil: solonchak (low alkaline sodic soil with a good structure) and solonetz (alkaline sodic soil with a precise B horizon). The southern subzone is characterized by highly sandy carbonaceous gray-brown and sierozem soils (isohumic soil poor in organic matter).

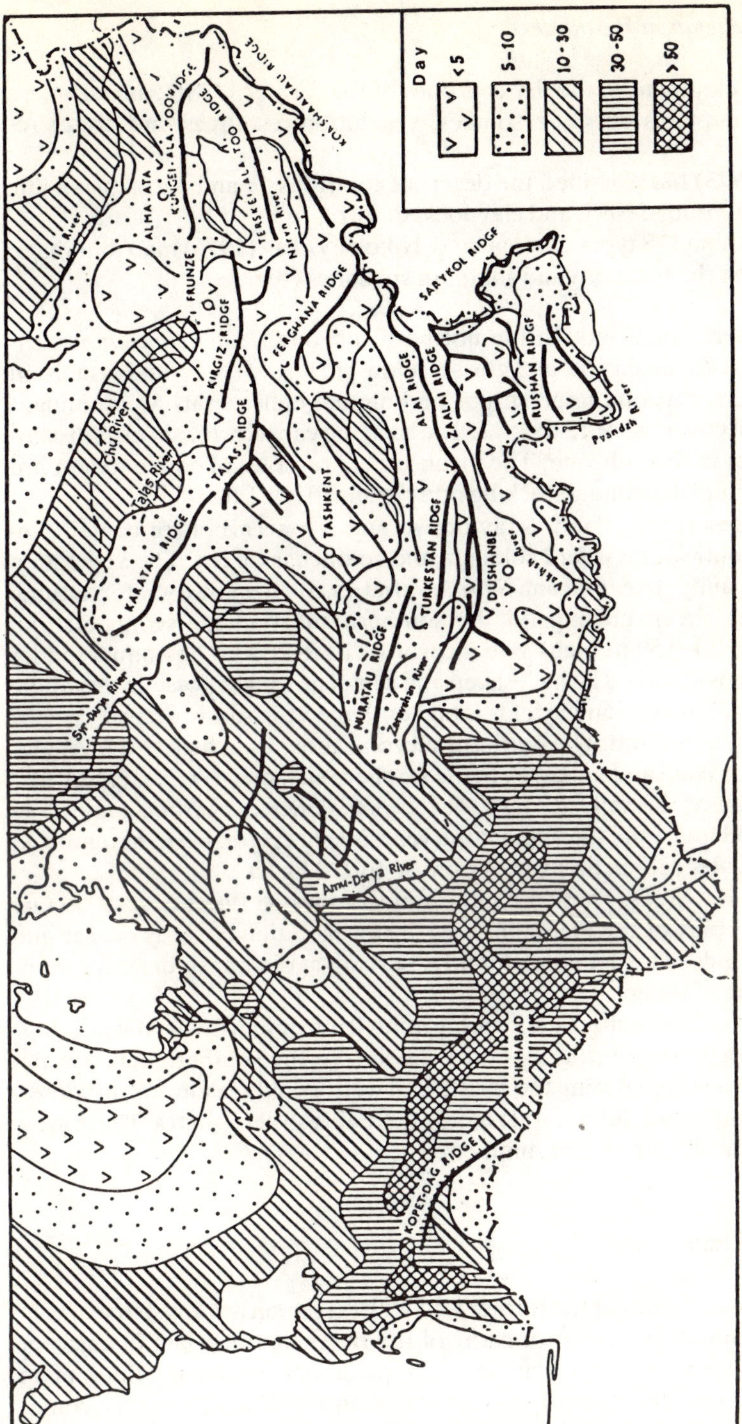

Fig. 33. Average annual number of days with dust storms. (Orlovsky, UNEPCOM 1982b)

3.5.2.3 The Vegetation Resources

Economic development in the dry ecozones of the former USSR, as elsewhere in the world, depends on soil resources, vegetation resources, and water resources.

Petrov (1973) has classified the desert of central Asia and Kazakhstan into sand deserts, gypsum desert, and clay-loess desert.

Distinguishing 178 types of rangeland, Nikolaev et al. (1977) map four major classes covering the northern and southern subzones (Fig. 34):

1. Sandy desert rangeland. In the northern subzone, *Haloxylon ammodendrum* and *Salsola*, shrubs, and grasses allow grazing of sheep and camels; in the southern subzone large shrubs and annual ephemerals are dominant with a total coverage of 12 to 25%. The carrying capacity for sheep and camels is higher in this subzone. The mean annual reserves of edible fodder are 100 kg ha^{-1} and the annual stocking rate per sheep is 9.4 ha.
2. Gypsum desert rangeland. Gray-brown and a low carbonate content, increasing southwards where solonetz and solonchack exist. The vegetation includes mainly *Artemisia* and *Salsola* varieties and may be used for autumn and winter sheep grazing. In the southern subzone sparse undershrub varieties (0.20–0.50 m high) give a cover of 10 to 20%. The annual fodder production is 60 to 208 kg ha^{-1}; sheep and camel grazing is possible the whole year round. Fluctuation in yields is higher than in the sandy desert.
3. Clay desert rangeland. The northern subzone abounds in high shrub vegetation (black saxaul and tamarix). In the southern subzone the fine texture and high salinity of the soils account for the prevalence of *Salsola*. The whole zone is used for camel throughout the year, for sheep grazing in autumn and winter. Fodder production varies from 80 to 220 kg ha^{-1}.
4. The piedmont areas. These are more stony or loessic. Loess deserts occupy the greater part of the piedmont area. The vegetation cover is denser and the rangelands of the loess-piedmont desert are at their best for spring-summer grazing of sheep.

   In the northern subzone the spring rainfall maximum favors the development of vegetation during the warm season. In the southern subzone the vegetation develops during the milder and wetter cold season, i.e., in winter. In the south, the hot, dry summer is only favorable to xerophytes. These grow also in the north, but in late summer.

3.5.2.4 Water Resources

In the central Asian states of the former USSR the river network is sparse. As in central Asia (China), the water of many of the rivers has its origin in adjacent mountains. Flowing down, they run over the piedmont, producing flood plain areas with their specific types of soils and vegetation. When the water comes onto the desert plains, some is lost to evaporation, some is absorbed by the soil, thus replenishing the groundwater, while the rest is used for irrigation.

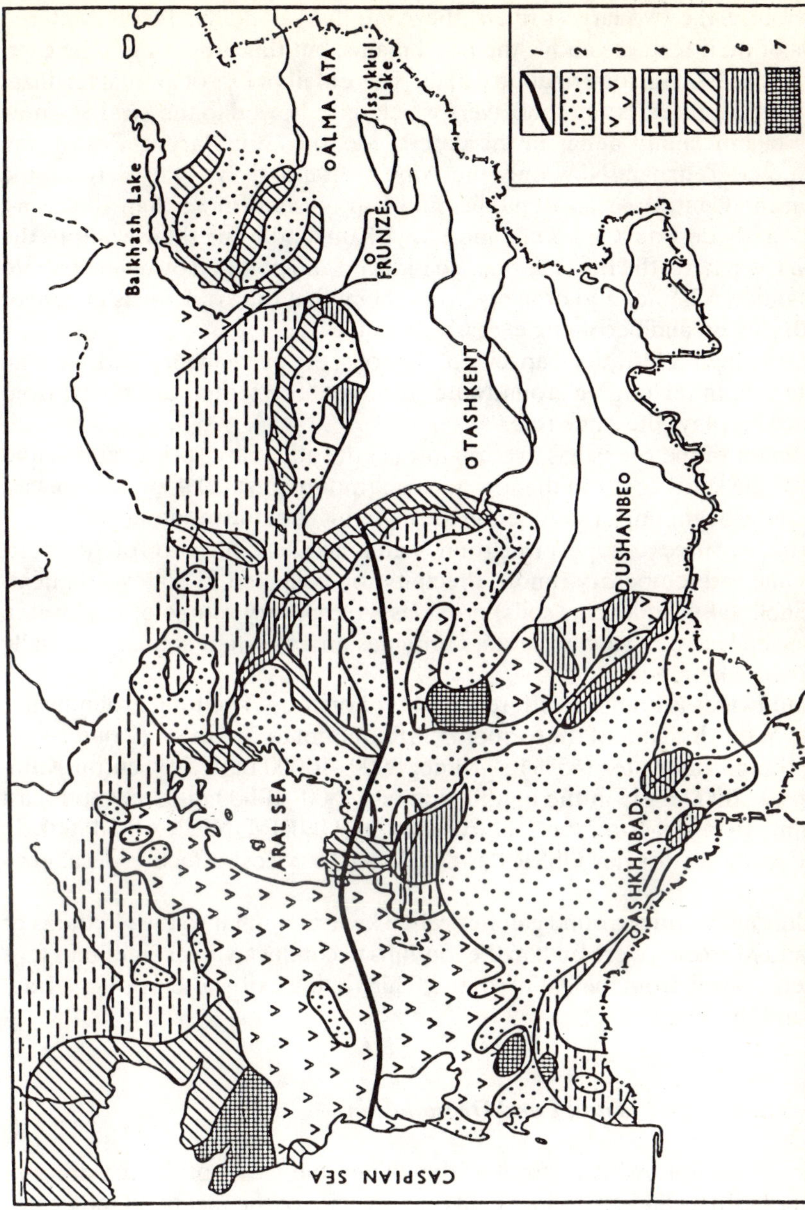

Fig. 34. Types of rangelands. (Orlovsky, UNEPCOM 1982b). *1* Subzone boundaries; *2* sand-desert rangelands; *3* gypseous-desert rangelands; *4* clay-desert rangelands; *5* rangelands in river valleys; *6* oases; *7* mountain area

Previously, the two largest rivers, the Amu Darya and Syr Darya, watered the fields of their terraces above the flood plains, but this is no longer the case, because of the 1000 km of irrigation canals. This entails a loss of annual fertilization for these terraces. Numerous rivers which once flowed to the Aral Sea now have dry channels and deltas in the desert, e.g., the Chu, Sarysu, Zeravshan, Murghab, and Tedzhan. Now only the Amu Darya and Syr Darya rivers, the longest in the Central Asian Republics and Kazakhstan (more than 1000 km), cross the sandy deserts Kara-Kum and Kyzyl-Kum and bring their water to the Aral Sea. Even here, the flow has been so reduced and polluted by diversion into giant irrigation schemes and drainage from them, that the Aral Sea is in danger of both drying up and becoming excessively saline.

With the high infiltration capacity of the soils, the permanent rivers have an important role in recharging groundwater reserves. The temporary runoff from the mountains plays the same role.

The water of the rivers has two origins: (1) the abundant spring rainfall and thawing of the snow cover in the low adjacent mountains; (2) snow and glacier thawing at the beginning of summer on the high ridges of the mountains.

The main sources of supply for groundwater include: infiltration of the rivers, main canals, and temporary runoff. Precipitation collected in hollows on near-impermeable takyr soils (clay soils) provides a temporary source of freshwater. The water collected in such reservoirs provides an important source especially for the livestock breeders.

Groundwater is found in different aquifer horizons. The total groundwater resources in the drylands of the former USSR amount to 26 $km^3 a^{-1}$ (Zonn 1986). Most of the groundwater (65%) is saline: 3000–10000 mgl^{-1} in Central Kara-Kum, up to 30000 mgl^{-1} in the marginal areas, 1500–10000 mgl^{-1} in south-east Kara-Kum. In Kyzyl-Kum the salinity is lower (UNEP-UNEPCOM 1980). In the last decade the former USSR has developed very cost-effective desalinization processes.

Traditionally underground galeries called kyarizs, similar to the ghettaras of Algeria and Morocco, were built in the foothills to conduct water sometimes tens of kilometers away from the initial source. This system exists also in China, Iran, Yemen, and the Maghreb.

3.5.3 Human Activities and Land Degradation

The previous section, which describes the vulnerable facets of the dry areas in the former USSR, can help to highlight that, in spite of the fact that land degradation depends mainly on natural conditions, in most cases the decisive factor is the impact of man on the environment.

Land degradation will be analyzed in connection with four types of human activities: old and modern oases; grazing; irrigated agriculture; industrialization, urban development, and communications.

3.5.3.1 Old Oases

At the western end of the Himalayas and on their northern foot, the former USSR has, like China, a set of oases (the ancient piedmont oases) the development of which is based on the use of surface water or shallow wells. The modern ones using deep water drilling (Ashkabad, Tashkent) are based on the development of huge irrigation schemes as on the Amu Darya.

When considering the processes by which the ancient oases were abandoned, the natural causes are superseded by social factors.

In the valleys of Khorezm, Zeravshan (main tributary of the Amu Darya river), and Sukhandarya there was formerly a flourishing strip of oases. The evolution of the ancient Khorezm oases described by Babaev et al. (1984) confirms that the dynamics of degradation was social and political.

Three thousand years ago an irrigated farming system had developed in the Khorezm oases. Iron tools and sedentary living appeared during the first millenium B.C. and marked the beginning of man's influence on his environment in this region. In the southern regions of Central Asia the first large irrigation systems were built on alluvial plains and in the catchment areas of the major rivers, the Murgab and the Atrek, and later the Amu Darya and its tributary the Zeravshan. All these systems were destroyed in the middle of the first millenium A.D., when the state suffered social upheavals. The extent of irrigated farming decreased and several cities were deserted. More than 1000 years passed before a second prosperous phase of development occurred in the 7th–8th centuries which lasted until the invasion in 712 by the Arab leader Kuteiba. During the third phase of prosperity (12th–13th century), vast previously abandoned areas on the left bank of Amu Darya werre cultivated. The density of population was high (80 to 150 persons km^{-2}). The Tatar-Mongol invasion broke this phase of economic growth, and the oases turned into desert. In the two following centuries a slow revival occurred until Tamerlane's invasion at the end of the 14th century, which destroyed the town and the walls of Urgench. Tamerlane ordered barley to be sown on the site of the city. Between the 15th and 17th centuries the irrigated area was transferred from the central Amu Darya valley to its delta. Irrigated agriculture continued with interruptions until the 19th century and was never abandoned.

This description shows that the overriding factors were the invasions, but certainly there was also another social factor: over-use of water. It is probable that these piedmont communities have attained the maximum of possible development, which caused soil salinization.

Other examples also analyzed by historical and archeological methods have shown that in southern Turkmenistan there were populations occupying these areas in the 6th to 4th millenia B.C. In the 5th to 3rd millenia B.C., oases were created, accompanied by drastic deforestation, which resulted in abandonment of the major part of the region (charcoal has been found made from juniper, which later disappeared from the area) and habitation became largely restricted to residual areas, the oases.

In the oldest human settlements only the peripheral parts of the desert were occupied by human activities. The flourishing oases were enclosed from the hostile steppes by long walls. Such walls have been found around the Murgab oasis and the Bukhara oasis. It was only later, with the development of the iron age, that human activities penetrated a little farther towards the heart of the desert. In southern Turkmenistan at the archeological site of Misrian (Dekhistan), a very large irrigation system was developed in the 1st and 2nd millenia B.C. which made penetration into the desert possible.

3.5.3.2 Land Degradation in Desert Rangelands

According to UNEP-UNEPCOM (1982) (UNEPCOM = Soviet branch of UNEP), natural rangelands in the former Soviet Union cover 370 million ha, excluding tundra and forest-tundra ranges. A third of this total area (about 122 million ha) is in the dry regions of central Asia and southern Kazakhstan: 95% in SW Turkmenistan, 84% in Uzbekistan, 89% southern Kazakhstan (Nechaeva and Pelt 1963), extending from north to south for over 20° latitude and from west to east over 36° longitude.

In the former Soviet Union, the dry rangelands, as previously seen are classified into three pastoral subzones with different climatic parameters: southern, northern, and foothills. Central Asian stock raising is divided into three types: desertic, oases, and mountainous, with three types of livestock: sheep, goat, and camel. The north was the domain of horse breeding, the oases the domain of cattle and camel, the mountains of sheep and goats based on a nomadic way of life which is nowadays replaced by a pastoral economy run by the state, with collective farms based on transhumance.

"On the whole, the total pastoral territory of Central Asia and southern Kazakhstan makes up around 129 million ha, with nearly 122 million ha situated on the desert range and the remaining area in the mountains. Given that the whole range area were devoted to livestock husbandry, it should be able to support about 34 million sheep in an average year for productivity. By comparison, it can be noted that in recent years this pastoral area has supported over 17 million sheep and almost 1 million cattle." (Nikolaev 1982) (Table 18).

Table 18. Potential grazing of rangelands in the desert regions of Central Asia and southern Kazakhstan (in terms of Karakul sheep population. (Nikolaev, UNEPCOM 1982)

Republic	Area, Mha		Mean annual forage yields, t ha^{-1}	Total forage, Mt	Grazing capacity, M head
	Total	Range			
Turkmenistan	49.1	41.0	0.13	5.41	5.7
Uzbekistan	45.1	33.5	0.29	9.74	10.3
Kazakhstan (south regions)	74.4	54.6	0.31	17.25	18.2
Total	168.6	129.1	0.25	32.40	34.2

The Karakul sheep is the most prevalent in rangeland activities, producing meat, wool, and skin. Thanks to the high quality of range forage, the lowest production costs of mutton and wool have been reached, but at the expense of land degradation as a negative consequence.

The most decisive causes of land degradation are: overgrazing, felling of trees and shrubs, and construction operations (road building, reclamation constructions). The consequences directly affect either vegetative cover or soils. In the vegetative cover there is a substitution of annual plants with a shallow root system for perennial plants and a reduction of valuable grass varieties, which are replaced by weeds or nonedible varieties.

The threshold of overgrazing is less often reached by overconsumption of the fodder than by inadequate water supply and poor water use of rangeland. The concentration of cattle around water points and wells leads to soil break up and the resultant emergence of wind erosion: these causes produce the same consequences in all the dry ecosystems of the planet. Morozova (1959), Kurochkina (1978), and Nechaeva (1979a,b) have contributed to the analysis of the vegetation depletion rings which (as in China and Africa) arise around water points:

1. Ring one (directly surrounding the water point, 0.5 km wide): vegetation is very sparse and residual in interdunal depressions, only 12 varieties are left. The loosened material is organized into active dunes.
2. Ring two (0.5 to 1.5 km from the water point); this is an area of moving sand with active dunes forming. A complete change of vegetation leads to the increase of shrubs and undershrub, to the disappearance of perennial grasses and a scarcity of *Calligonum rubens*. There are optimum conditions for the development of *Aristida karelinii*, and a restriction of floristic abundance, with only 25 varieties remaining on 20% of the surface, with dominance of the low-food-value grass *Aristida*.
3. Ring three (2 to 3.5 km): 70% of annual grasses have disappeared and loosening sand results in deflation. Aeolian blowouts appear, damaging the original vegetation and supplying sand for the development of active dunes. The main association consists of *Calligonum* associated with annual grasses, scattered trees, and high shrubs. *Carex physodes* and *Mausolea eriocarpa* become scarce, *Calligonum, Aristida karelinii*, and *Turnefortia* more numerous. The number of varieties decreases to 37 on 50% of the surface. The diversity of the vegetative cover is lost, which is probably the main indicator of rangeland degradation.
4. The remote ring (4 to 6 km from the water point): "The area around a water point, mostly affected by overgrazing, is equivalent to the radius of a daily migration of sheep: 5 to 6 km" (UNEP-UNEPCOM 1982). In this last ring the original vegetation (mostly white saxaul) is preserved, with nevertheless some substitution in the vegetative cover and moderate sand loosening.

The previous schematic description does not reflect all types of vegetation depletion but in the case of sand loosening and deflation the scheme is the same:

relatively flat fixed sands are turned into moving sands along with the appearance of well-differentiated hillocks and active dunes. Nechaeva (1979b) gives a precise time scale in rangelands affected by overgrazing: degradation takes 5 to 8 years to develop. Amelioration can take 6 years on sandy soils and takyrs, but if the degradation is more severe and seeds of certain trees and shrubs are rare, almost two decades are necessary.

The consequences of overgrazing also affects the soils: compaction by cattle decreases infiltration and augments runoff, worsening the infiltration/runoff ratio, decreasing the moisture content of the soil, and intensifying water erosion. The albedo is increased; wind erosion is also intensified. In the dry rangelands of the former USSR, just as in the African Sahel, "wind erosion is usually accompanied by a reduced content of organic matter, physical clay, nutrients, and the emergence of stones on the surface." (Orlovsky 1982a) (Table 19).

Scientists of the former USSR consider that tree and shrub felling is more detrimental than overgrazing. Animals cannot eat all the new vegetation that sprouts, while felling can totally eradicate shrubs and trees. This may become the first step towards desertification (UNEP-UNEPCOM 1982). In dry conditions *Haloxylon* and other trees combine with shrubs to create microclimates which are protective for other plants, so that the felling of one *Haloxylon* means

Table 19. Scheme of desertification induced by grazing

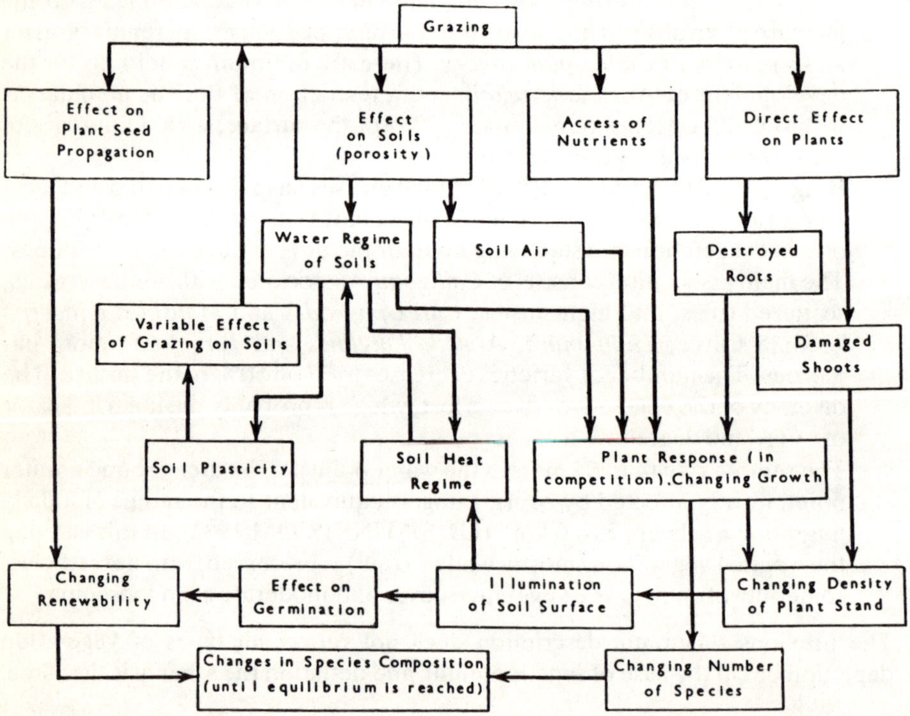

the disappearance of other plants living in the *Haloxylon* community. For cattle it also means the disappearance of natural shelter from cold winter winds and summer heat.

Petrov (1972), remembering the fuel deficit at the beginning of the century, describes how in the 1920's–1930's the dry areas were providing up to 100 000 tons of *Haloxylon* and other shrubs. This resulted in degradation belts 75 to 100 km wide around all human settlements, oases, and railways.

The unsound use of *Haloxylon* is continuing: in the Tau Kum the boundary of the *Haloxylon* forest has shifted 10 to 50 km to the north. In the Caspian Sea basin *Haloxylon aphyllum* and *Ammodendron* areas are also shrinking (Victorov 1970).

3.5.3.3 Ecological Implication of Irrigation in Drylands

Overgrazing and *Haloxylon* collection for firewood are not the sole factors concerned with environmental deterioration in the dry ecosystems in the former USSR. As in Australia, industrial development with all its train of new settlements, roads, gas and oil fields, the arrival of machinery, transport movements, and irrigation facilities, has also negative impacts and is the initial source of deflation in sandy areas.

The three main targets of irrigation are:

1. to compensate for water deficiency,
2. to reduce the danger of "desertification" in farming and grazing areas,
3. to convert drylands into highly productive lands.

Irrigation can initiate desertification processes. In the dry areas of the former USSR the main negative impact linked with irrigation can be classified under two headings:

1. Changes in the landscape: newly formed sand edifices, blowouts and deflated areas, suffosion[9] processes which lead to the genesis of holes and gullies, cutting of river flow, and accumulation effects.
2. Changes in the water quality: Chembarisov (1974 in UNEP-UNEPCOM 1982) reveals that sustained irrigation has brought with it a rise in the water salinity of the Amu Darya, the Syr Darya, and the Zeravshan rivers. These water quality changes have been neglected. Normally, the chemical composition of irrigation water changes seasonally with rainfall or increased evaporation including the effect of the return flow.

The return flow, the main constituents of which are seepage from the canals, surface runoff, and water losses through deep infiltration, will go back into the river network or aquifers and perhaps be used again for irrigation. It can be responsible for the removal of salts, fertilizers, herbicides, insecticides, fungicides, or other chemical or organic remnants, bacteria, pathogens, microele-

[9]Suffosion means collapse of the surface because of underground dissolution.

ments, and mineral particles from the irrigation system and their transfer via the drainage system to the river. This can totally change the composition of the river water and presents a danger for aquatic animals and in some cases even for terrestrial animals and man.

The effect of the irrigation water quality on environment can be so highly detrimental that Kovda (1977) proposed an in-depth study of the quality and chemical composition of these waters.

The effects of irrigation water are summarized in Table 20:

1. On soils in dry areas, an evaporation rate higher than the rainfall, induces the accumulation of salt at the surface or in the topsoil. This affects the quality and quantity of production.
2. Unwise use of irrigation water, and the extension of the irrigated area without respect for seasonal change in water availability and of the quality of the water intake leads to soil salinization.
3. They can produce great economic and social damage through sterilization of the land, destruction of agricultural activity, and the consequent necessity to re-settle the population.

In all the drylands the chief goal of agricultural development today is the intensification of production, which involves not only irrigation and drainage, but unfortunately the widespread and increasingly frequent application of chemicals, both fertilizers and biocides. These substances get into surface and groundwater supplies through the return flow from irrigation. The former USSR has experienced the ecological effects of pesticides on the local water supplies, i.e., short-term reduction of incoming oxygen, increase of carbon dioxide, decrease of pH values, changes of nutritional value, changes of plant growth, and alterations in the aquatic fauna.

The example of the Karakum canal demonstrates how river management for irrigation results in unexpected changes in the landscape and degradation.

Before the 1917 revolution a project was advanced for the regulation of the river flow and redistribution of the water towards water-deficit areas.

The description of the building and of the consequences of the Karakum canal by Grave and Grave (1986) gives an honest example of both its advantageous and detrimental impact on the environment.

The first stage (1959) was to bring the Amu Darya water through the Karakum canal to the Murgab oasis using an ancient river channel, the Kelif Uzboy, crossing the Amu Darya-Murgab interfluve (studied in 1985 by academician and geographer Obruchef). The major problem at the beginning was seepage. Water seeping from canals can have numerous detrimental impacts: rise of the water level of shallow groundwater, waterlogging in the case of poor natural drainage, and consequent soil salinization.

Nevertheless the bed of the canal became naturally and slowly lined by a sheet of fine silty material and the bedrock saturation and seepage decreased. This experience brought the knowledge that even a canal built in sand (300 on 397 km run through a sandy desert) can work without special measures to render the bed impermeable.

Table 20. Effect of irrigation on natural components. (After Zonn 1981; UNEPCOM 1982)

Natural components and their stability to irrigation	Changes in natural components induced by irrigation	Negative impacts pertaining changes in natural components
Lithogenic components, most stable	1. Creation of agro-irrigation relief	Formation of gullies, padings, blowouts
	2. Change of soil-formation conditions	Worsening of soil physical properties, decreased productivity
	3. Change of soil water properties	Mostly positive after-effects
	4. Change of salt content in the aeration	Soil salinization, decreased productivity fall out of agricultural areas in disuse
Hydroclimatic components medium stability	1. Change of river flow 2 Change of river water salinity due to diversion of return flow	Infringement of river flow, increased water salinity, changed vegetation in river basins, worsened habitat of hydrobionts
	3. General deterioration of water quality in rivers due to diversion of return flow	Increased biological and chemical requirements in oxygen, greater content of toxic elements in water, eutrophication, spread of disease-carriers, changed color and taste of water
	4. Change of groundwater table	Salinization, groundwater depletion
	5. Change of groundwater salinity	Soil salinization
	6. Change of microclimate, wind, moistening pattern in the horizon of aeration	Mostly positive after-effects
Biogeneous components, low stability	1. Change of natural vegetation and fauna	Mostly negative after-effects
	2. Change of cultivated vegetation	Mostly positive after-effects

The second stage took the canal from Murgab to Tedjen (138 km). In the same year (1960) the Haus-Khan water reservoir was built to retain the autumn-winter flood.

The third stage from Tedjen to Ashkabad (258 km) started in 1962 and finished in 1975, including several extensions.

The Karakum canal supplies water to the Murgab and Tedjen inland deltas, the western-Turkmen lowland, the cities and oilfields of western Turkmenistan. Its total length is 1100 km, 450 of which are navigable. The irrigated area is 550 000 ha with cotton and subtropical crops, and the desert rangelands in the western area of Amu Darya receives, over 3.5 million ha, 10 to 11 km^3 of water from the Amu Darya (Grave and Grave 1986). Nevertheless, Grave (1976) describes how along the Karakum canal the sand excavated for the construction of a pipeline and canals has been reshaped into moving dunes encroaching on the settlements of the workers. This process of degradation was aggravated by the combination of excavated sand, cattle grazing, road-building, and firewood collection.

Figure 35 shows the disturbances produced in the landscape by the operation of machinery:

— a belt of sands 0.5 km wide from where vegetation has disappeared;
— a second belt 1 to 1.5 km wide of semi-fixed sand, with marshes and seepage lakes where water-loving plants and weeds are now growing.

The hydrological impact of the canal involved more than 2000 km^2. The seepage losses had an effect on the water table, which rose by 10 to 15 m and even 20 m in places. Along the canal, in the interdunal depressions, the permanent seepage lakes combined with temporary lakes to give the area a swampy aspect. At the same time, however, the phenomenon of waterlogging had a positive aspect by attracting migratory birds: ducks, geese, seagulls, and cormorants. Wild boar settled in the cane growth; fish appeared in the canal and in its lakes.

Thanks to a new water distribution around the wells built along the canal, where in the early stages there had been areas of high degradation, new fodder reserves (alfalfa) farms and farms growing vegetables, melons, and even cotton were created.

In the Murgab and Tedjen oases the irrigated area reached respectively 212 000 and 190 000 ha. After the construction of the canal an efficient drainage system was established in the early 1970's. In 1970, 2 988 000 tons of salt were removed from the oasis of Murgab, i.e., 120% of the input. This means improve-

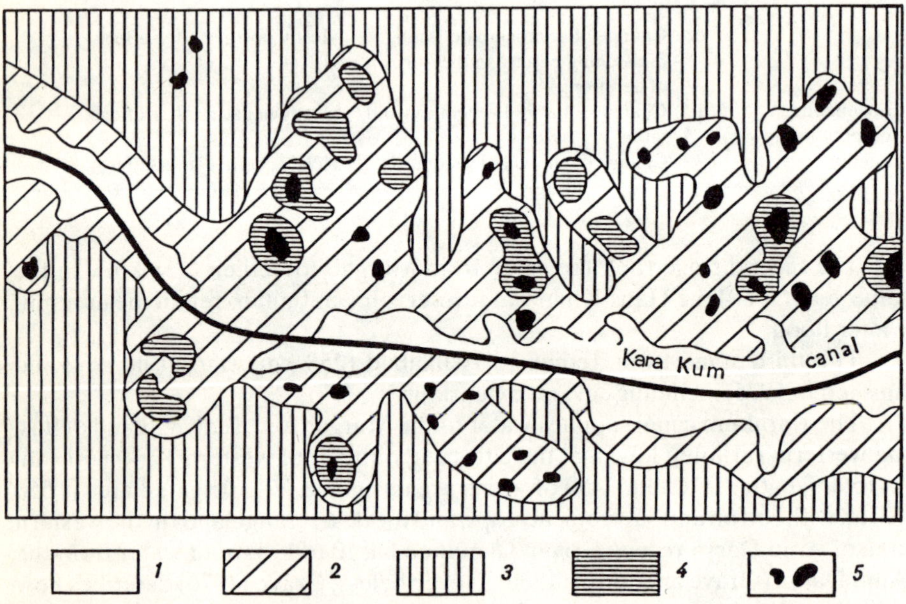

Fig. 35. Natural variation in conditions in the zone of the Kara Kum canal. (After Kharin 1976 in Orlovsky, UNEPCOM 1982a). *1* Destroyed sand along canal route; *2* semi-fixed sands; *3* fixed sands; *4* marshlands; *5* filtration lakes. This map covers 120 km from *left* to *right* and represents roughly the region of Tedjen

ment of the situation but not complete control of the salinization built up before the creation of the drainage system.

There are further projects for sand fixation, horticulture, and the installation of the Mary hydroelectric power station, the largest in Central Asia (1.2 million kw capacity), operating with canal water and natural gas.

The supply of Amu Darya water to the fertile piedmont plain of Kopet-Dag has made possible the increase of the irrigated area from 20000 ha (irrigated by local streams) to 100000 ha. A 300-km-long pipeline (Ashkabad to Gerbent, village NE of Ashkabad) brings water to farms and cattle raisers previously dependent on wells. The natural grassland was enriched by seeding fodder grass varieties in the irrigated areas.

The flood water is controlled and transferred to the Karakum dune areas, which will be developed for farming. This example shows how a planned desert land irrigation system can lead to the rehabilitation of degraded land, to the expansion of oases and to dryland pasture amelioration. Nevertheless, two main difficulties still persist in the drylands: the problem of salinization and the difficulty of combatting wind erosion.

Furthermore, along the Karakum canal, the rising water table over an area of about 1 million ha and the watering through water courses and lateral irrigation of another 4 million ha of grassland has permitted dispersal of the sheep population and decreased the trend towards increasing erosion around existing foci.

The land degradation linked with the Karakum Canal is only one facet of what became later on the ecological disaster of the Aral Basin in its whole and of the Aral Sea itself. In an attempt to evaluate the problems of the *Aral crisis, its origins* and *solutions*, and to prepare an applicable plan for rehabilitation, the former USSR organized two international symposia in Nukus, Karakalpak(Uzbekistan) in September and October 1990.

The present situation in the Aral Sea and its basin can be characterized by three interconnected problems: (1) the drying of the Aral Sea; (2) the health problem of the population in the basin; (3) the problems of overirrigation.

1. With respect to the drying up of the Aral Sea, several facts have to be considered. Between 1960 and 1981 the level of the sea dropped from 67 000 to 41 000 km^2 and its volume from 1064 to 404 km^3 (Glazowsky 1990). In 1989 the Aral divided into two parts. The cause of the reduction is mainly diminishing run off of water from Syr Daria and Amu Daria to the sea.

   "Withdrawal of water for irrigation resulted in changes of river runoff. Where in the period from 1911 to 1960 of the total 120 km^3 a year of river water originating in the region 56 km^3 reached the Aral Sea. In the mid-70's, due to the withdrawal of water for irrigation, it dropped to between 7 and 11 km^3, while in the 1980's during several years the inflow of river runoff to the Aral ceased almost completely" (Glazowsky 1990).

   The arising problems are:

a) The salinity of the Aral water increased from 10 to 24‰ and in some areas reached 28 to 30‰, and 38‰ in 1993. This increased salinity resulted in a complete change in the biological food chain and the species composition (Aladin and Khlebovich 1989), in a reduction in fishing and in economic fishery, which has completely ceased today. Formerly, fishing produced 40000 t fish annually; presently, some fishing exists on remaining small lakes and in some drainage collectors and lakes but the quality of the fish is unsatisfactory.

b) Huge amounts of seafloor are exposed with the effect of aeolian deflation of the formerly deposited salts and other particles in the whole region of Central Asia. This also results in blowouts and Solonchaks, which are most frequent in the eastern and southeastern parts of the Aral Sea.

c) The reduction of the Aral Sea waterbody has already a climatic and meteorological effect. The climate of the area is becoming more continental. Salt and sandstorms have increased in number. Over 25 years the number of days with dust storms in the coastal regions increased between 1.5 and 3.6 times (Molosnova et al. 1987). Amounts of salt blown out of the sea bottom increased: according to Glazowsky (1990), the more probable estimations are between 40 and 150 million t. The differences between minimal and maximal temperatures have increased by 4°C.

2. The following factors are connected with health hazards to the population:

a) High child mortality, high disease rate among children;

b) High incidence of intestinal cancer;

c) Low level of immunity of the population.

Deterioration of the environment due to irrigation and extensive use of chemicals in agriculture resulted in an increase in illness and death rates. Overpopulation is also a factor in the increasing problems in this area. From 1959 to 1987 the population growth in Uzbekistan was from 8.1 to 19 million, in Tadjikistan from 1.5 to 4.8 million, in Turkmenistan from 1.5 to 3.36 million.

3. The third set of problems deals with irrigated areas and agricultural questions. In many irrigated regions the groundwater table rose and secondary salinization is widespread. Rozanov (1984) has shown that 1 million ha of land is lost in Central Asia because of erroneous irrigation practices. According to Khakimov (1989), the percentage of moderate to severe salinized soils in irrigated areas reaches 60% in Uzbekistan, 80% in Turkmenistan, 35% in Tadjikistan, 40% in Kirghizistan, and 60 to 70% in Kazakhstan.

The beginning of the Aral ecological crisis may be related to the decision in the 1950's and 1960's to realize a large-scale irrigation project in the Central Asian areas of the former USSR and Kazakhastan with different objectives, as summarized by Glazowsky (1990):

— To increase the production of cotton and rice, to provide the population of the country with cotton cloth and ready-made goods, and to increase cotton export for earning hard currency;
— To enhance the production of vegetables and fruits turning the region into All-Union orchard;
— To provide the population with meat and rice;
— To raise population employment.

Since early 1960's the area of irrigated lands in Uzbekistan and Tadjikistan increased 1.5 times, in Kazakhstan 1.7 times and in Turkmenistan 2.4 times. Such expansion of irrigation was accompanied by considerable capital investments in agriculture. For instance, the main production stock of agriculture increased between 5 and 7 times, energy supply up to 6 times, the number of tractors 3.2 times and the capacity of the tractor motors 7.6 times. Simultaneously, there was from a 3.5 to 6 fold increase in the amounts of mineral fertilizers applied in the fields; the specific amount of fertilizers per hectare of plowland in the republics of Central Asia being 2.5 or 3 times as great as in the Russian Federation (Glazowsky 1990).

The decrease in river runoff resulted in biotic degradation in the delta areas of Amu Daria and Syr Daria. Reeds dried up in the shallow lakes. One million ha of food plains dried up (980 000 ha in lower Amu Daria and 200 000 in lower Syr Daria). The yields of reed communities diminished from 3 t to 7–130 kg/ha according to Novikova (1990) and to Zaletaev and Novikova (1990). By 1987 alluvial-meadow soils in the lower reaches of Syr Daria dried up and turned into Solonchaks. Gerasimov et al. (1983) estimated that about 730 000 ha were eliminated from agricultural rotation. Due to changes in vegetative cover the total fodder reserves decreased from 1.2 to 0.5 t and the productivity of grassland decreased three times.

A recent assessment of Khakimov (1989) shows that soil salinization results in a decrease in yield on irrigated land: 30% in Uzbekistan, 40% in Turkmenistan, 30 to 33% in Kazakhstan, 18% in Tadjikistan and 20% in Kirghizistan. The yields in general have decreased: in Turkmenistan in 1986 the cotton began to drop to the level of the 1980. In Kazakhstan and Kirghizistan the yield of raw cotton has not improved since 1980.

Specific expenditures, mainly for new irrigation constructions, are required for agriculture.

Great amounts of pesticides needed in the monoculture of cotton and large amounts of defoliants used for harvesting pollute the runoff waters and are often discharged back to the rivers. Finally, the increasing competition for water and for drainage canals has aggravated the social tensions between the Republics.

As summarized by Glazowsky (1990), the sources of the Aral ecological crisis are:

1. An erroneous strategy of development and location of productive forces;
2. Mistakes in the strategy of agricultural development: introduction of cotton as monoculture, excessive expansion of irrigated agriculture, wide use of herbicides, and development of low-productive lands difficult to ameliorate.
3. A low quality in designing, building and exploiting irrigation systems.
4. Too high and poorly substantiated irrigation standards.

3.5.3.4 Environmental Impacts of Industrial Development in Drylands of the Former USSR

Arid and semi-arid lands in the former USSR are areas of increasing industrial development in both the short and the long term.

According to Utkin (1986), industrialization increased from 1913 to 1967 by 137 times in the Kirghiz Republic, 72 times in the Tadzhik Republic, 35 times in the Uzbek Republic, and 37 times in the Turkmen Republic. The oldest industries in these republics were engaged in the processing of agricultural raw material: cotton, domestic spinning and weaving, tanneries, and small food-producing units. Later, larger plants were created for food industries and, in relation to mining, mechanical engineering, and energy production.

The drylands of the former USSR presently contribute about one third of the amount of natural gas and one seventh of the total amount of oil and coal produced by the country. They also produce large quantities of ferrous and non-ferrous metals, which places Kazakhstan at the highest rate for ferrous metallurgy and chemistry. This requires a parallel effort to develop modern urban centers and supportive agriculture. Central Asia is self-sufficient in energy for its urban centers, for mechanized irrigation, and for industrial processes, and there is even a surplus for export to other regions of the former USSR.

Industrial development, if not carefully organized, is a cause of land degradation: access roads, railways, and pipelines for transfer of oil and gas can also harm the environment.

In drylands of the former USSR hard experience has shown that there must be an integrated and permanent policy to protect all infrastructures against sand drifts. Behind this policy lies much technical and administrative progress, one of the most relevant points being to confine major work management to the sites of sand erosion or transport and not, as is generally done in Africa, to the accumulation sites. Shlikhter (1986) insists on the fact "that integrated protection of railways from sand drift is only possible through the mechanization of sand stabilization operations". In the Central Asian railroad hundreds of km of shelterbelts are necessary: over 600 km in length and 14.5 ha in surface have already been built. Roads are being protected by sand-arresting trenches dug by trenching machines.

In all cases of land degradation, the solutions exist but they are more expensive than was previously estimated.

3.5.4 *Aeolian Action and Land Degradation in the Sandy Deserts of the Former USSR*

Soviet scientists (UNEP-UNEPCOM 1986b) have classified sand surfaces in relation to their vulnerability to deflation:

1. Lack of deflation. The surface is swept by wind without the appearance of ripples: even with this first type there can be shifting sand mainly by creeping. It is often observed along the periphery of sand tracks.
2. Low deflation: there is evidence of ripples and sand accumulation without typical dune formation. The second type occurs on the outskirts of irrigated areas, shores of lakes, and solonchaks (salt deserts).

3. Moderate deflation: sand-ripples, ridges, and isolated small dunes appear in the same areas as in (1).
4. Severe deflation: a combination of active dunes and chains of active dunes (from 2 to 20–25 m high) alternating with hummocky sands with a cover of herbaceous-shrub vegetation.
5. Very severe total deflation: the whole landscape is shaped by active dunes; only some isolated plants persist in the interdunal corridors.

The sandy deserts can be divided into two parts in terms of intensity of aeolian actions and their natural potential:

a) areas with active dunes and intense aeolian processes,
b) low productive fixed sandy areas or vegetation-covered aeolian relief, but where the surface is affected by varying degrees of deflation.

According to their genesis, Petrov (1973) classified the sandy deserts into four subtypes:

1. old alluvial plains (Lowland Karakum, Sary Ishikotran, Muyunkum, Aral coastal Karakum),
2. coastal plains (North Caspian sands),
3. piedmont plains (small areas in the Kopet Dag piedmonts) with Cretaceous and Tertiary deposits,
4. structural plains (Trans-Unguz Karakum and Kyzylkum). The low plains of Central Asia are Quaternary. They were mainly built up under a continental desert regime and formed in recent valleys and deltas of the Amu Darya, Syr Darya, Murghab, and Tedzhen rivers. Following the drying-out of the ancient river system, the alluvial plain was subject to wind action and the present aeolian relief was shaped.

Shifting sands occupy 5 to 7% of the Central Asian deserts (Petrov 1950; Ivanov 1969). Their genesis is of two types:

1. Geological as in the west-Turkmenian Lowland at a time when the Caspian water was at a low level and the marine and alluvial deposits were exposed to strong deflation and winnowing. The shifting sands of Dardzha Kum and Kyzyl Kum have the same origin. The Kelkor dunal area results from the deflation of the new Caspian sediments in historical times (roughly in the last 120 years). The largest area of active dunes in Lowland Karakum, called Dzhilli Kum, was fed by the reworking of the Trans Unguz sediments.
2. Man-induced shifting sands are visible south of Dzhilli Kum. They evolved in the recent period as a result of agricultural over-use of the old Amu Darya delta and winnowing of the sandy-clayey alluvial material.

It is probable that wind erosion is maximum in the Karakum near watering points and because of vegetation cutting at or near all the sites where gas and oil pipelines are laid or around areas of geological exploration.

Dzhanpeisov and Dzhamalbekov (1978) give a very convincing and depressing example of land degradation induced by nonagricultural activities:

"Intensive geological surveys and transportation of drilling rigs are to be blamed for turning 18% of rangelands in the southern Mangyshlak plateau into temporarily unsuitable lands where from 70 to 80% of the vegetation is destroyed".

The migration of shifting sand has a threefold detrimental impact:

1. Burial by sand of production facilities, irrigated land, railways, and highways. In the mid-Amu Darya oases in the 1920–30's, sand encroached over irrigated land at the rate of several tens of hectares a year. On the same river, the town of Turtkul was almost completely overwhelmed. In the Bukhara oases thousands of hectares of irrigated lands were lost under shifting sand (Petrov 1950). Crop farming in the lower part of the Amu Darya and in other regions suffered the same fate.

   In drylands, transport systems require increased expenditure of financial, material, and manpower resources because they stand exposed to the threat of sand drifts and blowouts. Roads and railroads require sand fixation on their margins. Sand drifts, especially in areas of encroaching barchans or elongating linear dunes severely affect transport operations when the roads are cut by the drifting sand. Traffic can be totally interrupted for hours, even days, and shipment delayed. Sand and dust storms complicate aerial traffic, cause failure in fuel systems, electrical equipment, and lubrication, and increase the wear of the machinery;" airborne dust concentrations in excess of 10 mg m^{-3} shorten the service life of engines by a factor of 8 to 10." (Utkin 1986).
2. Blowouts, another constraining impact of shifting sand, can excavate the power transmission-line constructions, gas or oil pipelines, and the foundations of individual engineering structures.

   Already in the 1880's, control of shifting sands in Central Asia was necessary to protect the Transcaspian Railroad from encroaching sand. Clay and ballast cover, and matting of dry grass and bush were used.
3. Changes in the soil texture are another adverse consequence. In the reblown sands the percentage of clayey and silty particles is reduced to 1.5–2%. The percentage of grains over 0.250 mm is increased. As a result, the potential agricultural fertility of the soil is reduced.

3.5.5 Conclusion

The development in drylands is a difficult and challenging task. In these ecosystems the approach, closely linked with environment, entails strict attention to the protection of the ecological situation in line with the improvement of the economic situation. In dry ecosystems any enterprise is a violation of the ecological balance and this can result in land degradation. Irrigation may trigger waterlogging and salinization. Increasing water intake from the rivers results in a dramatic reduction of the discharge of the rivers into the sea or inland lakes.

Despite the dilemmas associated with irrigation, irrigated farming in the drylands remains the most efficient approach to combating land degradation. Compared with dryland farming, the return yield of cereals is sixfold higher and for industrial crops four- to fiveold higher. This explains why in developing countries the growth percentage of irrigated areas is four times higher than the growth of dryland cropping (UNEP-UNCOD 1978).

Overstocking of pastureland and the felling of trees and bushes may start foci of deflation and a reworking of the upper part of previously fixed sandy edifices producing tracts of shifting sands. In the former USSR, among the varied processes responsible for increasing land degradation, wind erosion in all its forms still remains the most important. Even with the sparing use of desert resources, combined with the expenditures of amounts of money devoted to agroreclamation and afforestation, neither the areas occupied by moving sands nor salinization are totally controlled in the former USSR. Nevertheless, the simultaneous study and development of the sandy territories has facilitated control solutions.

Several laboratories, the Institute of Deserts, Academy of Sciences, Turkmenistan; Central Asian Forestry Research Institute in Tashkent; National Scientific Research Institute of Agro-Reclamation and Afforestation in Volgograd, and the Institute of Railroad Engineers in Tashkent, to name but a few, have formulated the bases for sand fixation, studied the genesis of dune reactivation, of birth of sand areas, investigated the biology of sand-fixing plants, and followed the action and efficacy of chemical sand stabilizers and mechanical binders.

In the last few decades crop protection belts have been planted in areas with strong wind erosion: more than 6000 ha in Turkmenistan and 28 000 ha in Uzbekistan, mainly in the central part of the Ferghana valley in the Golodnaya and Karshinokaya steppes. Afforestation and sand fixation in Turkmenistan have reached 300 000 ha and 608 000 ha in Uzbekistan. From 1976 to 1980, 367 000 ha of forest was planted on irrigated sand and mountainous land and from 1981 to 1985, a further half a million ha were afforested (UNEP-UNEPCOM 1986b).

During the period 1976–1980 some 37.6 million ha (some source; the FAO says that all irrigated land in the former USSR was only about 20 million ha) pastures were irrigated. However, the deterioration of pastureland, which is another affliction of dryland ecosystems in the former USSR, is not totally prevented, even if irrigation and afforestation are well advanced.
also to be taken into account because it can result in silting up of the bed and in the breaking of dams. In recent years some 1480 ha of bank-protective forest belts were planted.

Along the banks of the Karakum canal alone, some 12 300 ha of hydrophytic trees and shrubs belts were planted.

All these processes of rehabilitation were applied after pilot research, but the speed of the therapy is insufficient and the economic cost too high to eradicate the land degradation in a short time. Nevertheless, the knowledge exists and "desertification" is not the right word to use because degradation in the former USSR can be reversed with manpower and money.

3.6 High Technological Capital-Intensive Pastoralism and Land Degradation or "Desertification" in a Dry Continent: Australia

3.6.1 Human Causes of Land Degradation in Arid Australia (Table 21)

Over 5 million km^2, about three-quarters of Australia, are arid. About 50% of these are sandy desert, mainly in the west central part and are unsuitable for grazing because of the quality of the vegetation which is impalatable and not nutritious (Perry 1984). The remainder is mainly limited to extensive grazing based on natural vegetation. "Australia is a pastoral nation" notes Graetz (1984). In fact, a third to a half of this area is too arid to be used for any agricultural activity. The dry rangeland is used on an extensive settled basis and is part of a monetary export market-oriented system, which is totally different from Chinese and African nomadism with herders and large animal migrations.

"The area began to be developed for grazing in the 1860's at about the same time as the western USA" (Perry 1984). Nowadays, of a total population of 15 million, only 2% live in these ecozones and use it for grazing: sheep (30 million) in the south, beef cattle (8 million) in the north. During the first 120 years of grazing the objective was maximum exploitation of the pastureland based on native vegetation without attention to carrying capacity. The land has paid the price: considerable land degradation has occurred, due to excessive animal population and grazing pressure. Overgrazing resulted in selective disappearance of the perennial and more palatable plants and their replacement by annual and less nutritious species, lowering the carrying capacity particularly during dry spells and droughts. Reduction in the density of the cover, aggravated at local points by trampling, has favored the mechanical processes of degradation: wind and water erosion.

In the arid areas of Australia, depletion of vegetation and soil erosion by wind and water caused by man's activities are the predominant processes of soil degradation.

Table 21. Status of the land according to land occupation in Australia. (Perry 1984)

	Total area (km^2)	% in good condition	% in need of restoration with changed management
Grazing	3 400 000	45	22
Nonarid grazing	1 300 000	63	20
Extensive cropping	443 000	32	
Intensive cropping	24 000	33	30
		Conditions which require land conservation treatment (%)	Treated areas (%)
Grazing		33	1
Nonarid grazing		17	
Extensive cropping		34	3
Intensive cropping		34	2

Whilst the areas of land are vast and the animal population high, the human population is low. These three characteristics together pose special problems of land management and land degradation, which are specific to Australia.

A government report estimates that of 3 400 000 km^2 used for grazing, 45% are still in good condition, 22% could be restored with changed management, and 33% require land treatment; 1% only has been treated (Perry 1984).

The problem for Australian graziers is to keep the balance between sustainable production at a commercially profitable level, against a background of variable rainfall and fluctuating world market prices, without causing irreversible damage to the natural resources of the ecosystem. The difficulty is that of satisfying a heavily capitalized enterprise while avoiding deterioration of rangeland resources. In the face of these difficulties, adjustments have been made: the state as leaseholder has laid down lease conditions, has improved communication for a better transfer of cattle to different grazing areas, has arranged for more efficient selling off in dry season of the surplus, and has institutionalized grazing control.

3.6.2 Grazing and Other Agricultural Activities in Semi-Arid Australia

In the semi-arid part of the country (rainfall 150–500 mm/year), all the land which is considered unsuitable for cropping is used for grazing (1300000 km^2): this includes steeply sloping areas, land with shallow soils, and areas which are too cold in winter or too hot in summer for cropping.

All the land was originally covered with forest (deep-rooted evergreen trees) and woodland. When cleared to increase the grazing surface, the water balance changed first, then water and land salinization appeared, and erosion came with overgrazing. One third is degraded (Table 21).

1. Extensive cropping areas in the temperate nonarid part of Australia occupy 443000 km^2, two-thirds of which are degraded (Table 21). The land used for cropping was also previously woodland. Different soils and a wide range of climatic conditions, often climatically marginal areas, are used for extensive cropping in a system including farming and grazing on the same farm.

   Because of poor assessment of the climatic risks and of the soil limitation under continuous cultivation, rainfed cropping is the main cause of land degradation in the semi-arid areas. The expressions of land deterioration are depletion of fertility followed by accelerated erosion: wind erosion mainly with sand shifting on light soils, water erosion following inappropriate tillage and finally the consequent enforced abandonment of farming land and the exodus of the rural population.
2. Intensive cropping areas follow the same steps of degradation as in the dry western USA. Intensive cropping produces high value crops, in Australia mainly sugar cane and horticultural crops. An area of 24 000 km^2 is used for intensive cropping. Two-thirds of this surface is degraded and needs changes in management practices or soil conservation measures but only 2% have been treated. Intensive cropping requires higher rainfall or

irrigation and tends to produce a rising water table, leading to waterlogging and salinization.

Mabbutt (1986) describes two areas affected by land degradation and salinization:

"The main area affected is the wheat belt of Western Australia, where the problem has long been recognized (Teakle 1938). Malcolm and Stoneman (1976) report that 167 000 ha of previoulsy productive farmland are now too saline to grow the preferred crops and pastures, and this area is steadily increasing. Although the area salinized is relatively small, amounting, for instance, to between 1.5 and 2.5% of the cleared areas, considerable parts of individual properties may be affected. Another semi-arid dry farming area affected is the Mallee region of northwestern Victoria, which has also been closely surveyed. Estimates of the area damaged by salinization here vary between 3600 ha (Rowan 1971) and 5000 ha (Northcote and Skene 1972). The problem also occurs in South Australia, where figures of 14 000 ha are quoted as being affected (Matheson 1968), and in Queensland (Pauli 1972), New South Wales (Van Dijk 1969); (Hamilton and Lang 1978), and Tasmania (Colclough 1973). For Australia as a whole, Northcote and Skene (1972) state that 197 000 ha that once produced satisfactory crops and pastures is now salinized. Estimates of areas affected vary markedly between observers, depending on the criteria used."

The salts are mainly sodium chloride, although salts of calcium and magnesium can also be present. The effects of salinization on crops through osmotic stress or soil changes which restrict the availability of water to the plants are the same as drought symptoms (physiological drought).

Land clearing for agriculture involving the replacement of deep-rooted evergreen trees and shrubs by alloctonous cereal crops is the main cause of salinization in semi-arid Australia. The cereals have been selected to grow with winter rains. The consequences are reduction in evapotranspiration, excess water for runoff, and seepage on lower slopes, even waterlogging and salinization in adjacent valley floors.

Mabbutt (1986) gives an exhaustive list of the surface aspects of land salinization:

1. Areas that may be devoid of vegetation with dead trees or shrubs or colonized by salt-tolerant plants.
2. Surface salinity expressed by salt crystals, crusts or cracking, puffy areas.
3. Surface often moist but dry in summer with salt efflorescences.
4. In the affected areas salt scalds may appear. (The word scald is used for areal erosion by wind or water). On the margin of the scalds, rills and gullies may appear, maximizing water erosion, while the cracking puffy surfaces over the loose topsoil increase wind erosion.
5. The saline and sodic soils which are very extensive in Australia, can, with agricultural development, become sources of salts in soil water which can result in impervious subsoil layers and perched soil water tables responsible for increased runoff, seepage, and waterlogging downslope.
6. Salt accumulation and clay dispersion in the top soils can produce crusted surfaces which increase runoff and lead to flooding of valley floors.

7. The hard surface seals maintain evaporation losses, add to surface salinization, and hinder natural revegetation of bare ground.
8. The increasing extent of wetted horizons above shallow saline water tables increases the capillary rise of salts to the surface in valley floors.

3.6.3 Land Degradation in Irrigated Areas

As a result of the transformation of the hydrological cycle under irrigation, waterlogging and salinization of soils can occur when drainage is technically poor.

Mabbutt (1986) gives a very realistic description of the Murray Valley where 80% of the 1 200 000 ha of irrigated land in Australia is situated (Fig. 36).

"Irrigation was introduced into the Kerang Region of northwest Victoria in the late 1880's and into the lower Goulburn Valley in the early 1890's. In parts of the Kerang Region water tables had already risen to within capillary reach of the surface by the turn of the century and some lands had gone out of production through consequent salinization. In New South Wales, where irrigation had commenced in the Murrumbidgee Irrigation Area (MIA) in 1912, first reports of rising water tables appeared in 1915, but the dimensions of the problem first became apparent following the heavy rainfalls of 1930–1931 and recurred with increasing severity in the wet winters of 1939, 1942, and 1956. Further west, the main problem is the increasing salinity of the water of the Murray River, as a source of irrigation water. The two problems are interrelated in that, in the areas of shallow water tables upriver, saline groundwater seeps into natural and artificial drainage systems or directly into the Murray River, which is the ultimate drainage collector of the whole valley.
A recent survey of salinity in the Murray Valley (Gutteridge et al. 1970) gave estimates of the areas of occurrence of shallow and saline groundwater in the Riverine Plains zone of the Murray Valley. This shows the Kerang Region to be the only area seriously affected by salinization at present, and that in the Deniliquin area the problem is in an early stage of development only. Nevertheless, predictions based on calculated accessions to the water table, supported where possible by the known rate of rise, indicate that the area now affected (640 000 ha) could increase by 150% by the end of the century, including 40% of the Deniliquin Region. Existing estimated annual losses of production of a $24 million could increase fivefold by the year 2000, if no action was taken." Mabbutt (1986).

3.6.4 Land Degradation in Nonagricultural Areas

Land degradation can result from all types of land uses: communication, mining, industry, urban settlement, recreation, tourism, but on a smaller scale, compared with agriculture. The degradation of water supplies by agriculture and industry is probably the major cause of environmental deterioration in Australia.

3.6.4.1 Impact of Urban Settlements

The older towns of the arid areas in Australia which originated as mining settlements formerly had larger populations. Perry and McAlpine (1976) describe the concentric zones of environmental impact:

1. The built-up area which formerly included industrial or mining plants. Degradation of soils can result in severe dust problems which affect urban

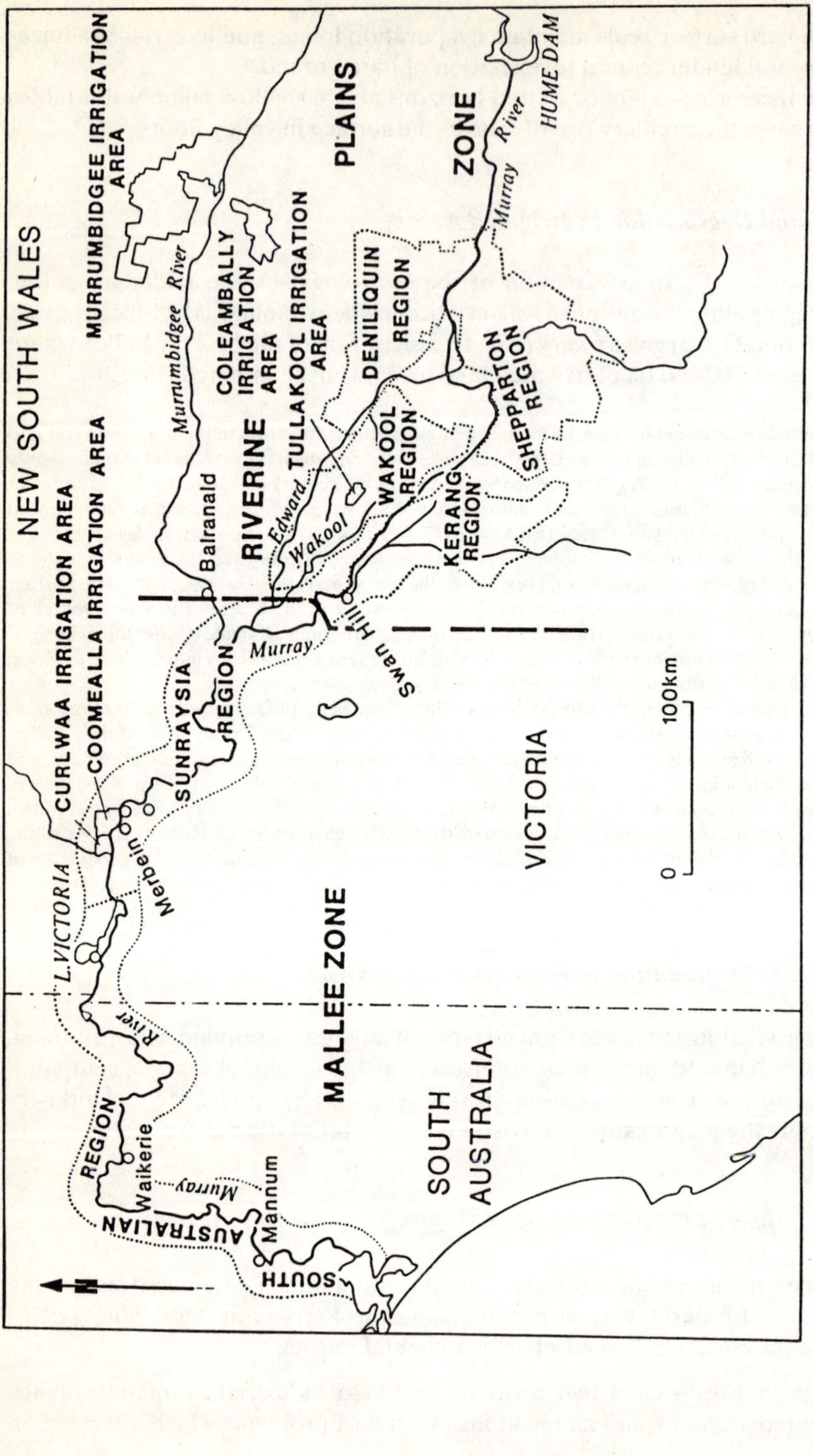

Fig. 36. Irrigated regions of the Murray valley. (Mabbutt 1986)

living conditions. "Many of these older arid zone settlements are indeed urban deserts rather than oases." (Mabbutt 1986).

2. The surrounding area depending on the needs and activities of the urban area: construction of the town requires timber, construction material, dumping of industrial wastes, needs for industrial water supplies, material for industrial communications, and items for the life support system of the town inhabitants, domestic water and energy supplies, grazing for domestic livestock, and material for road and railroad building.

In arid areas, rural activities, i.e., uncontrolled and offroad traffic, hunting, and unplanned settlements, have also environmental consequences. Varying degrees of degradation of the vegetative cover occur by removal of trees, grazing out of edible species, introduction of weeds and exotics; or complete devegetation leading to increased wind erosion with the classic loss of top soil, sand drift, and dust storms. Denudation and compaction of soils by vehicular and pedestrian traffic and sport can accelerate water erosion. In the neighborhood area of industrial and mining centers, pollution of soils, groundwater, and surface water can occur. Urbanization which does not take into account the natural surface drainage disorganizes the area and renders it unable to cope with periods of heavy rainfall. The flood in the town of Djibouti (East Africa) in April 1989 offers a comparable dramatic example of this type of mismanagement in an arid zone. However, it is not necessary to go to arid countries to deplore such a tragedy: autumn 1988 offered a similar example of flooding in Nimes, in the south of France.

The consequences of such pressures are heightened in a dry environment, where top soils are commonly loosely structures, where the vegetation is sparse and slow-growing, and disturbed soils in the hinterland are difficult to revegetate. The main impact on the hinterland of these older towns is a surrounding belt of some 30 to 100 km diameter where trees have been removed for fuel and timber for housing and mining.

3.6.4.2 Impact of Mining

The direct effects of mining (destruction of vegetation and disturbance of the lands by open cast mining, pipe lines, tracks, mine dumps and dust, noxious wastes and fumes, excessive pumping of water, discharge of water and solutions, and seepage) are not specific to desert operations, but in dry ecozones the greater fragility of the environment increases the difficulty of rehabilitation and calls for considerable investment and a high level of technological skills. Specific care must be given to the stabilization of mine residues because they are a source of very abrasive dust when wind-carried (rock flour). The stabilization of dump hills is another problem. Several techniques have been created, including stabilization by revegetation.

3.6.5 Conclusion

Today, thanks to new approaches and mainly by the utilization of satellite image data for monitoring land degradation within the rangelands, a good quantitative and qualitative estimation of degradation can be obtained. "Unfortunately the loss of productive land still continues" (Graetz 1984). In the future, Australia has an obligation to continue to use its dry rangelands but also to improve its conservation strategy.

The case study of Australia demonstrates that land degradation is a corollary of human activities even without human overpopulation. Here, animal overpopulation is the major cause. From the standpoint of the areas affected, land degradation in Australia is a problem of overgrazing of rangelands. In smaller areas in the southern parts of the continent, degradation is linked with dryland or irrigated agriculture, the economic impact of which probably exceeds that resulting from overgrazing. Attempts at recovery were undertaken in the 1930's and 1940's with soil conservation legislation, but degradation is still expanding. "In general the position is not desperate. There now exists an educated public alive to questions of environmental degradation." (Mabbutt 1986).

The recognition of the complexity of land degradation and the need to adopt a multidisciplinary approach to the solution of the problems, is a well-established behavior in Australia. The book *Land Degradation: Problems and Policies* (Chisholm and Dumsday 1987) concentrates almost exclusively on Australia and has brought together leading scientists and protectioners in the field of soil conservation and rehabilitation. In the last chapter doubt is raised whether land degradation in Australia is a real or an apparent problem and whether conservation programs are adequate or not adaptable.

3.7 High Technological Capital-Intensive Agriculture and Soil Degradation or "Desertification" in North America

In the excellent book *Desertification of the United States* written for the Council on Environmental Quality, David Sheridan (1981), in the introduction, says "The term desertification sounds odd in an American context". After analyzing desertification, in the conclusion Sheridan writes:

"Desertification in the arid United States is flagrant. Groundwater supplies beneath vast stretches of land are dropping precipitously. Whole river systems have dried up; others are choked with sediment washed from denuded land. Hundreds of thousands of acres of previously irrigated cropland have been abandoned to wind or weeds. Salts are building up steadily in some of the nation's most productive irrigated soils. Several million acres of natural grassland are, as a result of cultivation or overgrazing, eroding at unnaturally high rates. Soils from the Great Plains are ending up in the Atlantic Ocean.
All total, about 225 million acres[10] of land in the United States are undergoing severe desertification – an area roughly the size of the 13 original states."

[10] 91 million ha.

In the United States I have chosen to speak of dryland degradation more than of desertification because the richest country in the world has the technical and financial solutions for the rehabilitation of its degraded areas. When degradation has reached such an extent that reclamation needs too high an investment, the government has opted voluntarily for desertification. The word desertification can be used in this last case because the abandoned agricultural land returns to a state which is worse than before cultivation.

The publication of statistics including an abundance of soil survey data of land degradation, plus the contrast between the highest level of technology, and possibly the highest rhythm of degradation, leads me to accord special importance this chapter concerning land degradation in the USA.

There are two questions to be answered: which are the degraded areas in the USA? What forms and what extent of degradation can be observed?

3.7.1 Location of Land Degradation in the Drylands of the USA

Much of the United States west of the ninety-eighth meridian is arid – that is its annual average precipitation is less than 500 mm (Sheridan 1986). By arid, Sheridan means drylands including arid, semi-arid, and subhumid ecozones with three major deserts, the Sonoran, Mojave, which are arid ecosystems and not climatic deserts, as is the Great Basin, and six major river basins, Columbia, Snake, Sacramento, San Joaquin, Colorado, and Rio Grande (Fig. 37).

Dregne (1977) calculated that 1.1 million square miles (about 2.85 million km^2) (36.8% of the continent's arid land) are prone to severe land degradation. Some 10 500 square miles (about 27 200 km^2) have undergone very severe land degradation. Dregne compares this situation with that of Africa, where he estimates the total of severe degradation to be some 4 million square miles (10 360 million km^2), but considers that very severe degradation is more extended in North America, with 5500 square miles (about 14 245 km^2).

On the whole, according to Dregne (in Sheridan 1981), about 225 million acres (910545 km^2) of land have experienced severe or very severe desertification (Fig. 38). These desertification values are for the status (present conditions) not for risks (prone to) (Dregne 1983).

When Dregne speaks of severe desertification in North America, he means:

a) Substitution of edible grasses by forbs and shrubs not or less edible.
b) Badly controlled drainage and leaching resulting in salinity with 50% reduction in crops yield.

For Dregne, very severe desertification means:

a) Large, shifting barren dunes.
b) Large, deep and dense gullies.
c) Salt crusts nearly impermeable to infiltration which have developed on irrigated soil.

Where is land degradation observed in the USA? If we compare Dregne's (1977) map of the states of desertification with the map of the North American Deserts

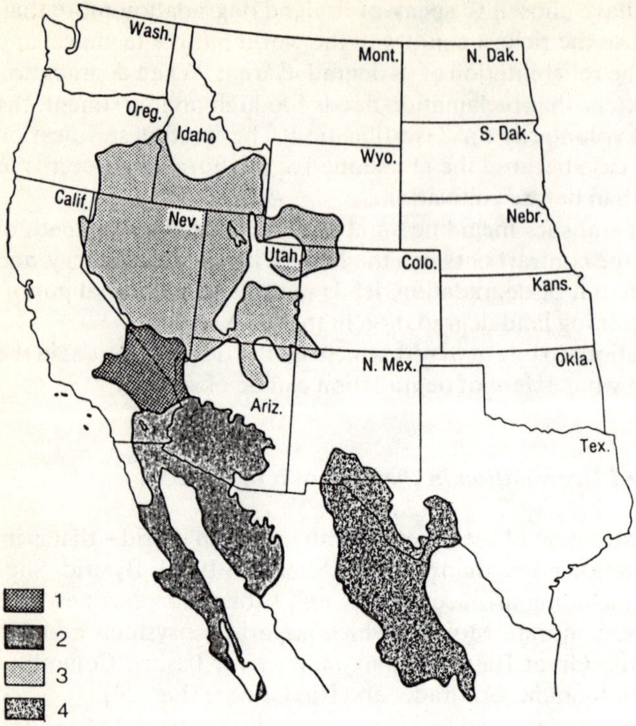

Fig. 37. North American drylands. (Sheridan 1981). *1* Mojave Desert; *2* Sonoran Desert; *3* Great Basin Desert; *4* Chihuahuan Desert

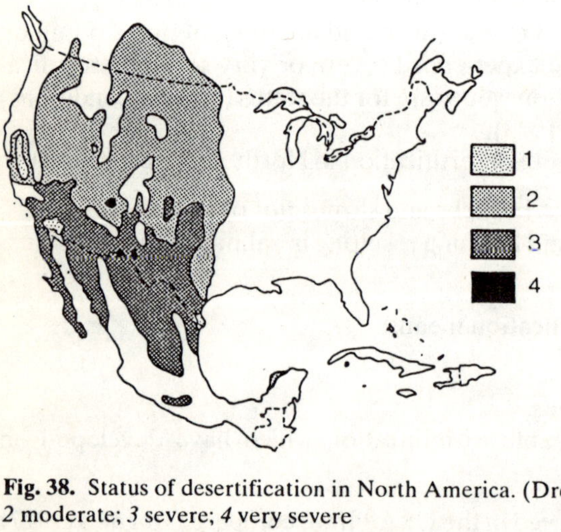

Fig. 38. Status of desertification in North America. (Dregne 1977, in Sheridan 1981). *1* Slight; *2* moderate; *3* severe; *4* very severe

(Mohave, Sonoran, Great Basin, Chihuahuan) we see the overlapping of drylands and degraded areas.

El-Swaifi and Dangler (1982), discussing the limits of "tolerable soil loss," quote Young (1981), who estimated a loss of 11 t ha^{-1}, and assumes an accelerated rate of top soil removal on cultivated land of 1 cm per 30 years. The Soil Conservation Service, USDA, has adopted an average rate of 1 cm per 60 to 300 years.

According to Sheridan, the degradation-prone areas in the USA are those which receive less than or around 50 mm rainfall per year. Fifteen states are totally or partly concerned (Fig. 39) (Geraghty et al. 1973).

3.7.2 *The Causes of Land Degradation*

3.7.2.1 Excessive Water Consumption: First Cause of Land Degradation

Thanks to new, sophisticated, and expensive technology, it is increasingly easy to obtain water. Three main techniques exist:

1. Pumping of groundwater, facilitated on a large scale by the centrifugal motor pumps which can suck water in greater volume, at greater speed, and from greater depths.

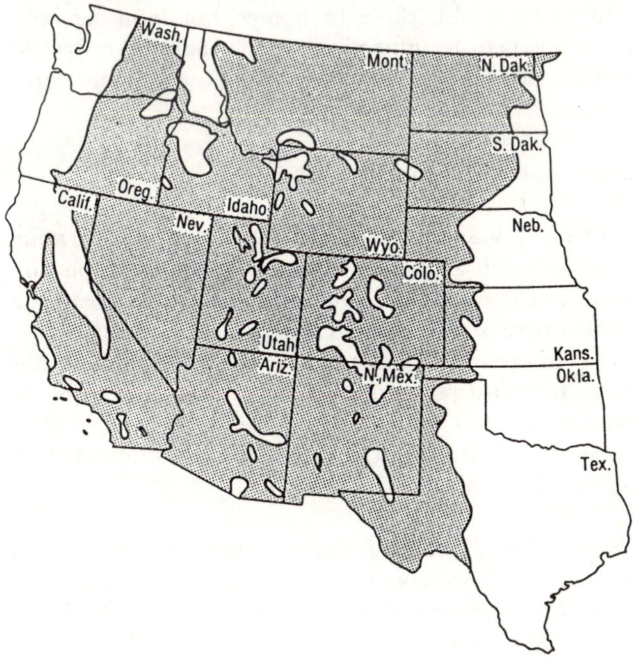

Fig. 39. Areas with less than 20 inches average annual precipitation. (After Geraghty et al. 1973 in Sheridan 1981)

2. Damming of the rivers to make water available on demand.
3. Long-distance transport of water from areas where it is abundant to areas where it is scarce.

The first difficulty in the western states of the USA is the over-consumption of water: the ratio between pumping out and natural replenishment is out of balance; the users, mainly irrigation and municipal needs, are drawing on groundwater reservoirs faster than they can be recharged.

According to the US General Accounting Office (1977a) groundwater supplies account for 20% of the water consumed in the USA. By contrast, in Texas groundwater supplies 75% and in Arizona 62%.

Quinn (1968) reveals that: "One out of every five persons in the arid western States is served by a water supply system that is imported from a source a hundred miles or more away".

Overconsumption of water results in depletion of groundwater resources, which is a cause of desertification because it leads to the abandonment of irrigable farmlands. During the phase of depletion of the water of an aquifer, the total energy necessary for pumping rises. The energy costs are proportional to the depth of the well.

Another deleterious consequence of pumping can be land subsidence, especially when the aquifer is in unconsolidated sediments. Thus, 13 468 km^2 of land have sunk in the San Joaquin valley (east of San Francisco) as of 1972 and about 10 878 km^2 more than 0.3048 m. The damage of subsidence to irrigation and drainage canals and pipes between June 1975 and September 1976 was $ 3.7 million (Sheridan 1981). It increases the work of farmers, who have to realign their irrigation systems. Damage is also done to homes, communities, and buildings. The long-term consequence of compaction of mined aquifers is the loss of storage capacity, which is a serious problem in a dry area needing irrigation.

3.7.2.2 Overgrazing: Second Cause of Land Degradation

Even today, just as in Africa, the USA drylands do not respect the carrying capacity of their rangeland (Johnson 1978). Until recently, the USA had no general or systematic rule of grazing limitation; the policy is opportunist. The productivity of the land for livestock has decreased since the last century.

Sheridan (1981) writes: "Improvident pasturage or overgrazing, as it has come to be known, has been the most potent desertification force, in terms of total acreage affected within the United States."

According to Dregne, in North America the three areas most affected by very severe desertification have been plagued by overgrazing.

Case Study One: Navajo Agricultural Products Industries. In the 6 million ha of the Navajo Agricultural Products Industry, overgrazing resulted in severe water erosion and vegetation substitution: noxious shrubs replaced edible grass. What are the causes? According to Billy (1978):

- demographic explosion about 3% increase per year;
- no alternative nongrazing economic possibility;
- 63% unemployment;
- reluctance to reduce cattle and sheep to the carrying capacity;
- fencing has not been introduced because it is in contradiction to Navajo tradition, whereby grazing land is not private property but can be used by all members of the clan;
- water and timber including firewood are also communal property.

Federal efforts to control overgrazing have been made since 1937. According to Billy (1978) in 1937 the number of sheep on the reservation was 1 300 000 on rangeland with a carrying capacity of 600 000. Livestock reduction was imposed but met with strong opposition. In 1978, on Navajo rangeland there were 2 170 300, which means three times the estimated carrying capacity: how the land is able to sustain this will only be seen in the future.

Case Study Two: Challis Area (Central Idaho) also Analyzed by Sheridan (1981). This further confirms that overgrazing is a very severe degradation force in the dry west of United States.

The US Bureau of Land Management (BLM) manages 170 million acres (68.8 million ha) of rangeland in the Challis area of central Idaho and the Rio Puerco basin of New Mexico, 40 miles (64.36 km) northwest of Albuquerque.

In the Challis Unit, overgrazing led to substitution of mixed sagebrush and grassland by sagebrush and an increasing invasion of rabbit-brush and snakeweed not edible by livestock. In some areas, the topsoil was completely eroded and in the grassland, pedestals with plants, rills, and gullies covered the whole topography.

The BLM was requested in December 1974 by the US District Court (for the district of Columbia, which established in Natural Resources Defense Council v. Morton that BLM was in violation of the National Environmental Policy Act) to reduce livestock use on 40% of the allotments to allow rehabilitation of the very poor range grass conditions, to reduce blowing dust, to avoid a larger formation of sand dunes and invasion of mesquite (*Prosopis juliflora*), which puts down a very deep taproot (15 to 30 m) and thus it succeeds at the expense of shallow-rooted species.

Case Study Three: Rio Puerco Basin (New Mexico). In the Rio Puerco Basin the situation is even worse. Sheridan, quoting Dortignac, writes that overgrazing has reached such a point that water erosion in the area results in a soil loss of 2 to 8.7 tons per acre (0.8 to 3.3 t/ha) annually with an annual regeneration of 1 t maximum per acre (0.4 t/ha) per year. The suspended sediment concentration in the water of Rio Puerco downstream reaches 267,000 mg l^{-1}. (Fish suffer at about 80 mg l^{-1}).

The annual rainfall of the Rio Puerco varies between 230 and 350 mm per year and corresponds to the dry ecosystems of Africa in the Sahelian zone.

Since the mid-19th century, numerous small farms have flourished along the alluvial plain of the river Rio Puerco, using ditch irrigation. The farms became so flourishing near the turn of the century that the region was known as "the bread basket of New Mexico". But around 1880 a drought revealed erosion features which even today have not healed. Water tables began to drop, arroyo erosion reached a maximum, large amounts of sediment were exported and, according to Dortignac, between 1885 and 1962, 1.1 to 1.5 billion tons (1 to 1.36 billion t) of soil was washed into the Rio Grande from its tributary, the Rio Puerco. This resulted in the rise of the bed of the Rio Grande: (1.5 m) between 1931 and 1938 (Murray and Lowry 1976). Downriver irrigation systems suffered severe siltation, since the Rio Puerco brings less than 10% of the water of Rio Grande but 50% of its sediment load (Sheridan 1986). The agricultural villages along the Rio Puerco were abandoned during the first half of the 20th century. The arroyos behaved like gullies progressively eroding good alluvial deposits.

Dortignac (1963) observes:

> "In the main drainages, the channels have cut down to bedrock, or to a point where downcutting and resistive forces are offsetting. In places, the channel bottom has been lowered as much as (15 m). Presently, the main channels are widening. As a steep, raw bank is undercut and falls into the channel, the flow is diverted and begins to undercut another bank, which falls in its turn. This process will continue until the channel is so wide that the water flows will lose the erosive forces needed to undercut the banks."

Other arroyos in the dry southwest of the United States have experienced similar cutting back and increased runoff, also attributed to overgrazing. Some scientists think that climatic deterioration also contributes, but in general this cause is totally rejected.

A pertinent observation by the paleobotanist Bohrer (1979) states: "Historic overgrazing has created extremely dry conditions for plants due to the removal of litter, loss of soil cover, and the trampling of the ground that prohibits rainfall from reaching the roots of plants". This last observation is very valuable, because overgrazing is usually considered in terms of overconsumption of forage rather than of compaction by trampling or destruction of the structure of the top soil by cutting hoofs. Those of goats are the most dangerous. The soil compaction hinders the return of annual vegetation.

Government efforts have been made to halt soil and vegetation degradation:

1. mainly by reducing the numbers of cattle;
2. by initiating a rest-rotation system with a rest of 1 year on one third of the rangeland surface,
3. range improvement by exclosure of some areas with fences;
4. digging of wells,
5. establishing water catchments,
6. seeding of grass in areas where the dominant snakeweed was burned or cut down,
7. creation of a team of cattle guards which did not lead to an unequivocal success. It was suggested that the only solution is to close the whole area entirely to grazing.

Such a region is difficult to rehabilitate because:

1. The area is dry, often less than 300 mm rainfall per year.
2. The soil erosion started more than 100 years ago, resulting in low content of organic matter and disappearance of the top soil, which bring unfertile conditions for plant life.
3. The edible species have disappeared, leaving unpalatable species, mainly snakeweed, which does not disappear under present conditions.

Bohrer (in Sheridan 1981) makes very precise observations:

1. The absence of cool season perennial grasses like mutton bluegrass (*Poa fendleriana*) once widespread in the Rio Puerco Basin. They now reproduce only in refuges inaccessible to sheep and cattle and not in grazed areas.
2. The vine mesquite grass (*Panicum obtusum*), indicator of recovering from overgrazing, is rarely present in the Rio Puerco Basin.
3. The diminishing rate of Indian ricegrass (*Oryzopsis hymenoides*) is also an indicator of permanent overgrazing.
4. Almost one-third of the Rio Puerco Basin has fine sandy loam soils especially vulnerable to wind erosion. Wind erosion exports annually around 2 to 4 tons of top soil per acre (0.7 to 1.4 t/ha) and 10 tons (3.7 t/ha) on the most fragile soils.

There are also human reasons for the slow rates of rehabilitation; the strong political opposition of the ranchers to the reduction of grazing on public rangeland; the ranchers' point of view on carrying capacity differs from that of range scientists. The ranchers, from their practical experience, disagree with the range scientists, who compare the present state of plants and animal life to their estimate of what the pregrazing conditions were, knowledge which is difficult to verify and easy to challenge.

To halt the downward spiral of vegetation and soil degradation in the Rio Puerco and in other grazing areas, Congress in 1934 passed the Taylor Grazing Act, which failed. The Taylor Grazing Act, which intended to end overgrazing on the public rangeland, failed because of lack of an adequate monitoring system, incapacity to reduce the number of livestock on the land, and disagreement between rangers and scientists on the carrying capacity. No painless solution exists.

In 1978 Congress passed the Public Rangeland Improvement Act because forage production was still below the potential and the state of the rangelands was still declining or unsatisfactory.

Case Study Four: Santa Cruz and San Pedro River Basins (South Arizona). In less than one century, just as in the Rio Puerco Basin (New Mexico), livestock overgrazing, exacerbated by climatic fluctuations, has metamorphosed the two floodplains of the Santa Cruz and the San Pedro River Basins of South Arizona.

The climatic change, according to Hastings and Turner (in Sheridan 1981), has, since 1900, been a downward trend in rainfall of about 2.5 mm every 30 years with winter precipitations dropping markedly and summer ones slightly, and an upward trend of the mean annual temperature of about 1.5 to 2 °C.

The livestock increased in Arizona from 5,000 cattle in 1870 to 1 095 000 in 1890. This demographic explosion of cattle weakened the grass cover, affected the ratio between runoff and infiltration in favor of the former, and increased arroyo cutting. Sheridan (1981) explains:

— "Overgrazing is no longer the area's major desertification force, but the land still suffers the aftereffects of the 1870-91 cattle orgy.
— The upward migration of plant species from drier, hotter conditions to moister, cooler ones continues, and there is no sign that the pace has slackened.
— The invasion of mesquite and tamarisk into riparian habitats, at the expense of grasses and traditional tree species, has accelerated.
— Human overdraft of groundwater is now the major desertification force at work in this area, and one of its chief consequences — the abandonment of irrigated cropland — is spreading."

The 8.1 million acres (3.27 million ha) of rangeland of the river basins produce less fodder than 100 years ago. The density of vegetation is lower and the vegetation itself has changed. Mesquite (*Prosopis juliflora*), tamarisk (*Tamarix pentandra*), acacias, noxious weeds, and shrubs are the invaders replacing former grassland. Overgrazing and soil erosion, and their natural corollary, groundwater depletion by human overdraft, are the main problems.

The growth of the city of Tucson is one of the causes of water overdraft. Tucson, with 275 mm of rainfall per year, relies entirely on groundwater, which it pumps out at five times the rate at which natural replenishment can occur.

Land subsidence accompanied by soil cracks and landslides caused by groundwater overdraft, as in the San Joaquin case study, is another problem in Arizona.

Agriculture in Arizona uses 90% of the water, so the Government has tried to take steps to decrease water consumption: the Government buys up farmland and water mining companies to decrease the surface of irrigated agriculture (pecan nuts and cotton) and has begun recycling water.

In 1980, a water management and conservation law was passed, demonstrating that, after a laissez-faire system, the State of Arizona has entered an era of awareness: groundwater is a finite resource and should be saved.

3.7.2.3 Salinization: Third Cause of Land Degradation (after Sheridan 1981)

Case Study Five: Wellton-Mohawk District (Arizona). Salinization is the long-term problem in the Wellton-Mohawk irrigation district in Arizona east of Yuma.

Water was diverted from the Colorado River 30 miles east to irrigate 60 000 acres (24 281 ha) of desert land (rainfall is 10 mm per year) along the Gila river (Holburt 1978, Curry 1979).

Because of an impermeable layer which blocks downward drainage of water, saline water underlies a large percentage of the irrigated lands. The problem is aggravated by the fact that the Colorado River receives the water of the Gila River which is greatly over-utilized. Thus the water of the Colorado river had an average salinity of 800 mg l^{-1} in 1961, which rose to 1500 mg l^{-1} in 1962, and the water from the Wellton-Mohawk's drainage system had 6000 mg l^{-1} of salt when reaching the Colorado. The drainage wells built in the 1960's were insufficient to absorb all the saline water.

The result was that the water of the Colorado which was released into Mexico was too saline. In 1973, the United States agreed to release to Mexico water with salinity not more than 115 mg l^{-1} (Sheridan 1981). An expensive desalinization plan was established in Yuma ($178 million to construct and $12 million to operate) to desalinate about 120 000 acrè/foot (48 562 ha per 0.30 m) of water per year. One acre-foot is enough water to cover 1 acre (0.404686 ha) to a depth of 1 foot (0.3048 m). The prices were so high that the Government decided to purchase and withdraw 5000 acres (2023 ha) of land from production. This solution is also too expensive, because the farmers do not want to sell and when they accept, are asking 3000 dollars for one acre (about 7400 dollars for one ha).

The solution lies in greater irrigation efficiency through better water management using the Israeli irrigation model. In the Negev, citrus needs five times less water than in Wellton-Mohawk.

To develop solar salt ponds can also be efficient and can replace the project of a desalting plan. It appears that salinity control is a very difficult question which requires not so much desalinization facilities as complete socio-economic restructuring. In the case of Wellton-Mohawk the question to solve is twofold: water salinity and soil salinity, both linked with irrigation. Irrigation, which is the most expensive of all development schemes, must be established more cautiously. The economically logical solution might be to withdraw all the irrigated land now under cultivation and let it return to desert.

3.7.2.4 Physical Soil Erosion: Fourth Cause of Land Degradation

Case Study Six of Kiowa, Crowley Counties in Colorado, and Gaines County in Texas. Kiowa and Crowley Counties in Colorado and Gaines County in Texas are the areas where the magnitude of soil erosion reaches the level of the Sahel in Niger. The problem is not new, as the three counties figured in the dust storms of the 1930's. What are the causes, the mechanisms, and the consequences?

The causes are related to soil texture in southeastern Colorado. The soils are inherited aeolian silts and fine sands, highly susceptible to wind erosion. In Gaines County the fine sand, which does not hold moisture well, is also vulnerable to wind. Fragile soil texture combined with low rainfall, around 40 mm per year, agriculture with very little plant cover in autumn and winter, repeated plowing in order to control weeds, and economic parameters (high wheat prices in 1973–74, for example) resulted in heavy wind erosion.

In the three Counties wind erosion can be considered as the mechanism of degradation with a serious problem of groundwater depletion in Gaines. The

Table 22. Soil erosion in six counties in the arid west of the USA. (Sheridan 1981)

County	Farms sampled	Estimated annual soil loss (tons per acre)				
		0 to 5	5.1 to 10	10.1 to 20	20.1 to 40	Over 40
Gaines, Texas	39	1	0	2	5	31
Roosevelt, New Mexico	28	2	7	9	10	0
Finney, Kansas	35	1	23	2	9	0
Benton, Washington	20	0	11	8	0	0
Whitman, Washington	30	5	14	11	0	0
Burleigh, N. Dakota	11	7	4	0	0	0

rates of wind erosion as shown by Table 22 are at a maximum in Gaines County in Texas, where the export of particles reaches a total of 40 tons per acre (16.19 t ha^{-1}).

Gaines County is mining its underground water at twice the rate of natural recharge. The aquifer exploited is the Ogallala Aquifer, which stretches under parts of seven states (Fig. 40).

Gaines County is at the southern end. The groundwater has depleted drastically and the cost of its pumping, because of the increased depth of the aquifer and increased prices of energy, has risen for the Gaines County farmers from about $1.5 in the 1970's to $60 per acre-foot 10 years later (Bell and Morisson 1979).

The rising energy costs, coupled with the lowering of the water level, can only lead to abandonment of irrigated cropland. This fallow crop land will return to desert with some weeds and increased dust storm potential. Sheridan (1981) notes that The Texas Department of Water Resources sounds a note of warning:

"If this overdraft continues, the aquifer ultimately will be depleted to the point that it may not be economically feasible to produce water for irrigation.... The actions of the water users will determine whether the projections of this study come to pass...."

Given such problems, the continued nonregulation of groundwater in Texas is difficult to fathom. The only method used to regulate the amount of groundwater pumped on the Texas High Plains is well-spacing (Sheridan 1981).

Graves observes:

"Texas law continues to regard most groundwater as a mysterious blessing... legitimately subject to capture and use in unlimited quantities by any property owner who digs or drives a well."

As a consequence, there exists what economists call a "negative incentive to conserve the resource", and he continues:

"Finally, it should be recalled that the federal government has subsidized the rapid depletion of the Ogallala Aquifer -- first by price supports for commodities such as cotton, then by crop disaster payments, the various cost-share "soil conservation" programs, as well as by the low-interest loans of

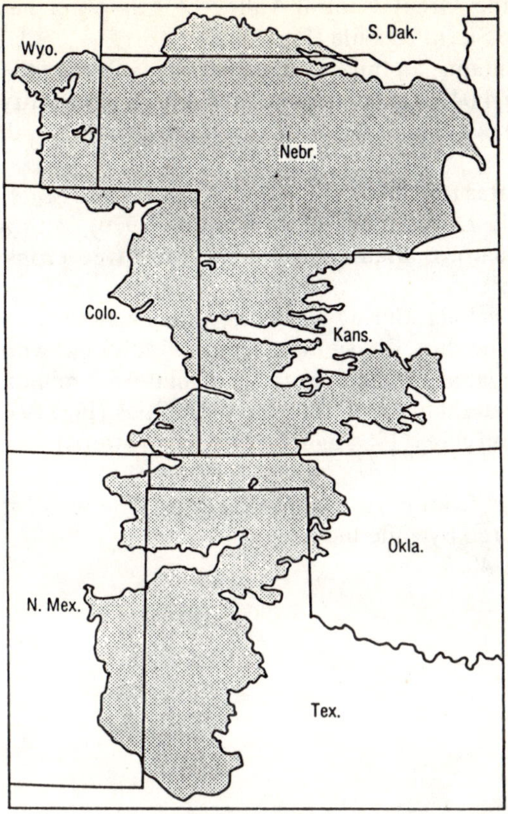

Fig. 40. The Ogallala aquifer in western America. (After Gutentag and Weeks 1980 in Sheridan 1981)

the SBA and the FmHA. In addition, federal tax policy encourages the depletion of this resource. High Plains farmers are granted a *depletion allowance* on pumped groundwater, thereby enjoying a tax break similar to that which the oil industry enjoyed for many years and which the mineral extraction industry currently enjoys. The more water they consume, the less tax they pay."

3.7.2.5 Combined Causes and Mechanisms of Land Degradation

Case Study Seven: San Joaquin Valley (California). The previous case studies have placed the emphasis on land degradation caused by single factors, e.g., overgrazing, salinization, or soil erosion.

The San Joaquin Valley case study is analyzed by Sheridan (1981) to demonstrate that land degradation can have multiple causes.

Five major causes of desertification are detectable in the San Joaquin Basin: poor drainage of irrigated land, overgrazing, crop cultivation of highly erodible soils, waterlogging, and damage caused to vegetation and soil by mechanized agriculture.

The San Joaquin Valley in the Great Central Valley of California lies between the coastal ranges and the Sierra Nevada (Fig. 41).

The different landscapes, mountains, foothills, and the flat valley floor of the basin cover, 7 365 540 ha of which 4 047 000 ha are flat land. The region is dry with rainfall of 14 inches (355 mm) per year in the north, decreasing in the south to 5 inches (127 mm) per year.

Despite its aridity, the Basin rates fourth in production in the United States after Iowa, Texas, and Illinois (US Department of Agriculture 1979), with a production of cotton, grapes, tomatoes, alfalfa, sugar beets, and tree crops: walnuts, almonds, oranges, and apricots.

The growing season is 8 months long, during which the valley is frost-free. The climate is Mediterranean in type: hot, dry summers and mild, humid winters. All these conditions entail irrigated agriculture in almost all the 5.5 million acres (2.2 million ha) of the cultivated land and drainage difficulties (Fig. 42). Salinity and sodicity affect 2.2 million acres (1984 figure) (Dregne in litteris).

1. From outside the basin, mostly from north California: 20%. The water is stored behind dams and delivered by aqueducts.
2. From the aquifers of the basin: 40%.
3. From the rivers: 40%.

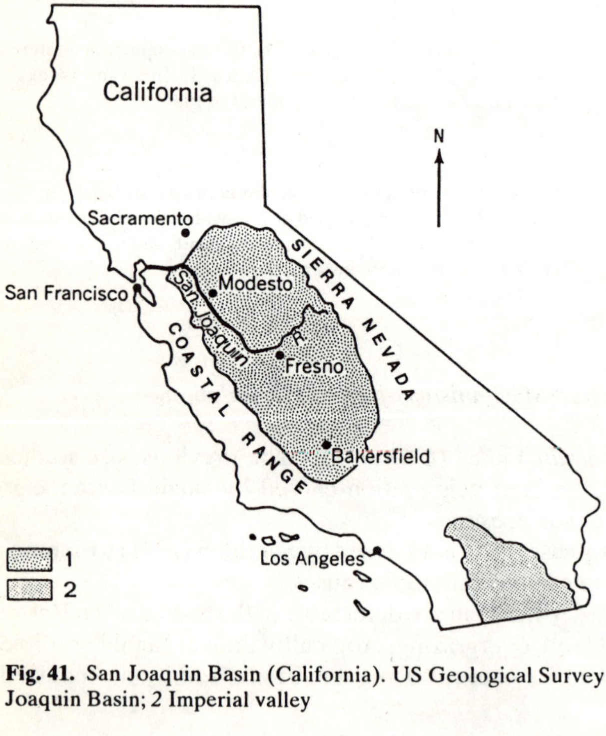

Fig. 41. San Joaquin Basin (California). US Geological Survey 1974, in Sheridan 1981). *1* San Joaquin Basin; *2* Imperial valley

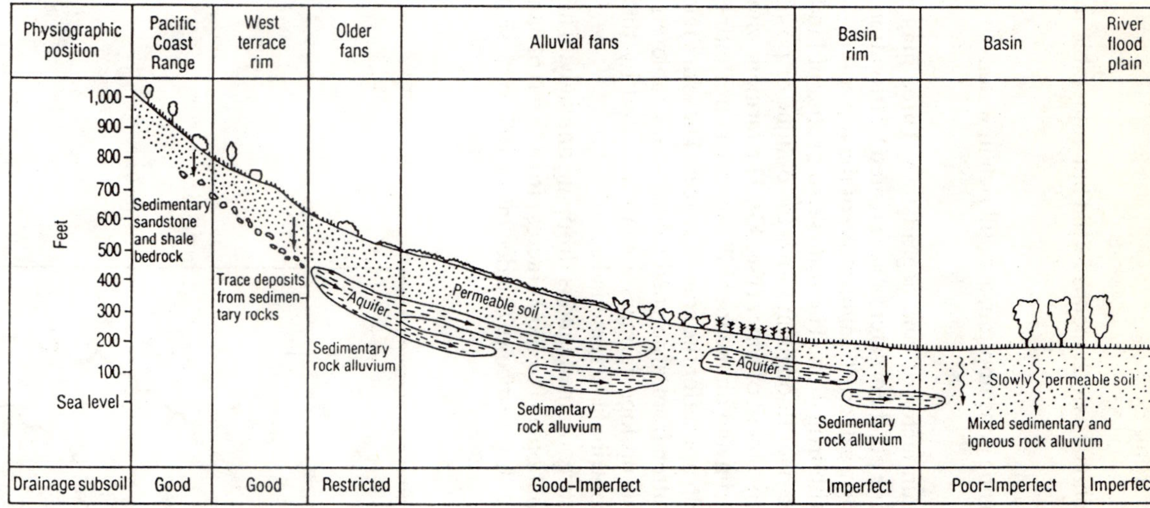

Fig. 42. San Joaquin Basin (California). Area with drainage problems (present and potential). (US Bureau of Reclamation and California Department of Water Resources, Agricultural drainage and salt management in the San Joaquin valley, in Sheridan 1981)

The prosperity of the region comes from:

1. The soils, which are colluvio-alluvial, which means fertile but vulnerable to wind and water erosion.
2. The length of the growing season.
3. The aquifers, in spite of increasing consumption, are still rich in water.

Salinization. Salinization is the main factor of land degradation. Three major aspects can be identified: highly saline perched water tables created by irrigation; a few hundred acres of land wearing a salt crust; salinization of the top soil.

These difficulties are linked with irrigation. The transverse profile of land from the Pacific Coast Range to the river flood plain shows the oblique disposal of the aquifers with a down slope movement of water (Fig. 43). The farmers who are established on the west terrace rim and on the foot hill fans are contributing to the problem of salinization, but do not suffer the consequences. These are felt by the downslope farmers, who suffer from a perched saline water table endangering their crops. Because the upslope farmers are not affected by this salinization, they are reluctant to contribute to remedial action.

The agricultural yield from poorly drained land is dropping annually. The evaluation of the Bureau of Reclamation on the 400000 acres (161875 ha) of San Joaquin farmland needing drainage shows since 1970 a decline of 10% or $ 31.2 million.

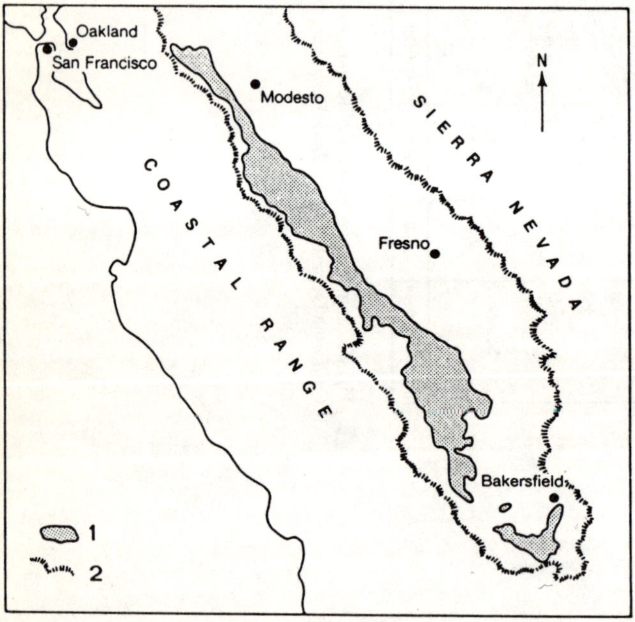

Fig. 43. San Joaquin Basin (California): profile of a drainage problem area. (US Department of Agriculture). Soil conservation service, in Sheridan 1981)

Without action, by the end of the millenium the degraded land will reach 700 000 acres (283 280 ha) with annual cropyield loss of $321.3 million writes Sheridan (1981).

At the beginning of the 1980's, the rich San Joaquin Valley was losing 14000 acres (5665 ha), of prime farmland each year to urbanization.

Local and federal administrations are looking for solutions. One of the solutions consists in the establishment by each farmer of an on-farm drainage system to collect the perched saline underground water. Next, the unused irrigation water can be drained through wells or deep ditches beneath the root system and beneath the waterproof ground layer which blocks infiltration. Another solution is to collect the saline water in an on-farm sump, then pump it along a lined ditch to a master drain from where it will be expelled to a natural depression, a salt lake, or in an evaporation dump.

The Pacific Ocean can play the role of the natural sink. For this purpose, a concrete-lined ditch of some 470 km along the natural valley would drain the saline water to the joint San Joaquin and Sacramento delta. A quarter of this drain (the San Luis drain) has already been built. There has been opposition from environmentalists and residents of the Bay Delta area and several other problems are still not solved: mainly the impact on the delta of the arrival of salty drainage water in the oceanic water of Suisun Bay. The questions asked are: (1) Will the freshwater consumed by the inhabitants of the delta area be endangered? (2) Will arsenic and mercury reach at toxic level? (3) What will be the effects of the project on the last remaining large wetland of California?

More costly would be to discharge the water directly further west into the Pacific Ocean. Other alternatives exist, but are limited by the cost among which is the use of desalinization plants.

Developing a local disposal system with local evaporation ponds is another alternative to the drainage problem. The difficulties attached to this local approach are the large surface area needed by all the ponds and the cost of building the initial drainage system.

The last question to solve is: who will pay for the projects?: farmers, irrigators, State of California, Federal Government? The Bureau of Reclamation has calculated the cost of the master drain project: $258 million from the Federal Government, which would be repaid in 40 to 50 years by the farmers, and $89.1 million from the State of California. To this figure must be added the on-farm drainage systems and the cost of collecting water from the farms to the main drain, which is the responsibility of local administration in the water districts and drainage districts.

The local alternative appears more practical than a valley-wide project. The best solution to reduce the negative outcome of irrigation lies in more efficient water use, so as to decrease the overall costs of irrigated agriculture, to decrease energy costs, and prevent the build-up of saline groundwater.

It is surprising, but all the surveys agree that lack of technical information is the main reason for the failure of more efficient water use. The irrigators have to be educated in scientifically precise irrigation techniques.

1. Not all irrigators know the most efficient irrigation method, whether drip, sprinkler, or surface flooding for their specific soils and particular crops.
2. They must be told the water requirement for each crop cultivated.
3. They must be trained in reading the root zone moisture to avoid overirrigation.
4. They need advice on the optimum time of irrigation so as to obtain the minimum evaporation.
5. They have to be educated on field leveling for better distribution of water during surface flooding.

All these techniques require extension services to provide farmers with technical information on water and soil conservation. Farmers' remedies are short-term remedies: replacement of deep-rooted, salt-sensitive crops by shallow-rooted, salt-tolerant crops or converting agricultural land to pasture; but in all cases, irrigation is necessary so that the level of the shallow aquifers rises. Few farmers are using highly salinized fields as evaporation pods, instead they simply abandon the salt-encrusted land.

Another question remains: has cost-benefit analysis really considered environmental factors adequately? Many would argue that this technique is far from adequate.

Overgrazing and Mechanical Soil Erosion. Overgrazing is the second drastic form of deterioration of the San Joaquin Valley. In some places in the 338000 acres (136767 ha) of Basin rangeland, the vegetation is so degraded that it cannot regenerate in a short time. In other places woody and noxious plants replace good forage vegetation. In these cases the threshold of irreversible desertification on the human scale is reached. Overgrazing causes a chain of degradation with substitution of the vegetation species or reduction of the vegetation density. This first stage leads to soil erosion, to lower soil fertility, to water or wind erosion, and to floods.

In the San Joaquin Basin moderate to severe water erosion by rillwash, sheet-wash and gullying caused deterioration in the foothills and lower mountains slopes which were earlier the chosen areas for cattle and sheep-raising, but then were later transformed into orchards and vineyards also favorable to erosion. Because of the absence of slope, water erosion on the flat valley floor is negligible, even where the grassland nicely dotted with oaks was replaced by food and fiber fields. Rangeland is the first victim of erosion, with an annual loss of about $1.2 million, according to the USDA.

Water erosion consists not only of ablation but also of transport and sedimentation reducing the capacity of streams and reservoirs and causing flooding. Floods destroy croplands and accumulate sediments which are expensive to remove.

Overgrazing is in this area also a cause of wind erosion. The USGS has made a very precise description of the dramatic windstorm which occurred on December 20th 1977 in the Bakersfield vicinity of the southern San Joaquin Valley. Early in the morning a windstorm began to blow from the southern border of the valley and moved more than 25 000 000 tons (22 679 500 t) of soil in an area of 373 square miles (966 km^2). In some foothills strips 23 inches (584 mm) wide appeared. Recently plowed fields were scrubbed. A plume of dust formed over the San Joaquin and extended for 360 miles (579 km). The wind removed 167 tons (151.56 t) of soil per acre (67 t ha^{-1}) from the deteriorated rangeland. The causes were the high wind velocity 186 miles (299 km) per hour and poor vegetative cover. The loss of nutrients was estimated at about $24 million. Crops growing on the valley floor suffered from sand blasting; immature vegetables, potatoes, onions, and carrots suffered from root exposure; alfalfa, vineyards, and orchards were sand blasted or buried. According to the USDA, the damage caused by wind erosion is $8.6 million annually, but this windstorm did more damage to agriculture in 24 h. The intensity of the windstorm was less damaging than the sand removed from plowed fields, road cuts and shoulders, channelized stream beds, construction sites, oil fields, recreational over-used areas, and degraded pastureland.

In addition, powerline pylons and other structures were also damaged. The windstorm also carried spores of a valley fever disease, coccidioidomycosis, which caused high fever in inhabitants of the San Francisco Bay and Sacramento areas, and sometimes even death.

The most severely eroded land and storm-affected areas were those with a high degree of overgrazing. After the windstorm, heavy rainfall continued for 3 months, causing floods.

Urbanization and Land Degradation. The San Joaquin sector has three fast-growing urban areas: Modesto, Fresno, and Bakersfield. Between 1972 and 2000 urbanization is projected to take over 407 100 acres (164 748 ha) of irrigable land. Urbanization results in the loss of good agricultural land. Over the past 5 years, California has lost 55 000 acres (22 258 ha) per year (California Department of Conservation).

The Williamson Act (1965) allows local government to contract with land owners to keep land in agricultural production, by lowering property taxes, but the Act was a failure in the areas surrounding the towns and useless in the areas far from towns.

3.7.3 Conclusion

Today the arid West accounts for about one-fifth of the USA's total agricultural production in food and fiber; but the arid West can be considered as marginal land for intensive agriculture. The development of the arid West illustrates the conflict between the human struggle against aridity and the ever-growing need for energy and water which results in massive groundwater overdraft.

A variety of solutions are suggested for the different problems:

1. Stage one: in-state projects such as the Central Valley project in California authorized by Congress in 1933. The Central Arizona Project will divert annually 1.2 million acre-feet (485 623 ha per 0.30 m) of water from the Colorado river and transport the water to urban and agricultural areas by means of a 250-mile (402 km) long pipe.

a) San Joaquin project in California
b) Santa Cruz and Avra Valley Basin (in Arizona) to satisfy the water consumption of Tucson.

2. Stage two: intra basins projects

a) Today the Missouri and upper Colorado Basins are major energy export areas for New Mexico and Utah.
b) The Texas Water Plan is a very ambitious interbasin water transfer project. According to Graves (1971), the water would be extracted from Mississippi and, transported across Louisiana, before being delivered to the Texas water system. These high-technology solutions require high electric generating capacity and a tremendous budget. The estimates at the beginning of the 1970's, according to Graves, were $10 to 14 billion. The citizens of the lower Mississippi Valley show no enthusiasm for sending so much water to the high plains of West Texas.
c) Similarly the citizens of the Columbia River Basin refused to transfer the water from the Colombia river to California.

The five major causes of land degradation in the USA are:(1) deforestation (in the 19th century), (2) overgrazing by cattle and sheep, (3) dryland cultivation of highly erodable soil, (4) overpumping of groundwater[11], (5) poor drainage of irrigated soils. It is surprising to find a country with such a high level of technology allowing its land to be degraded, but nevertheless degradation in the USA is a reality.

The second surprise for those who know the dramatic level of land degradation in Africa, the most affected continent, is to observe that the causes and mechanisms of degradation are almost similar on both continents, i.e., the ignoring of conservation techniques and its corollary, mismanagement by the land-users. These reasons for the misuse of land are common to both undeveloped Africa and sophisticated North America. The techniques which move the most rapidly are those promising reduction in land exploitation and, more slowly, those involving dissemination of conservation technologies.

[11]Dregne (1992, in litteris) writes: "I fear that you over-emphasize the damage already occuring in over-mining of groundwater. That is a problem almost exclusively confined to the states of Arizona, New-Mexico and Texas, at present. The area on which irrigation is no longer practiced is only a few hundred thousand acres, at most. In the 21st century, the number will increase to all six states in the Great Plains underlain by the Ogallala aquifer and to California."

3.8 General Conclusion, Chapter 3

The aim of Chapter 3 was to assess the different degrees of deterioration (land degradation) and the existence of irreversible degradation (desertification) on different continents.

This attempt to verify the existence of desertification leads to an impression of homogeneity of causes and manifestations. Land degradation is a phenomenon shared by all the continents in their arid, semi-arid, and subhumid ecozones. It can occur even in the wettest tropics: Brazil, where they burn off the trees and plant sorghum for a few years until the soil is exhausted; Cameroon in the overgrazed Sudano-Guinean ecosystem, and Mali where the maximum degradation has occurred in the Sudanian ecosystem, for example.

The drylands of Latin America have not so far been considered because land degradation is not the primary problem for this continent. I am, however, aware of the deterioration of the land in the north and northwest part of Mexico, the northern fringe of the continent of South America, the northeast of Brazil, the Pacific coastal deserts of Peru and Chile and the northwest of Argentina. I have also omitted those parts of eastern Asia which have semi-arid areas, since they cover only a small percentage of the region.

Even if land degradation is a common phenomenon, and even if its causes are man-induced, these causes can differ in detail. If we try to summarize them in one sentence: for Australia the causes are animal overpopulation, for China human overpopulation, for Egypt also human overpopulation exacerbated by the interception of Nile silt by the High Dam, for the USA overuse of groundwater and highly vulnerable soils, for the USSR overuse of vegetation, erodible soils and irrigation; in short, all these causes can be summed up in one word: mismanagement.

Too much or too little rainfall only serves to reveal or enhance the effect of socio-economic malfunctioning. We have seen that without any exception, all these different results depend on human activities.

However, this is not the most important contribution of these analyses; the essential is the confidence that everywhere remedies exist. Can one then speak of desertification? Yes, it occurs when economic conditions push countries to abandon territories without a struggle because of limiting financial resources.

Several other lessons emerge from these case studies which can be classified under three headings: one concerning water, the second grazing areas and the third soil erosion, mainly by wind action.

Even if water projects are essential in the development of arid lands and even if there have been agricultural activities based on irrigation almost since the beginning of agriculture, in these regions unresolved issues are numerous.

"New water projects are needed to sustain the continued growth of the US arid region. They will, however, require enormous capital investment and even a nation as rich as the United States has a finite capital supply and must be concerned about its most efficient allocation. How does the return on a new water project compare with a capital investment elsewhere in a nation? The analytical tools available to us to answer such a question are woefully inadequate. Yes, certain costs and benefits of the proposed project can be estimated, i.e., quantified. Others, however, cannot. Our ability to quantify potential environmental costs, for instance, is quite limited." (Sheridan 1986).

Beside the question of cost-benefit ratio are the questions of who will pay for the projects? In the USA, this could be the society as the whole, or the beneficiaries of the project: flood-control, water supply users, hydro-electricity users. In the latter case, on what basis would payment be made? What will be the rate of interest?

In developing countries, where will the money come from? Tragically, international aid has not provided the solution. What is certain is that neither governments nor users have the money to pay.

Concerning the allocation of the water resources, can we agree with the economists, who are convinced that the water must go to the "highest value use"? For them the "highest value use" is certainly not environment conservation or land rehabilitation. And the "highest value use" is certainly not the same in the USA and in Africa. Food and energy self-sufficiency should be considered as the "highest value use" in almost all developing countries. The irrigated agriculture in the American West competes with industry for water use because industrial productivity is greater per m^3 of water. In Central Arizona, the copper industry (mining and refining) and its growing light industries are the main rivals of agriculture.

This is, however, not the main aspect of water problems. Environmental degradation linked with water consumption must be the major concern: irrigation and over-mining of groundwater, and irrigation and salinization, are certainly the two pairs which merit the highest attention.

The opinions of experts are very different concerning water mining. Some consider that groundwater is simply a mineral which must be exploited as long as it is economically efficient: but water is the only mineral that is drunk. Other experts consider that pumping must be limited to the natural capacity for recharging the aquifers. A third group considers that new or more cost-effective technologies (precipitation enhancement through bombardment of cumulus clouds with silver iodide, hauling of icebergs from polar zones, desalinization of marine water) can solve future water supply problems.

To avoid the pollution of soils by salts, irrigation must be the last-resort proposal for development and if it is decided on, then adequate drainage facilities should be built in order to avoid environmental degradation and later increased costs.

However imperfect, cost-benefit analysis has demonstrated that despite the high costs of irrigation schemes, the bigger the project the greater seems to be the ratio of costs to benefits, both financially and environmentally.

Arid grassland degradation presents special problems. In connection with his study of desertification in India, Sdasyuk (1986) writes: "Grazing lands are in the worst condition, as a result of the excessive pressure of numerous cattle herds. Notwithstanding the official classification of 13 million ha as permanent rangelands, they are practically devoid of any vegetation due to overgrazing and other uses (Indian Sixth Five-Year Plan 1981). Rangelands in India are sources of soil and vegetation degradation, affecting adjacent territories as well."

Despite numerous research programs, nobody is able to offer the livestock breeder a reliable criterion or guide to a simple definition of overgrazing or carrying capacity in relation to vegetation resources, soil, and climatic data (rainfall, temperature, evaporation). On all the continents, rangeland users have over-estimated the carrying capacity, especially during drought periods. Nobody can be sure that drylands can be rehabilitated after they have deteriorated through dryland farming or overgrazing and in particular when the native flora and the topsoil have both undergone severe loss.

The most advanced projects of rehabilitation were organized in the Rajastan desert with aerial reseeding. According to Sheridan (1986), it can cost from $ US 60 to 3000 ha^{-1} depending on the amount of rainfall, the topography, and the quality and depth of soil for reclamation. When the cost of reclamation is high, its implementation is questionable. In general, rehabilitation of soil must be envisaged on a geological time scale, even if in their arid regions the Chinese have demonstrated in Shapotu (South of Tengger Desert) that clever brains can make it happen, on a human time scale.

That the awareness of the erosional force of wind is confirmed in all the case studies is certainly not a negligible result. This awareness is underlined by Sheridan (1986) when he writes:

"We can no longer afford to underestimate wind as an erosional force. A study of a windstorm which scoured 2000 km^2 in the agriculturally rich San Joaquin Valley of California on December 20, 1977, reveals that this single storm mobilized approximately 50 million tons of soil (Wilshire 1981). A number of factors contributed to the severity of the erosion. A long drought had preceded the storm, leaving the land desiccated. Moreover, the rangelands in the affected area were vulnerable to wind erosion because they had been overgrazed for years. And the area's croplands had just been plowed in preparation for seeding."

Wilshire notes:

"Observation of modern dust storms has led to the surprising conclusion that, within the continental masses, wind is transporting an amount of sediment more nearly comparable to that transported by water than was previously believed. For example, Pimental et al. (1976) estimate conservatively that of 5 billion tons of soil lost annually by erosion in the United States, fully 1 billion tons are moved by wind. In some parts of the world windborne sediment far exceeds that carried by water". (Wilshire 1981).

Fryrear (1981) demonstrates how dryland cropping of arid lands decreases soil productivity when wind erosion removes fine particles eliminating nutrients attached to these particles.

We cannot conclude the third chapter of this book on a pessimistic note. Everyday mismanagement and deterioration of the environment bring a new lesson: we know that the irreversible threshold of land degradation has not been reached technologically. It is more of a socio-economic problem. Nevertheless, even from a technical point of view, scientific efforts for a better knowledge of the mechanisms of land degradation and the threshold of irreversibility must be developed.

In the next part of the book I shall try to restate the main processes of land degradation so as to understand what is happening in the drylands. Only then shall I focus my attention on the solutions.

Chapter 4 Processes Leading to Soil Degradation and "Desertification"

4.1 Introduction

At the beginning of 1989, the depletion of the ozone layer was considered the most critical environmental hazard. In the meantime, soil degradation is forgotten. The first danger seems more frightening than the second; it has the charm of novelty. Soil degradation is defined as decrease of soil quality and consequently of potential productivity; it is well demonstrated by Jackson (1989).

Soil degradation research is essentially down-to-earth, time- and capital-consuming, also it is neither new, nor glamorous, even though the acceleration of soil erosion is widely recognized, and disturbing statistics are published. According to Judson (1981) and Brown and Wolf (1984), fluvial sediments carried into the oceans increased from 10 billion tons a year before intensive agriculture, grazing and other technological activities to between 25 to 50 billion tons thereafter.

As quoted in Lal (1988), in the last decade the economic aspects of soil degradation have been highlighted. In chronological order, according to a first estimation of the UNEP 1980 (United Nations Environment Programme) each year crop productivity on about 20 million ha is reduced to zero or becomes uneconomic because of soil erosion. Ten years later this assessment seems unduly overestimated.

Dudal (1981), 1 year later, estimated that soil erosion is leading to an irreversible loss in productivity on about 6 million ha of fertile agricultural land per year; even this value seems pessimistic.

Buringh (1981) quoted an annual global loss of agricultural land as 3 million ha due to desertification: this represents another decrease in assessment.

Higgins et al. (1982) report that crop yields of rainfed agriculture may decrease by one-third over the next 20 years, which is also probably an overestimate. Kovda (1983) considers that since the beginning of settled agriculture, soil erosion has degraded 430 million ha of agricultural land. Larson et al. (1983) estimate that about 60% of the total annual sediment load of 1 billion tons carried by rivers from the United States is eroded from agricultural land. Clark (1985) reports that the off-site damage caused by sediments in the USA amount to $6 billion per year, one-tenth being for dredging rivers, harbors, and reservoirs (Pimental et al. 1987).

Considering soil deterioration as the most critical component of land degradation and, in the framework of irreversible land degradation, as the main factor of desertification, I wish in Chapter 4 of this book to recall only the physicochemical, chemical, and physical mechanisms of land degradation.

Lester Brown, president of the Worldwatch Institute, writes:

"Because of the shortsighted way in which one-third to one-half of the world's cropland is being managed, the soils on this land have been converted from a renewable resource to a nonrenewable one. Assuming an average depth of remaining topsoil of 7 inches (19.56 cm), or 1120 tons per acre (411.16 t ha^{-1}), and a total of 3.1 billion acres (1.255 billion ha) of cropland, there are 3.8 trillion tons (3.45 trilliont) of topsoil with which to produce food, feed, and fiber. At the current rate of excessive erosion, this soil resource is being depleted at the rate of 0.7% a year – 7% each decade. In effect, the world is mining much of its cropland, treating it as a depletable resource not unlike oil". (Foster in Lal 1988).

Soil erosion can be defined as the physical detachment, entrainment (or transport), and deposition (or sedimentation) of a soil particle. The term soil loss is the quantity of material exported at a designated point at the outlet of an area (plot, field, water catchment); it can be expressed in tons per square kilometre per year (ton km^{-2} a^{-1}), but the processes responsible for soil deterioration are not only external and physical, they are also internal, biological, and chemical. As we have seen, salinization, which is a chemical process of degradation, is a universal plague.

Physical deterioration is the rearrangement of soil particles owing to the removal of finer particles. This results in compaction, ablation of the topsoil, damage to soil structure, and detrimental changes in texture. Texture refers to the proportion of the various particle size divisions of the mineral element of soil. Soil is made up of grains of many different sizes but it is usual to find a predominance of a particular size range. Soil texture is normally classified according to the following size categories:

	Diameter
Clay	less than 2 μm
Silt	2–50 μm
Fine sand	50–200 μm
Medium sand	200–500 μm
Coarse sand	500–2000 μm
Gravel	2–20 mm

Biological degradation is the reduction in the quantity of organic matter and living organisms and in the rate of vegetational decomposition in the soil.

Chemical degradation includes salinity, alkalinity, or acidity. A high concentration of salts in the soil gives rise to saline or alkaline soils. This is often the result of irrigation without adequate drainage and water with a high mineral content: the water is transpired while the minerals increase in concentration. If the soil solution becomes more concentrated than the solute concentration of the plant cells, water is unable to pass into the plant and a reverse movement of water and nutrients can occur, causing wilting and death to the plant.

Another form of degradation is leaching, which is the washing out of minerals, particularly potassium and nitrates, from the soil.

Toxicity in the soil can also occur as a result of pollution, pesticides, radioactivity, and other waste products from cities and industries, which can wash chemical toxins into the soil.

The analysis of the processes of soil degradation has in this book two objectives:

1. To provide a general outline of the types of erosion and to analyze their interference with desertification. The processes responsible for soil degradation are primarily those linked with water and wind actions.
2. To contribute to the protection of "soil capital" as the agricultural production capital and to the conservation of the downstream infrastructures: dams, irrigation devices, farmland.

4.2 Physicochemical Processes of Degradation

4.2.1 Potential Modifications and Decay of Soil Structure

The soil surface including the top layer of the soil at the interface between earth and atmosphere is the place where atmospheric attack (also called weathering) is maximum. Consequently, the surface and the top layer of the soil can be subjected to permanent natural modifications of their properties.

Between the pedosphere (soil) and atmosphere there is the intermediary phytosphere which has, according to its density, a protective role. The processes of soil degradation are accelerated when the vegetation is destroyed by human activities.

The vulnerability of a soil is mainly dependent on its structure. This structure is the special arrangement of the mineral particles and their bonds by organic material and/or iron or aluminum hydroxides.

Organic matter is present as humus particles of colloidal dimensions or as unstable sticky material or thin films on the mineral particles. If the mineral particles are clayey (smaller than 2 μm), another bond occurs. The clays have a plate shape and bear a negative charge. Two clay particles can be bridged by the divalent cation of two water molecules.

As a result of these different bonds, a top soil is a roomy edifice composed of soil aggregates separated by a porous space. The larger pores (> 8 μm) contain the soil atmosphere, the smaller ones (< 8 μm) in a moist soil are the water holders by capillary forces. Soil structure depends on soil pore space and the stability of the aggregates. The larger and the more solid the aggregates, the smaller the potential vulnerability of the soil. For instance, wind cannot blow away aggregates of more than 2.5 mm.

Physical aspects of soil fertility are determined by the soil structure. It must allow the plant to develop its root system in the pore space, the distribution of which must permit the roots to function: i.e., enough water combined with air to breathe. Soil structure determines also the environment of soil microflora and microfauna, which bring about the decomposition of vegetal and animal materials, thus liberating the plant nutrients.

We have seen that soil structure determines the physical fertility. The available plant nutrients in the soil determine in their turn the chemical aspect of soil fertility. The chemical properties of soil fertility generally alter more slowly than soil structure which can be changed almost instantaneously through human activities or climatic change. Some factors to be taken into consideration are: tillage increases soil pore space; heavy implements driven on the soil decrease the pore space, raindrops and soil splash compact the topsoil, runoff removal of particles on the slope results in colluvial transport and deposition.

4.2.2 Leaching

"Leaching is the removal from within and outside the soils of colloids and the chemical bases associated". (WMO technical note, no. 178, p. 33). (Colloids are the particles with less than 2 μm of clay minerals, iron, aluminum oxides, and grains of quartz.) Leaching leads to the loss of bases such as Ca^{2+}, Mg^{2+}, K^{+}, and Na^{+} and accumulation of Al^{3+} and Mn^{2+}, and can be associated with loss of the clay fraction.

Leached soils are distributed over most of the climatic zones of the world, excluding arid and semi-arid climates in which evapotranspiration is higher than rainfall for most of the years. Nevertheless, leaching can exist at a dramatic level in the dry ecozones under irrigation. Percolation of water into the soils is high and the profiles are heavily desaturated and acid.

Leaching consists in the migration of nonsolid substances from the upper horizons to the lower ones. The transfer can be vertical when the accumulation is in the deep levels of soil and oblique when leaching is produced on a slope or in a profile located downstream.

When leaching concerns the bases, the scientists speak of lixiviation. The bases are adsorbed on the clay mineral. Water is the agent of both leaching and lixiviation.

A relative excess of permanent or temporary stagnant water in one or several soil horizons can modify the soil morphology and its properties. Imbalance high enough to become degradation, may occur in the dynamics of the soil. Leaching is a common process which does not lead to degradation unless the phenomenon is speeded up and causes an irreversible drop in soil fertility.

It could be said that desertification is reached when a new structure and water regime of the soil is irreversible with a significant loss in fertility or in potential development of crop or vegetation. An excess quantity of water due to precipitation, irrigation, or runoff in relation to the speed of infiltration causes a stagnation at the surface. In the case of flat topography, a slow infiltration causes leaching in the depth of the soil and reprecipitation when water becomes stagnant.

The water regime of the soil is characterized by the balance between inflow and outflow:

1. Inflow at the surface results from the amounts of rainfall, condensation, irrigation, and runoff; at deeper levels it is due to lateral circulation, and/or rising groundwater.
2. Outflow depends on evaporation into the atmosphere and lateral and vertical percolation.

Water in the soil circulates as a fluid or as vapor according to:

1. the pedoclimate determined by the nature of the soil atmosphere, and
2. the permeability and the porosity of the horizons which determine the speed of water penetration.

4.2.2.1 Aspects of Degradation by Leaching in the Upper Horizons of the Soil

The horizons that are leached by losing their clay fraction lose their texture balance. Their granulometric distribution is modified in favor of enrichment in sands. The porosity and structure characteristics are modified.

The exchangeable bases adsorbed on the clay, which are partly nutrients (K^+ and NH_4^+), are also partly lost.

In this way the resistance to water erosion (runoff) is degreased due to the disappearance of a part of the clay and the topsoil becomes more fragile.

Potential fertility is reduced because of depletion of the mineral colloids, which are those able to fix the chemical bases in an exchangeable state, suitable for their assimilation by plants.

4.2.2.2 Aspects of Degradation by Leaching in the Lower Horizons of the Soil

The deplaced colloids are trapped in the lower horizons (or in clay pans), where colloid coating increases the size of aggregates. Porosity is reduced and because of the aggregate coating and secondary precipitation of silicium, aluminium and ferric hydroxides, a structural increase with reduction of permeability results in a constraint of root penetration. The root system becomes sparse and the usable depth of the soil is reduced. Waterlogging and hydromorphic conditions can appear and even iron concretions can occur.

Leaching is not a dangerous process as long as human activity does not modify the dynamics of soil water.

With irrigation, there is a modification of the conditions of water circulation in the soil, inflow is higher than outflow, and waterlogging results.

When deforestation is followed by crop planting, the amount of water that can percolate towards the deeper horizons increases. The water supply balance of the soil is modified and leaching can be intensified to a level which can be termed degradation due to adverse changes in the mineral composition of the soil. The rate of decline of exchangeable cations is less with mulch farming, frequent use of legumes, grasses, and planted fallows, and with agroforestry systems.

4.2.3 Waterlogging

Soil is a porous medium. When the porosity of the soil is entirely filled with water, the soil scientists speak of waterlogging or raising of the water table. Waterlogging affects one or several horizons of the soil, seasonally or permanently.

When waterlogging becomes excessive or "abnormal", it may result in degradation, mainly when the water brings salts towards the surface.

This soil degradation results from "abnormal" waterlogging, i.e., when excess water in a soil horizon limits the aerobic life. In the evolution towards an anaerobic medium the microorganisms responsible for biodegradation of organic material are inhibited or destroyed. An accumulation of organic matter occurs resulting from slowing-down of mineralization and humification of plant debris. Fulvic-type acids are freed, the medium becomes acidic and more favourable to clay and iron leaching. Below pH 4.5 there is a strong increase in exchangeable aluminum and the latter becomes toxic for vegetation (WMO 1983a).

Certain of these chemical or structural changes decrease stability and fertility and increase the vulnerability of the medium until irreversible degradation occurs.

The causes of waterlogging are:

1. Natural poor external drainage linked with the topography: depression or low-lying areas without natural outlets.

   Climatic factors may also be a determinant in areas of high seasonal rainfall in a short rainy season with excess water inflow in the case of exceptional flooding, and in the case of degradation of the plant cover, which can result in a large increase of the runoff rate.
2. Poor internal drainage may also induce waterlogging. A clayey texture, an impermeable horizon, compact iron or aluminium hydroxides – duricrusts – can be responsible.

   In the fixed dune system in Niger, north of a line Maradi-Zinder, in all the interdunal depressions, the very shallow soils composed largely of scarcely differentiated horizons have hydromorphic characteristics owing to poor drainage during the rainy season even though the rainfall is only between 300 and 500 mm/year.
3. Irrigation is the third cause of waterlogging when the drainage system is not well designed. In areas with a dry season salinization can occur after waterlogging.

   In the two decades 1968–88, periods of drought and statistically exceptional rainfall have been recorded in all ecosystems of the world: in Certao (NE Brazil), much of Africa (1968–73, 1982–83), Australia, and Melanesia. Numerous phenomena of surface flooding have occurred in the USA and in Latin America (Cuba, Ecuador, Peru, Bolivia), and Polynesia. In September-October 1988 dramatic flooding in Sudan, after drought and plant cover deterioration, brought a heavy increase of runoff and sudden flooding of the wadis.

4.2.4 Salinization and Alkalinization

4.2.4.1 Processes

Salinization is the accumulation of soluble salts (chlorides, sulfates, carbonates) of sodium, magnesium, or calcium.

Alkalinization or sodication involve an enrichment in sodium ions.

Salinization occurs mainly in arid and semi-arid ecozones, and coastal areas.

"Soil and salinity problems are more widespread and acute in arid than in temperate areas. They affect one-third of the 200 million ha of irrigated land in the world". (Armitage 1985)

According to the FAO, a soil is classified as saline when its content in soluble salt exceeds 1 to 2% in the 20 upper cm. In a more practical way, salinity conditions are said to have developed when the accumulation of salts has reached a level harmful to plant growth. With the increase in soluble salts in the soil, the osmotic pressure increases and simultaneously increases the difficulty for plants of utilizing water, a state of physiological drought, although some plants, e.g., barley, are more salt-tolerant than others, e.g., wheat, while some halophytes are extremely salt-tolerant, e.g., *Atriplex*, the salt bush.

"About 900 million ha are presently affected by excessive salts. Salt-affected soils are especially common in irrigated regions of Asia, Australia, and western USA." (Lal et al. 1989).
"The problem is particularly severe where the mean annual evaporation (PET) significantly exceeds the precipitation (P), typically where P/PET is less than 0.75. Soils with high concentration of salt in the root-zone are called saline soils if the sodium absorption (SAR) is less than 15

$$SAR = Na/\sqrt{\frac{(Ca+Mg)}{2}}$$

"Soils are called alkaline if the predominant cation accumulated is sodium and the SAR exceeds 15". (Lal et al. 1989).

If the salty perched water table comes within 1–1.6 m of the roots, some of the water can rise through capillarity and cause damage because it inhibits the plant's ability to absorb moisture. The plants become stunted or die, depending on the salt concentration. If the salty groundwater reaches the soil surface it evaporates and leaves salt crystals on the surface of the fields and if the quantity of salty water reaching the surface is great enough, a relatively impermeable salt crust will form over the soil, diminishing infiltration and natural leaching.

The origin of the salt or sodium ions are: marine, from seawater, paleolagoons or inland lakes, deposits or salt accumulated in sediments or in groundwater which rises up in artesian water. Salt can be brought by wind: two of the best examples are the "Pri-Aral" and the Chalbi desert (north Kenya), which can be considered as a natural edaphic semi-desert. This area is occupied in the center by an ancient sebkha which is seasonally dry. The evaporites of the sebkha accumulate on the spot and the surface efflorescences of the salty soils are scattered around the basin by winds. Around the sebkha the vegetation cannot grow normally because of aridity, but mainly because of the wind-scattered salts.

The accumulation of the soluble salts (salinization) of sodium ions produces soil degradation over the entire profile, resulting in such harmful effects as:

1. changes in structure, which becomes unstable, compacted,
2. changes in the morphology, with decrease in permeability and porosity,
3. changes in the biological activity, which is weakened; the content of organic matter tends to drop or does not decompose satisfactorily.
4. changes in the soil chemistry; the pH decreases or increases beyond 9 or 10,
5. changes in the characteristics of the vegetation which develops on them: reduction of cultivated or natural vegetation or even its total disappearance,
6. changes in water/soil/plant balance.

The soil profile can be modified by the appearance of saline efflorescences or even a surface crust. Such a case of salinization is partly responsible for the disease of the palm tree plantations in the oasis of Tafilalet (Morocco). Deep crusting with shallow saline groundwater can also occur. These crustal salts are exploited locally for salt, e.g., the Bilma (Niger) salt deposits.

The basic principles of reclaiming salt-affected soils or preventing further salinization are well known, but generally not applied because of high costs. Salt accumulation related to irrigated cropland will increase with expansion of irrigation. Szabolcs (1986) has evaluated the increase in the irrigation land areas from 8 million ha at the beginning of the 19th century to 235 million in 1980. In the arid, semi-arid, dry subhumid ecozones, the high productivity expected will call for supplemental irrigation. Therefore the potential dangers of excessive salt accumulation will increase.

4.2.4.2 Case Studies of Salinization: An Age-Old Irrigation Problem

West Africa is characterized by runs of below-average rainfall years. Each run has direct effects on river salinity:

1. Reduction of water availability because of low river levels.
2. In the dry season marine incursions of saline seawater progressing upstream from the mouth of the river. After a deficient rainy season the saline water cannot be flushed out.

The Casamance river in Senegal (after Farmer and Wigley 1985) is an example of this problem.

Figure 44 shows a curve of the normal situation, one of the low salinity of the river water after the good 1968 rainy season, and the 100-km upstream seawater intrusion in July 1969 after a poor rainy season. The year 1984 was preceded by a run of dry years. In May 1984 the river water over 200 km inland reached a salinity level of 120‰, an amount which is three times that of seawater. The

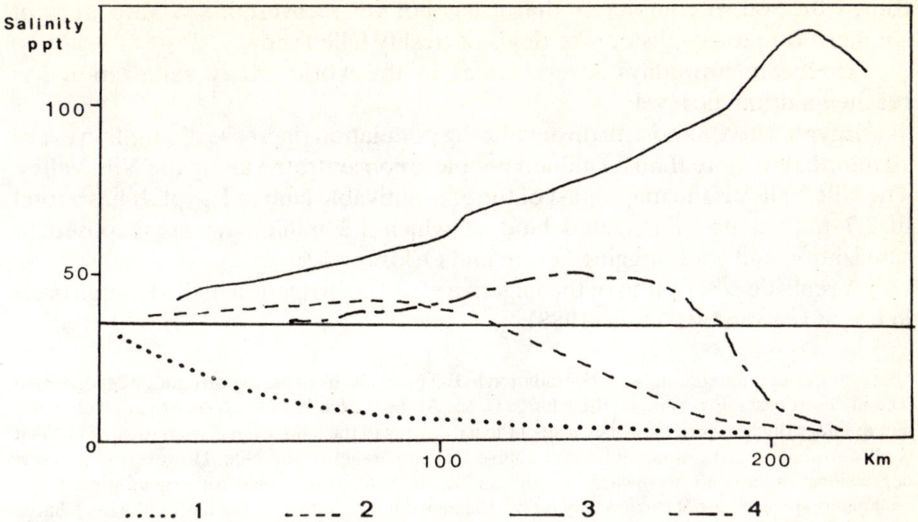

Fig. 44. Casamance River salinities as a function of distance (in km) upstream from the mouth. (Pages, CRODT, Dakar, Senegal, in Farmer and Wigley 1985). The ordinate is in parts per thousand (ppt). Seawater salinity (35 ppt) is shown as a *horizontal line*. *1* Oct. 1968; *2* July 1969; *3* May 1984; *4* Sept. 1984

salinity level hardly decreased after the rainy season in September 1984 and is now well above the low level of October 1968.

The situation is similar for the Senegal river. For the drought periods (e.g., 1974–1980) the river water has invariably a high level of salinity and for good runs of rainfall years (1953–59) results in a low salinity level. The year 1981 had the furthest saline water intrusion upstream. Lake Guiers, located on a tributary of the Senegal river 160 km inland and one of the sources of water for Dakar (25% of the water comes from the lake), is in danger of drying up. The other 75% of the water necessary for Dakar is pumped from a water table located beneath the town, where the danger of contamination by seawater has reached a critical stage.

In Libya the coastal aquifers are so depleted that seawater is coming in (e.g., in Benghazi and more recently in Tripoli). The attempt to stave off the evil moment by pumping fossil water from the sub-Saharan aquifer is only a palliative. This water is not being replaced and when it is exhausted, Libya runs the risk of being uninhabitable. The intrusion of a level of seawater under the continent is a worldwide phenomenon; the dry tropical areas are not alone. It is, for example, a plague in all the coastal areas of France from Normandy to "Côte d'Azur".

The excellent text of Eckholm (1976) *Losing Ground*, in which the author has written the following moving sentence, sounds the alarm about the almost irreversible danger of salinization: "The first recorded civilization, that of the Sumerians, was thriving in the southern Tigris-Euphrates Valley by the fourth millenium B.C. Over the course of 2000 years, Sumerian irrigation practices ruined the soil so completely that it has not yet recovered. . . . Vast areas of southern Iraq today glisten like fields of freshly fallen snow. . . ."

There are nowadays several areas in the world where salinization has reached a dramatic level:

Egypt, where almost all the increasing population (increase of 1 million every 10 months) of more than 60 million people is concentrated along the Nile Valley. The Nile Valley is the major part of the 3% cultivable land of Egypt. It has a total of 2.7 million ha of irrigated land of which 1.2 million ha are exposed to salinization and waterlogging (Zonn and Orlovsky 1986).

A realistic description of the negative effects of irrigation in the Kharga oasis in Egypt is given by El Baz (1988):

"An example of unintentional land degradation in Egypt results from the over-irrigation of reclaimed land in desert oases, for example, the Kharga Oasis. Along the banks of the Nile, irrigation channels and drainage ditches work easily because of the natural lay of the land; from Aswan to Alexandria it is all downhill, and the water follows a course that is parallel to the Nile. However, in a desert depressions, there is no northward downhill gradient. Water that is used for irrigation by desert reclamation projects is not carried very far by drainage ditches. It stays within basins like the Kharga depression and clogs the soil."

Egypt is not the only example; almost all the other irrigated areas in Africa have salinization difficulties:

— Mali in the irrigated area north of Segou,
— Morocco in its oases along Wadi Draa, Ziz, and Rheris,
— Senegal, as shown previously.

For the western United States, Sheridan's (1981) very precise analysis demonstrates the tragic level reached in this highly developed country. Three examples already analyzed will nevertheless be repeated to illustrate the degree of salinization: the San Joaquin basin (California), Wellton-Mohawk (Arizona), and Colorado River.

• In the San Joaquin basin (California) about 400 000 acres (161 874 ha) of irrigated farmland are affected by high brackish water tables. If this continues at the same rate, by the year 2080, 1.1 million acres (445 154 ha) will have become unproductive. In San Joaquin, in West Texas, as in the Tigris-Euphrates Valley, a shallow layer of fine clayey material keeps the salty unused irrigation water from infiltrating far below the level of crop roots.

• The problem in Wellton-Mohawk (Arizona) is also saline groundwater. The situation here is even worse because of the waterproof subsurface layer which blocks downward drainage. The Wellton-Mohawk project is built on the Gila river tributary of the Colorado river. This project is the primary cause of the increasing salinization level of the Colorado which has evolved from 800 mg l^{-1} in 1960 to 1500 mg l^{-1} in 1962.

The Colorado river salinity control was envisaged in the 1960's mainly because of the objections of Mexico to its increased salinity.

The US Salinity Laboratory has established a general classification of the salinity hazard to crops of irrigation water (in mg l^{-1}):

— Low: 100–250
— Medium: 250–750
— High: 750–2250
— Very high: > 2250

The salinity of the water was about 809 mg l^{-1} in 1979 at the Imperial Dam in the United States just before its release to Mexico. Therefore it was decided to build a desalinization plant near Yuma.

In the USA 27% of the irrigated land is affected by salinization, while 80 000 ha have been abandoned two decades after completion of an irrigation project in California (Zonn and Orlovsky 1986).

Africa and North America are not the only continents affected by salinization. In Iraq, nearly half of the irrigated area is affected by salinization and waterlogging. Along the Pakistani Indus, 15 million ha are irrigated, while waterlogging and salinization, to varying degrees, affect a total of 10 million ha (Zonn and Orlovsky 1986).

"In 1952–54, it was estimated that 20% of 11.4 million ha surveyed in Punjab was salt-affected, as was 56% of the 7.25 million ha surveyed in Sind and 5% of the 0.55 million ha covered in Northwest Frontier Province. Of the area surveyed in Punjab, 16% was waterlogged, as was 38% of that in Sind. It was estimated in 1976 that, of the 4.5 million ha of salt-affected lands, 87% was affected in this way by fossil, i.e., natural, salinity that had developed over very long periods. The remaining 15% were affected by secondary, man-induced salinity (Shah 1978). Whereas 16 000–20 000 ha were being added annually to the area affected by waterlogging and salinity in 1945–46, the rate had increased to 45 000 ha/year in 1962–63, two-thirds of it in the lower parts of the Indus Plain." (Khatib 1971 in Armitage 1985).

Armitage (1985) adds:

"The man-made share of these salinity problems arises principally from raising of the water table through the continuous, region-wide passage of massive quantities of water in unlined canals, excessive applications of water by farmers, and inadequate natural and artificial drainage. To be effective, reclamation and amelioration must be undertaken on a very large scale. This is, in fact, being done through an agency specially set up for the purpose that began operations in 1960. The plan involves the construction of thousands of kilometers of drains to take surplus water away, massive use of tube wells to lower the water table, followed by a series of soil amelioration practices that include leaching with sweet water. In the 1st year, 1964, the water table had dropped by an average of about 28 cm in the areas treated and hopes for success were rising by 1970. Progress is slowed by the extensive financing that is required but, in general, the country believes it is holding its own: broadly speaking, new losses of productivity to salinity and waterlogging are being balanced by gains due to reclamation (Eckholm 1976)."

As seen in Chapter 3 (3.5.3.3), the Aral sea basin ecological crisis is also the direct result of the socio-economic development of irrigation in Soviet Central Asia and Kazakstan with rapid salinization of soils because of excessive water application and insufficient drainage (Létolle and Mainguet 1993).

4.2.5 Conclusion

My own conclusion is that it is absolutely vital that irrigation, especially large projects, in mainly arid and semi-arid ecosystems, should be considered only as the last option of management and if no other solution for development exists.

The main reasons are:

1. The cost of all irrigation projects.
2. The difficulty of maintenance of the schemes.
3. The threat caused to the environment.
4. Salinization and other induced dangers summarized for the former USSR by Zonn and Orlovsky (1986) as:

"1. Propagation in irrigation systems of the vectors of viral and infectious diseases and intermediate hosts.
2. Propagation of microorganisms causing communicable diseases in man (diphteria, typhoid fever, cholera, etc.) in the lakes and rivers that receive the return effluent flow from irrigated territories or that serve as a source of drinking water.
3. Water contamination by chemicals where poorly treated wastewaters, industrial or agricultural, are used for irrigation.
4. Inadequate sanitation and hygienic measures to insure safety during the construction and maintenance of irrigation systems.
The most prevalent and dangerous of the common diseases in irrigated tropical regions are schistosomiasis and malaria."

The same authors report on the intensification of aquatic snail reproduction along irrigation schemes: schistosomiasis has affected 200 million people in 71 countries of the tropical and subtropical zone. A typical example is the Khashm-el-Girba irrigation project on the Atbara river in the Sudan.

4.3 Physical Processes of Land Degradation: Water Erosion

4.3.1 Soil Loss

Erosion depends on climatic, topographic, soil-formation, and land-use factors. Hudson (1971), in a paper on Soil Conservation, notes that: "Erosion is a joint result of the erosivity of the physical agencies and erodibility of the soil: Erosion = *f*[(erosivity) . (erodibility)]." (Fig. 45)

In a rainstorm, erosivity depends on the amount and peak of intensity of rainfall. In the United States climatic erosivity is higher in the southeast and lower in the west with a ratio about 20 to 1 (Lal 1988). In the US it is higher than in Europe but less than in the tropics (Dissmeyer and Foster 1980). It has a seasonal distribution which varies with the location.

Erodibility of soil can vary from 1 to 10 according to four main parameters: soil texture, organic matter content, soil structure (and soil permeability), and chemical content (mainly iron and aluminium oxides, and sodium ions). Topography is a major factor: slope steepness and profile affect erosion more than slope length. The erosion rate is higher at the base of a convex-shaped slope than on the end of a uniform slope. The trend of concave slopes is to reduce erosion because of sediment deposit at the foot and the reduction of erosion on the upper portion.

Soil erosion by water was defined by Ellison (1947) as "a process of detachment and transportation of soil material by erosive agents." He pointed out that

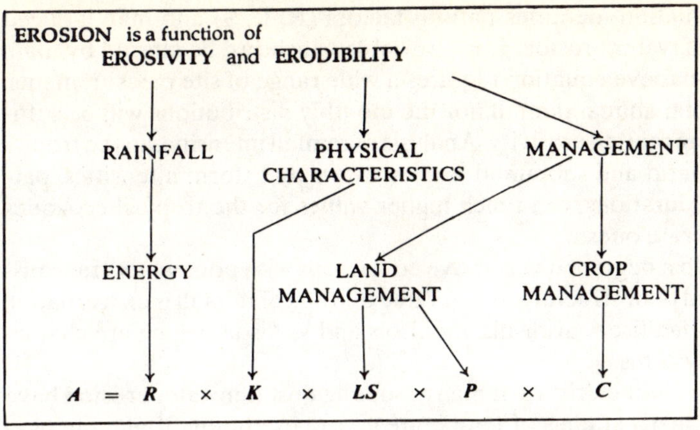

Fig. 45. Structure of the universal soil loss equation. (WMO no. 591 1983a)

for erosion by water these agents are rainfall and runoff. "Soil erodibility may be separated into its detachability and transportability components." (Wischmeier 1959; Meyer and Wischmeier 1969).

A mathematical model of four values — detachment by rainfall, detachment by runoff, transport capacity of rainfall, and transport capacity of runoff — was developed showing that "rainfall effects are dominant near the top of slopes but runoff effects are dominant beyond some points downslope and detachment capacity greater than transportation capacities on the upper portions of moderate slopes."

Later, the Universal Soil Loss Equation (USLE) of Wischmeier and Smith (1978) identified the parameters which control soil loss by water.

$$A = R \times K \times L \times S \times C \times P ,$$

where: A = total soil loss per unit area (acre) and per unit time (year); R = rainfall factor or rainfall erosion index expressed by EI, where E = energy of the rain storm; I = maximum intensity over a specified period (commonly 30 min) of rainfall in rate per hour; K = soil erodibility factor. The erosion rate per unit of erosion index for a specified soil in a cultivated continuous fallow on a 9% slope 72.6 feet (22.6 m) long. A predictive nomograph has been developed based upon five physical characteristics of the soil; L = slope length factor: the ratio of soil loss rate from the actual field slope length to that from a 72.6 feet (22.6 m) length of the same soil type and gradient; S = steepness or slope gradient factor: the ratio of soil loss from a gradient of the actual field to that from a 9% slope (of the same soil type and slope length; C = crop cover factor or crop management factor: the ratio of soil loss from a field with specific cropping or management to that from a fallow condition on which the factor K is evaluated; P = erosion control practice: the ratio of soil loss with contouring strip cultivation or terracing to that with straight row, up and down slope farming. This factor is linked to the reduction in erosion through soil management techniques.

The above equation identifies natural factors (R, L, S) and man-induced factors (K, C, P) of water erosion. L is a variable which can be altered by man. For its solution the above equation requires a wide range of site measurements.

Neither the total annual rainfall nor the monthly distributions will be sufficient for quantifying rainfall erosivity. Analysis of rainfall intensities in the tropics even in arid, semi-arid and subhumid tropics shows that storm intensities, particularly for short durations, can reach higher values for the tropical ecozones than for the temperate ones.

Combined with a degraded vegetative cover, soils with poor water transmission properties, and poor structure, the K value of tropical rainfall is exacerbated. This is the case with alfisols, aridisols, mollisols and vertisols, which are characteristic of dry ecosystems.

The protective characteristics of many crops against rainwater erosion have been identified in many studies of temperate ecozones: timing of planting and harvesting, cover densities and height, growth stages, rotations, and residue management have been linked with seasonal distribution of rainfall erosivity to define the C factor of the USLE. No similar information is available for tropical crops, but field observation has shown that cassava (*Manihot esculenta* Crantz) and millet provide little protection for the soil.

The P factor quantifies the role of supporting land management practices: tillage against the slope in contour strip cropping and terracing, equally well known in Europe and the USA, is only beginning to be widespread in wet tropical ecozones, although terracing has long been practiced on steep hillsides: e.g., the Nuba mountains (Sudan), the Mandara Hills (Nigeria), Nepal and China.

The USLE is difficult to apply and is really only applicable to most parts of the USA for which it was developed. The parameters must be checked carefully for noncultivated land. Wischmeier (1959) proposed a simplified expression for the calculation of rainfall erosivity:

$$R = K\,(A.B.C.),$$

where R = mean interannual erosive capacity index; K = regional coefficient; A = average interannual rainfall; B = maximum 24 h rainfall with a 2-year recurrent period; C = maximum 60 min intensity with a 2-year recurrent period. Temperature is not mentioned but is nevertheless important, as viscosity and density change appreciably with temperature. For its solution the above equation requires a wide range of site measurements.

Studies conducted by Cormay and Masson (in Babau 1983) indicate that for Tunisia $K = 35 \times 10^{-5}$. This value is only valid for the region where it has been established.

At a given site erosion by water is enhanced by natural and man-induced conditions:

1. The natural conditions are:
   — high rainfall intensity and long duration,
   — high rates of overland flow from adjacent topographical uplands,
   — long and steep slopes.

2. The natural and man-induced conditions are:
   — soil with natural or man-induced poor structure,
   — low infiltration rates,
   — scarce or absent vegetation and/or organic residues.

3. Man-induced conditions are:
   — land cultivation,
   — tillage operations parallel to the slope.

4.3.2 Deteriorating Mechanisms of Water Action

Throughout the world, the earth is subject to water erosion as a normal mechanism of topographic shaping. When accelerated by human activities, water action causes deterioration.

Processes of water action can be classified as:

1. surface attack at the top of the soil,
2. action throughout the depth of the soil profiles.

Surface attack is divided into two types:

a) raindrop attack,
b) runoff action.

Farmers are more concerned with the cultivable areas located between stream channels and by splash erosion, rillwash, gullying, and ephemeral channels than by continuously flowing rivers.

4.3.3 Raindrop and Splash Effect

"A raindrop with a diameter of about 4 mm, falling in still air, reaches a top speed (terminal velocity) of approximately 34.29 cm s^{-1} and so a maximum kinetic energy of roughly 203 dynes. Theoretically it would be possible with this energy to lift a sand layer (d = 2.65 g cm^{-3}) of 1 cm^2 and 1.3 mm thickness up to about 6 cm. This example may illustrate that large raindrops obtain energy sufficient to disturb the structure of the top-soil." (WMO No. 591 1983a)

Field observations, laboratory research, and field experimentation have brought ample evidence that splash is the most important process of water erosion. The pre-eminent role of raindrops on the soil surface was already recognized at the end of the last century by the German soil scientist, Wollny (1877), who analyzed the role of plants, and the role of surface mulch in intercepting rainfall and preventing the physical deterioration of the soil structures.

During the first decades of the 20th century, attention was more concentrated on runoff and its scouring and capacity of transport.

In 1944 Ellison led the first experimental study of the mechanical action of raindrops and demonstrated that a falling raindrop is a complete erosive agent involving the three phases of detaching, transportation, and deposit of soil particles. These three phases are the result of splash.

Hudson (1963) has given the best estimate of the kinetic effect of raindrops on two similar experimental plots, in which the second was protected by a gauze stretched 10 cm above the surface in order to reduce the kinetic energy of the rain without any change in the quantity of rain reaching the soil. On an unprotected plot he measured between 1953 to 1956 an erosion of 2900×10^2 kg ha^{-1} a^{-1}. On the protected plot erosion was reduced to 24×10^2 kg ha^{-1} a^{-1}.

The erosiveness of a raindrop is proportional to its kinetic energy, which can be calculated from the size of the raindrops, their density, and their distribution by classes of size.

Hence the intensity of the attack on the soil structure by a raindrop is a function of its diameter and its speed, e.g., its kinetic energy. The kinetic energy of a water drop is:

$$\frac{1}{2}mV^2,$$

where m = mass of the drop (proportional to the diameter of the drop). The maximum diameter of a drop is 9 mm and 5 to 6 mm in a turbulent atmosphere. The surface tension of water is insufficient to maintain a bigger drop intact. V = speed of the end of fall (of an equivalent sphere of the same mass).

The velocity increases up to the terminal velocity "which increases with the diameter of the raindrop from 4 to 9.3 m s^{-1} when the diameter increases from 1 to 6 mm (Laws 1941; Gunn and Kinzer 1949).

Hudson (1963) observed that falling velocity is affected by wind and proposes a formula:

$$V1 = Vv/\cos\alpha,$$

where V1 = real velocity of the drop falling obliquely, Vv = vertical velocity, α = angle of the drop-falling direction and the vertical.

Hudson, in the same paper, proposes taking the median diameter. He observed that this median diameter stabilizes at around the 100 mm h^{-1} intensity and decreases slightly afterwards.

The splash effect is quadruple:

1. Compression of the top soil.
2. Decrease of the soil permeability: the jumping water droplets, into which a raindrop is split when falling on the surface, plug soil pores, reducing the infiltration rates. The rainwater then inundates the surface since a layer of closely packed particles builds a thin smooth crusted soil surface where silt has clogged the pores of the soil, and keeps the rain out.

This thin layer of closely packed particles, called crusted surface, is in fact a waterproof one. In Mali and in Niger on the fixed dunes of Hausaland the phenomenon is common. The wind exports the loose A horizon but leaves the top of the B horizon which, because of splash erosion, emerges smooth, compact, and waterproof. The Bambara call this scalded surface Wala-Wala. Wala-Wala is an advanced stage of degradation, favorable to runoff. Water erosion takes the place of wind erosion and the soil becomes more and more unproductive.

3. Explosion of soil aggregates when the aggregates are saturated and the enclosed air is compressed.
4. Splash erosion of soil particles suspended in the splashing drops. Raindrops, thanks to their kinetic energy, detach, from the surface of the moistened aggregates, particles which are carried away in suspension by bouncing back after ejection (elastic impact) and by being fragmented into several droplets.

On a slope, the splashing droplets result in a downhill movement of the soil particles. In fact, the soil particles splashed by the falling droplets do not fall symmetrically. Those which are splashing down the slope meet the soil at a greater distance from the spot where the raindrop has hit the surface than those which splash up the slope.

This whole mechanism of particle loosening, and short downslope colluvial transport under the impact of raindrops produces splash erosion.

4.3.4 Runoff Action

Runoff occurs when rainfall intensity exceeds the infiltration capacity of the top soil. As soon as runoff appears, the second phase of the processes of attack of the soil surface, the removal phase, is initiated. Runoff can also have an abrasive effect by using the solid load.

The most severe water erosion problem occurs in the Dali catchment. The Dali river is a tributary of the middle reaches of the Yellow River (Huang He). It drains the gullied loess region in North Shaanxi in the Peoples' Republic of China with an export of 25 600 t km^{-2} a^{-1} from a basin area of 96.1 km^2 and 16 300 t km^2 from a 3893 km^2 basin (Mou and Meng 1980). One of the highest water erosion rates is 34 500 t km^{-2} a^{-1} in the Chinese loess (Derbyshire 1978).

Running water on the soil surface has a triple erosive effect: detachment of particles, rill colluvial entraining, and deposition of soil particles.

The different forms of water-induced erosion include (1) geologic water erosion, called also natural or normal water erosion; (2) accelerated or man-induced forms of erosion: sheet erosion, sheet-wash (inter-rill), rill and rillwash, gully and gullying, tunnel, pedestal, pinnacle, puddle erosion, streambank erosion, valley trenching, and vertical argillic migration.

Sheet erosion and rills (rillwash) represent, from the agricultural point of view, the most damaging and also the most dangerous, as they are very insidious forms of water erosion.

The analysis of the physical processes of water erosion, as of all other physical processes of soil degradation, has two objectives: to contribute to the preservation of the soil capital, but also to protect the infrastructure of the lower part of a drainage basin from the silting up of dams and lakes; the accumulation of sediments in irrigation devices and on farmland where the deposition of sediments can be harmful, and excessive deltaic deposits.

El-Swaifi and Dangler (1982) quote Carter (1977):

"Soil erosion is perhaps the most serious form of land degradation throughout the world. Even in the United States, and after more than 40 years of intensive research, the problem still persists and recent information indicates that soil loss rates may indeed again be on the rise." (Carter 1977).

In order for water erosion to start, loose particles must be available, and carried to a lower point on a field scale.

The detachment of elementary particles from the soil mass involves an initial input of energy which is provided by kinetic energy of rainfall. Once separated from the main soil mass, this elementary particle requires another form of energy to carry it from one point to another: this second form of energy is runoff. Both are linked to the characteristics of rainfall but also to slope factors: length, angle, and roughness of the surface, depending on soil characteristics, plant cover, and cultivation practices.

"During agricultural operations, cultivation, row direction in planting, and water management practices designed to convey runoff without allowing excessive erosion all change the natural rill system. In row crop agriculture, rills generally form in tillage paths, especially where the crop is cultivated. For up-and-down-hill rows, the rills enlarge as runoff amounts and rill erosion increases. For contour rows, rill enlargement is slower because of lower runoff velocities. However, the capacity of rills parallel to the rows is often exceeded under large storms, and rills parallel to the slope are formed. Terraces built across a slope interrupt downslope rill patterns and keep rill formation at the elementary stage." (Mutchler et al. in Lal 1988)

In the initial stages of water erosion the runoff remains diffuse and sheet erosion (or sheet-wash) appears without concentration at any particular point. Once a certain threshold is exceeded, the runoff tends to concentrate and rillwash appears with rivulets, rills, and gullies. These forms are not directly linked to kinetic rainfall energy but are dependent on the force of runoff. Bank erosion is another form of water erosion, linked not to runoff but to river current or wadi flow. In thc areas affected by desertification, the transport in river beds is negligible in comparison with runoff erosion.

All water erosion processes follow a series of steps starting with sheet erosion and finishing with river bed and bench erosion.

All comparative studies in both tropical and temperate ecozones lead to the same observation: rainfall in the tropics is six times more energetic than in the temperate zones.

The loss of arable soil corresponds to a loss of capital and therefore must be measured. Assessment of soil loss was initiated in the USA: in 1907 the US

Department of Agriculture declared an official land policy. In Utah, the Forest Service in 1915 achieved the first quantitative experiment on erosion followed in 1917 by similar experiments in Missouri.

4.3.5 Conclusion

Water erosion must be considered as a single complex system from falling rain drops to flowing rivers which is well summarized in the book *Soil Erosion Research Methods* (Lal 1988):

"Raindrop impact erodes the land surface between rills and also initiates transport of detached soil particles to the rills. The complete transport system involves the initial movement of soil from the point of raindrop impact to small rills, to larger rills, to ephemeral channels, and to permanent flowing rivers. Erosion and sedimentation can occur at any point in the system. This process is called geologic erosion when it occurs without human influence. Alternatively, the process is called accelerated erosion when activity by humans causes increased erosion, such as land disturbed for crop production."

It is neither difficult to diagnose qualitatively the mechanisms responsible for degradation, nor to evaluate quantitatively the amount of degradation, but the difficulty begins when assessing whether or not ongoing erosion represents a problem. The question which is difficult to answer arises as to the amount of soil loss which is normal and can therefore be tolerated. We have not found in the literature one single paper which answers this question for areas threatened by desertification. It was analyzed in the USA but not in tropical areas.

The complex nature of the problem of soil loss tolerance is highlighted in the paper of El Swaifi and Dangler (1982):

"Certain special considerations must be emphasized in formulating erosion loss tolerance for tropical soils. These include:

a) The extreme variability among soils which may necessitate formulation of many different tolerance limits.

b) Even for "typical" soils (e.g., Oxisols), the characteristically high weathering (regeneration) rates are probably more than off-set by the low initial nutrient content, and the fact that subsurface horizons are often characterized by high acidity and unfavorable structure or indurated layers.

c) The unusually high turbidities caused by specific concentrations of erosional sediments from oxidic soils and subsequent impact on the quality of destination water bodies (Ekern 1977).

d) The very limited resources available to the typical subsistence farmers in developing tropical countries for correcting the detrimental effects of erosion.

On the one hand, these combined factors suggest that soil loss tolerance limits for tropical soils should be lower than their temperate counterparts. On the other hand, realistic determinations of tolerance limits must consider the socio-economic elements pertinent to a specific environment; population pressures may modify the "technically feasible" tolerances upward."

The magnitude and importance of water erosion, particularly when it is exacerbated in areas where vegetation is scarce or absent, is well known. Nevertheless, not all erosion is deleterious. Rich material can be transported from areas where it is of little agricultural value to regions where it is of great agricultural importance: e.g., erosion of Ethiopian mountains produces silt deposited in the Nile plain, loess winnowed by wind from the deserts and fertilizes the Great Plains of China. But water erosion is the not sole physical process of land degradation. Wind erosion, so often overlooked, can have the same adverse economic effects.

"Equally ruinous are the adverse effects of wind erosion. Although wind erosion is less than water erosion on a world scale, the problem is severe in many semi-arid and arid regions. However, even less research data is available for wind erosion than for water erosion. The basic principles governing wind erosion processes and erosion control are similar to those for water erosion. Nonetheless, the specific cause-and-effect relationships, the magnitude of wind erosion in different ecologies, and the effectiveness of erosion control practices on management systems have not been investigated widely." (Lal 1988).

4.4 Physical Processes of Land Degradation: Wind Erosion; Where Is the Danger?

Natural erosion is a continuing, inevitable, and even beneficial process for the creation of soils. Prehistorical and historical acceleration until a speed is reached, where the threshold removal is higher than the rate of soil formation, is damaging and can become irreversible in human terms and results in so-called desertification.

The steps of damaging processes by wind erosion will be analyzed. It is commonly understood that wind in drylands is mainly dangerous due to its effect on the formation and migration of sand dunes.

Heathcote (1983) is right when he notes that "the advancing sand dunes are in fact a very special and localized case and a popular misconception of desertification."

It should be stressed that dunes inexorably moving forwards are only the end product of a chain of degrading wind processes and the final stage in aeolian land degradation. The dunes, often (too often) used as an indicator of desertification, are even less than the visible part of the iceberg. Very often in the popular press (for example the French newspaper *Liberation*, 6 March 1989) or even in scientific papers, when the word "desertification" is written it is accompanied by a photograph of active dunes, without any indication of the location or type of dunes. Unfortunately, the real important wind effects resulting in major soil loss are not so easily visible, are not spectacular on a photograph, and so thus, not very interesting for the media.

Wind erosion as a factor of "desertification" has more dangerous effects than burying some fields or buildings under moving dunes. This more dangerous effect is deflation.

What is deflation? It is the capacity of wind to sweep away the top soil and to impoverish it in organic matter and in fine particles. This deflation effect is stealthy and insidious, it can affect great expanses of soil, destroying texture by increasing the proportion of sand and degrading the soil structure. The soil becomes loose and easily exportable.

Millions of tons of topsoil can be removed by a single dust storm or by the repeated blowing away of fine material. It is this fine material which contains the majority of soil nutrients and is able to retain soil moisture.

Soil loss by wind erosion in ten Great Plains States of the USA, for example, amounts annually to 1.4×109 tons, 61% of the loss being from cropland and 39% from rangeland. The annual average wind erosion rate for cropland exceeds 11.2 tons ha^{-1} in three of the ten States. (Diderriksen 1981).

Wind erosion is economically very costly; for example, the total annual off-site costs from all sources of wind erosion in New Mexico are estimated to be US$ 465 million or US$ 358 per person (Huszar and Piper 1986).

The severity of wind erosion depends on a number of factors, such as climate (aeolian dynamics, marine dynamics in the case of coastal dunes), vegetative cover, soil erodibility, and human activities.

4.4.1 What is Wind?

When pressure is the same throughout a system, the air is said to be in a state of equilibrium. Under normal circumstances, air, like any gas, behaves like a turbulent fluid and continuous variations in pressure occur owing to temperature differences and other meteorological variables or topographical features. Wind is therefore the air movement resulting from pressure differences between two points: the airflow goes from a region of high pressure to a region of low pressure, and the greater the pressure differences the stronger will be the resultant wind. It can be classified according to its direction, its orientation, and its speed, which is usually given in m s^{-1}, km h^{-1} or knots (nautical miles h^{-1}). On a global scale, horizontal air movements are more frequent than vertical ones. There are two types of airflow:

— A turbulent flow is characterized by vortices in which there is movement of particles around an axis. The speed is inversely proportional to the distance from the axis and it can reach five times the speed of the undisturbed wind. Vortices can move or be stationary. The moving ones are usually temporary and erratic in their migration, while stationary ones are more permanent and form in the lee of obstacles such as escarpments and the crests of dunes or isolated rocks. When air passes over the ground the resulting friction generates heat, which, together with the natural radiation of heat from the earth's surface, causes a rapid exchange of gas molecules. The resulting turbulence causes a disruption of the smooth air currents.

— A laminar flow occurs at slower speeds when the airflow is made up of parallel streamlines. Those at the soil atmosphere interface are slower than higher ones. There is both molecular exchange and radiant heat transfer between adjacent streamlines; the fastest streamlines push the slower ones and give them their energy, and the net result is unstable conditions.

4.4.2 The Influence of Topography on Wind Conditions

Wind speed decreases near the ground as a result of surface roughness, whether caused by plants, natural or man-made obstacles, or just uneven ground. The resulting air turbulence consists of convergent, divergent, and ascending air currents. Where the ground is vegetated, wind speed starts to decrease at ten times the height of the vegetation. Bernoulli's application of Boyle's law of fluids (that the product of pressure and velocity is constant) gave rise to the Venturi principle. This states that the mechanical energy of fluid motion is also constant. When applied to airflow, the result is that any increase in windspeed is associated with a corresponding decrease in air pressure.

The model of air flow is by the Bernoulli rule:

$$P + \varphi \frac{V^2}{2} = \text{constant},$$

where P = pressure, V = wind speed, φ = specific air mass.

Such a phenomenon occurs when streamlines converge, the air behaving as if it were flowing through a nozzle, with airspeed increasing and pressure falling. When streamlines diverge, as a diffuser (Fig. 46), the opposite occurs, i.e., a rise in air pressure and a fall in airspeed.

This effect depends on the profile of the obstacle and its permeability to air. For example, an obstacle with gentle slopes will give rise to the pattern of wind dynamics shown in Fig. 47.

In contrast, an obstacle with steep slopes generates a different pattern of air flow (Fig. 48).

1. The windspeed accelerates on approaching the obstacle due to the compression of the streamlines and so the pressure decreases. At the base of the obstacle on the windward face is an area of deflation or erosion.
2. A deposit appears to leeward of the obstacle as a result of the decrease in airspeed as the streamlines expand and pressure increases; this is the limit of the turbulent sector where airspeed is almost nil.

   The maximum influence on airspeed is produced by a vertical impermeable obstacle such as a wall or palisade that is perpendicular to the wind. A permeable obstacle, e.g., a shelterbelt of trees, with a porosity of 50%, causes an even decrease in airspeed without affecting the streamline pattern. The pattern of sand sedimentation leeward and windward of the obstacle depends on its porosity and shape. For example, if the porosity is greater than 50%, the sedimentation profile will be long and low.

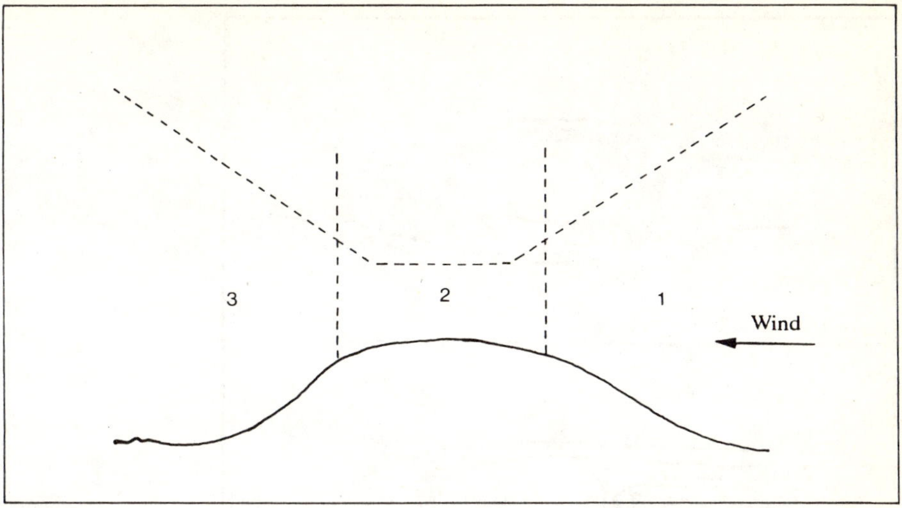

Fig. 46. The relationship between streamline patterns, pressures, wind velocity, and sand dynamics. *1* Upwind side, nozzle or convergent: low pressure, high wind speed, erosion area; *2* throat: transport area; *3* leeward side, diffuser or divergent: high pressure, low wind speed, accumulation area

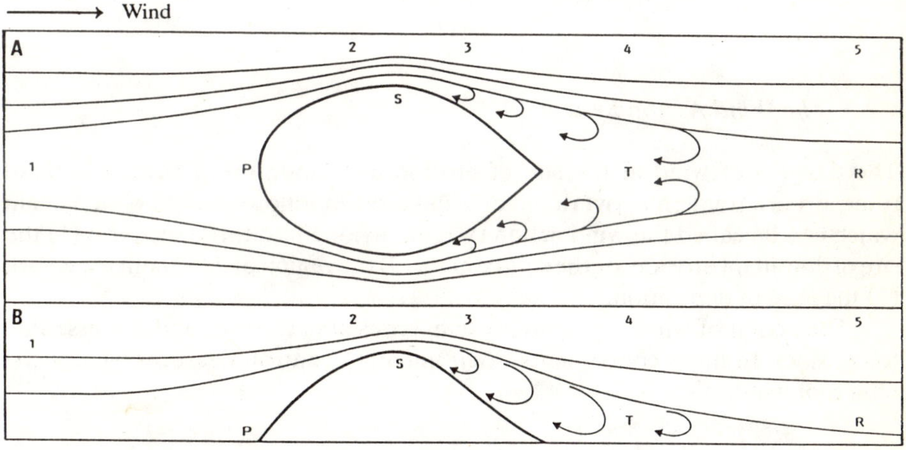

Fig. 47a,b. The effect of a gentle sloped obstacle on the dynamic behavior of wind streamlines and particle load. *A* Plan; *B* cross-section. *1* Upwind section of the obstacle: the pressure is highest and the air speed is lowest, even nil at a point *P*, at the bottom of the obstacle. Point *P* is an area of discharge and deposition. *2* Nozzle section: where the streamlines around the obstacle become dense on the sides. The airspeed increases and the pressure falls. This area is a transport area where wind erosion becomes very active. *3* Section of separation: the streamlines diverge from the obstacle at point *S*. The boundary layer separating from the surface is affected by vortexes which produce a turbulent drag with higher wind speed. Deposition occurs at the outer limit of the drag area (point *T*). *4* The drag area behaves as if it were itself the obstacle. Its length and width do not depend on air speed but on the form of the obstacle. Leeward of this section in R, the streamlines are no longer affected by the obstacle and they rejoin. *5* The section of rejoining is an area of sand deposition

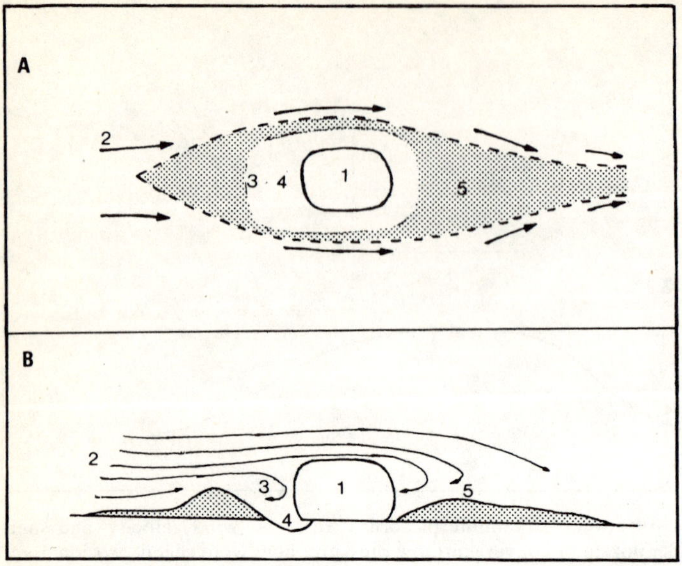

Fig. 48A,B. Pattern of airflow windward and leeward of an obstacle with steep slopes. *1* Steep-sided obstacle; *2* streamlines (direction of the wind flow); *3* high pressure and inversion of the airflow; *4* erosion; *5* accumulation and return flow. *A* Plan; *B* cross-section

4.4.3 The Wind Action System

The dangers of wind at the site of erosion and along the pathways of sand transport are frequently overlooked. Wind erosion follows a degradation chain which can be called the wind action system, divided into three sections: (1) the site of dominant erosion, or the source area; (2) the region of dominant transport; (3) the area of deposition.

The action of wind is potentially dangerous at each stage and it is essential to consider all three components when initiating control measures against the effects of wind.

4.4.3.1 Source Area

Threshold velocity is the minimum air speed necessary for a particle to be lifted off a surface, and initiates sand or soil movement. It varies with the square of the diameter of particles. The velocity increases as the particle size increases.

The threshold velocity is lowest for the quartz particles of 80–100 μm (200 μm in the water); above and below this size the air speed must be higher in order to produce particle movement (Fig. 49).

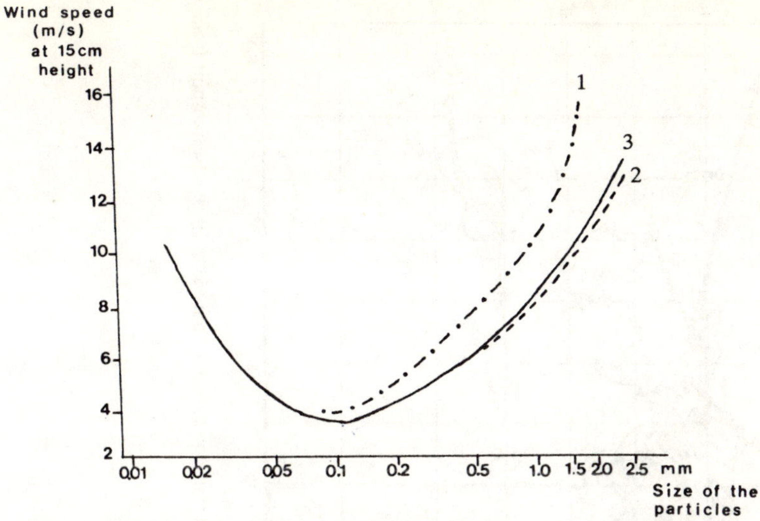

Fig. 49. Relation between the diameter of particles and threshold velocity. (After Chepil 1945; Yakubow 1959; Guyot 1987). *1* Aeolian sand; *2* silty sand; *3* fine silty sand

Deflation of particles by wind in function of their size was analyzed (Lyles and Takarko 1986) in ten soil sites in western Kansas, sampled in 1948, and re-sampled in 1984, to compare the particle size distribution (texture) and organic matter content in the top 10 cm. Except at one site, the sand fraction increased; this increase ranged from 0.9 to 23.3%. The greatest changes occurred in the moderately-coarse and coarse-textured (sandy) soils. Overall changes in particle distribution were 6.5, 7.2, and 0.7% for sand, silt, and clay, respectively, indicating that silt was being removed through sorting by wind. Organic matter declined at eight of the ten sites, averaging about 19% overall or about 0.01% per year. Wind erosion is causing a slow decline in silt content and organic matter in these soils, with potentially detrimental effects on soil structure, nutrient availability, and water-holding capacity.

Threshold equilibrium occurs when shear forces equal friction forces. The actual transport of sand by wind depends on the weight of the particle and the wind speed cubed.

A decrease in wind speed causes a decrease in the ability of the air to lift the particle, and therefore sedimentation of the particle occurs.

When wind reaches a sufficient speed, the first manifestation of wind erosion in source areas is dust storms, which consist of small particles of clay and silt with a diameter of between 2 and 50 μm. All continents experience them and the more powerful can have plumes 500–600 km wide and 2500 km long (Péwé 1981) (Fig. 50).

Meteorological surface observations from three sites in the western USA show an increase in the number of dust storms over the past 30 years, while the

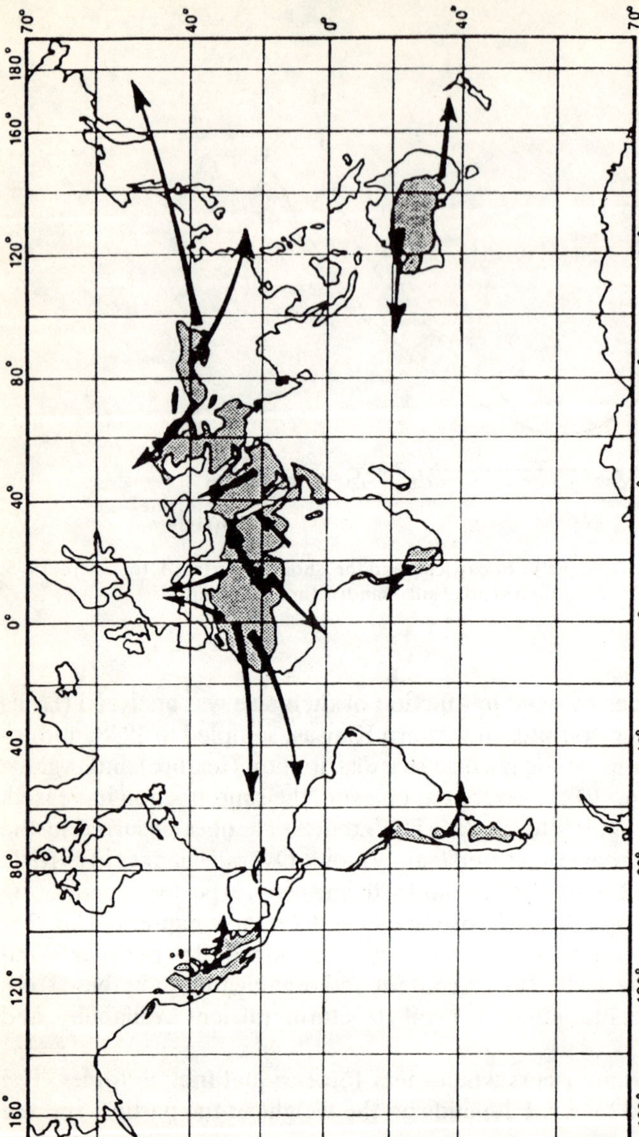

Fig. 50. Deserts of the earth and major directions and distances of dust transport (deserts from Meigs 1953: dust transport from all available published and unpublished sources). (After Péwé 1981)

mean wind speed associated with dust storms has decreased in growing urban areas. Dust is reported more often in developing areas (such as Denver and Tucson) than in less disturbed areas (Winslow, Arizona). Vehicle-generated dust from unpaved roads can limit visibilities to one-third or less of the molecular scattering limit. Dust devils may occasionally contribute large amounts of dust, and a possible correlation between dust devils and human activity is noted (Hall 1981).

Dust is carried from the Sahara towards northern and central Europe, Greenland, the Atlantic Ocean, the Caribbean, or even towards Kazakstan in the former USSR. Mineral dust is transported by the Harmattan (dry and hot, continental *trade-winds*) from arid regions of North Africa into the tropical north Atlantic in the summer months. The dust often reaches 5–7 km altitude, spreads over hundreds of km in latitude, and extends to the Caribbean Sea and the southeast United States. In winter, large quantities of soil dust are transported primarily to South America. The annual mass of Saharan dust transported over the North Atlantic Ocean is estimated to be about 260 million tons (Jeanicke 1979; Prospero et al. 1981).

All mineral aerosol size distributions indicated a maximum between 0.06 and 0.08 μm in size (Almeida and Schutz 1983).

The severe drought currently afflicting the Sudan-Sahelian zone to the south of the Sahara desert has been suggested as being instrumental in producing an increased input of soil-derived aerosols into the atmosphere from the region. During the very dry period 1972–74 aerosol concentrations at Barbados, West Indies, affected by the African dust plume, were three times those of pre-drought levels (before 1968). A marked increase in the frequency of severe dust occurrences in northern Nigeria was also noted during 1972 and 1973. Data from selected meteorological stations show that dust-storm activity in the east and west of the Sudano-Sahelian belt has dramatically increased during the drought years (by a factor of 6 in Mauritania and up to a factor of 5 in the Sudan). Egypt is affected by the occurrence of about ten dust storms per year; there are about 20 per year in West Africa, 30 per year in China, and as many as 60 per year in both Mexico City and Kazakhstan (Middleton 1985).

4.4.3.2 Transport Area

Soil erosion by wind is a two-step process: detachment and transport. For particle detachment, a higher speed is necessary than for transport: about 3 to 4 m s^{-1}.

Three principal means of aeolian transport are recognized: suspension, saltation, and creeping. Desert dust is carried by suspension; fine and medium-textured sand is transported by saltation while coarse sand moves by a process of creeping.

1. *Suspension.* Dust particles have a diameter of less than 60 to 80 μm and can be of organic or mineral origin. When the speed of rising air currents is greater than the speed of the descending dust particles, the latter remain suspended even

when the wind speed has decreased. This airborne dust is known as an aerosol. An extreme example of suspension is the occurrence of dust storms (called simoon in the Middle East, harmattan in Nigeria and haboob in the Sudan), by which dust may be transported up to altitudes of between 200 and 2500 m and at speeds up to 20 m s^{-1}. One index of the severity of dust storms is based on visibility, which may be less than 800–1000 m.

Desert dust can be transported by wind for thousands of km and be responsible for the origin and fertility of many soils. It may be deposited on the margins of deserts or further afield: for example, the soils of the tropical rainforests along the Gulf of Guinea for example, the soils along the Nile valley owe their fertility largely to wind deposition and not just to silt deposition, as in the Nile river. The Juba forest in Sudan receives dust probably from Kenya via the low corridor separating the Ethiopian Highlands in the north from the Kenya Highlands in the south, where a wind current can be detected according to aeolian field indicators of transport.

One of the most efficient ways in which silt and clay-size particles are blown upwards is in a spiraling dust devil which may carry the dust up to a height of several hundred meters. These are particularly common on sparsely vegetated surfaces where there are no obstacles such as trees to slow down or break up the vortices. Around the margins of soda lakes I have observed in Kenya strings of dust devils in areas of degraded savannah woodlands, despite the abundance of trees. This is the result of a phenomenon known as Salzprengung by which the prevalence of airborne salts from the lake causes the complete breakdown of soil structure and even rock fragments. A fine powder is produced which is particularly vulnerable to wind erosion. Wind erosion frequently suspends enough particles to create visibility hazards; 100 to 400 mg m^{-3} are needed to reduce daytime visibility to 200 m (Hagen and Skidmore 1977).

2. *Saltation.* Particles between 0.1 and 0.6 mm are lifted almost vertically in successive bounds because the air speed is reduced to almost zero near the soil surface. Grains then fall by their own weight along a curved trajectory at an oblique angle to the soil (Fig. 51).

Unlike fluids, granular material requires a finite shear force to overcome Colomb (dry) friction between grains. For normal aeolian sand, an airspeed of 7 m s^{-1} is necessary to overcome this friction, i.e., the point where shear force exceeds the friction force. At this threshold velocity, saltation occurs. Threshold velocities vary with the particle diameter and they are described by an asymmetrical parabolic function (Fig. 49). The lowest threshold velocity occurs in particles with a diameter of 0.08–0.1 mm. Particles either larger or smaller than this require higher threshold velocities. Once initiated, the wind speed required to lift the same size of particles is only 3 m s^{-1}.

On Ciscaucasian coarse clay and clay loam chernozems containing more than 50% of aggregates of 1 mm in diameter in their upper 5 cm, and at wind speeds of not less than 9–10 m s^{-1} at 50 cm above the soil surface, soil particle transport ranges from 1 mm at wind speeds of 9–12 m s^{-1} to 2–3 mm at wind speeds of 15

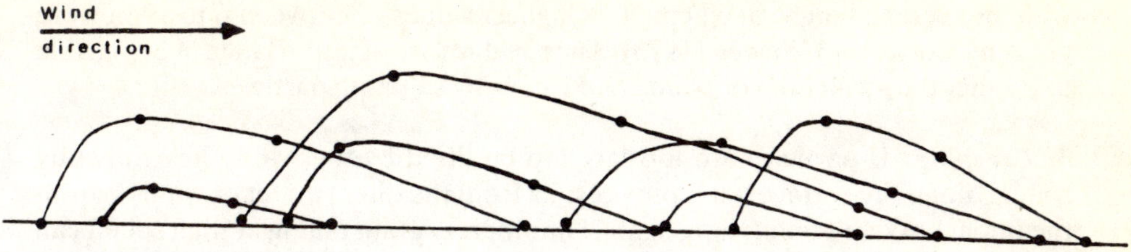

Fig. 51. Saltation of particles of sand

m s^{-1}. The transport of soil particles begins when the moisture content falls to 5–7%. Losses of the humus layer average 1.0 to 2.3 cm a^{-1} (Vasil'yev et al. 1978).

The height to which grains ascend depends on the initial air speed. Once the threshold velocity is reached, wind can transport more sand on a hard coherent surface than on a crumbly surface. The higher a grain is lifted, the more it is exposed to wind energy and the greater its impact on the soil. In the same wind, grains are lifted to different heights. Saltation is initiated either by direct wind pressure beginning when threshold velocity is reached or by the impact of saltating grains colliding with grains on the surface and then rebounding, a process attenuated by the sand. The sand movement increases with the textural heterogeneity of the grains. After winnowing, saltation is increased because the top layer is coarser.

A grain lifted vertically by moving air encounters air resistance in the opposite direction to the direction of air flow. During saltation, the grain is rotating and the angle α of fall is given by Bagnold's equation:

$$\frac{V^*}{V} = \tan \alpha,$$

where V^* = terminal velocity of the grain (m s^{-1}), V = air speed (m s^{-1}), tan α = tangent α.

For a grain with a diameter of 0.3 mm, $V^* = 2.7$ m s^{-1}. In a wind of 12.7 m s^{-1}, the tangent is equal to 0.216 or tan 12° 25′.

At a windspeed of 5 m s^{-1}, saltating grains are usually in a layer of about 10 cm thick, and occasionally as much as 1 m thick, but this depends on the surface. The maximum density of saltating sand on a pebble surface is at a height of 2 m but on a sand-sheet only 9–60 cm. This has important implications for vegetation establishment and growth owing to the damage that can be caused to leaves and seedlings. For example, short grass will be unaffected by saltation if it is growing on a pebbly surface, while young trees 1–2 m high would be susceptible to damage. In contrast, the opposite effect would be seen on a smooth sand surface. During a sandstorm in the Ubari Sand Sea in Libya, I observed saltation bounds reaching 4 m on a substratum of very coarse sand.

Authors vary in their estimates of load height: Chepil (1945) and Sharp (1964) reported that 90% of the load circulates below 31 cm over a stony surface with an average height of 63 cm. The highest values of between 6 to 19 m were given by Cooke and Warren (1973). These values are of importance in being able to predict the scratching of plants and leaves by saltating particles.

3. *Creeping.* If particles are too large to be lifted by the wind, they move by rolling along the surface. Creeping results from the direct effect of wind or from the impact of one grain falling against another. A grain falling at high speed can move another grain which is six times its diameter (more than 100 times its mass). The relative proportion of saltation and creeping is proportional to the diameter of the particles (Birot 1981) and is shown in Table 23.

Bagnold (1941) estimated that one-quarter of the wind load is transported by creeping, while Chepil (1945) found that it depended on grain size. He found that 15% of grains between 150 and 250 μm and 25% of grains between 250 to 839 μm are transported by creeping. Horikava and Shen (1960) suggested a figure of 20%, while Tshihara and Iwagaki (1973) gave lower values of between 6.5 and 16.6%.

Table 23. Saltation and creeping with respect to the diameter of the particles. (Birot 1981)

Diameter (μm)	Creeping (%)	Saltation (%)
150–250	16%	84%
250–830	25%	75%

4.4.3.3 Deposit Area and Dune Formation

There are many types of dunes, and two in particular, barchans and seifs, have been selected for further attention in this section. It is important to make a fundamental point about sand dunes, which is that any dune covered with vegetation is effectively almost stabilized. However, once this vegetation is lost, the sand can rapidly become reactivated and the sand reservoir, previously an area of deposit, becomes predominantly an area of erosion. This phenomenon occurs at a regional scale in the Sahel south of the Sahara. The dunes and sand sheets, previously fixed areas of deposition destabilized by human activities, have first lost their vegetative cover and then, because of strong wind erosion, their top soil. This is real aeolian land degradation which could eventually be called desertification if it becomes irreversible.

While dust is exported for long distances, sand is carried for much shorter distances and can form into dunes in the transport or in the deposit area. A reduction in windspeed can cause moving air to deposit its load. This would include topographical obstacles, new roughness introduced by a vegetative cover, roughness introduced by a sandy cover.

After depositing its load, wind has increased energy, which allows it to take a further load, but of finer sand. This phenomenon can be called load substitution.

In the source area, sand particles are removed from an exposed sandy area by the surface drag of wind and by pressure exerted by the wind on the surface of the soil. This results in one of three types of sand migration, shifting by saltation and creeping, shifting of sand along a linear dune, migration of sand as whole dunes, particularly barchans.

The quantity of particles removed and transported depends on their size and upon the wind speed. Above a threshold speed of 4 m s^{-1} sand will become airborne, whilst as wind speed drops below this level, sand deposition ensues.

The geomorphology, dynamics, and genesis of the two most important types of sand dunes, barchans, and seifs, will be described prior to showing the various methods for their control in Chapter 5.

Barchans

The most dangerous form of sand dune is the barchan, and it is of sufficient importance to warrant extensive description. Barchans, in fact, represent a family of dunes (Fig. 52). Some of them are stationary and therefore do not constitute an immediate threat, whilst others, including the true barchan and the barchanic dihedron, are dangerous because they migrate.

Geomorphology. Barchans are dihedral crescent-shaped dunes with a convex windward face. They are characterized by two oblique horns, or wings, which trail leeward and which are extended in strong winds (Fig. 53).

A perfectly formed barchan is rare, as it is usual to find horns of slightly different lengths. The dunes can vary in length from 50 cm to more than 1 km and are between 0.3 and 50 m high. The two faces of a barchan are topographically and dynamically different: the windward slope is a face both of deposition and removal, and it has an angle varying from 6° to 12°. The leeward side of the dune (the slip face) is much steeper, between 22° and 33°, and it is a slope of deposition and a slip face.

There are several pre-conditions for barchan formation:

1. A wind regime with one prevailing wind direction.
2. Flat topography.
3. Availability of aeolian sand with a diameter between approximately 0.125 and 0.35 mm.
4. A cohesive substratum of coarse sand and gravel.
5. A dry surface without vegetation.

Barchans or barchanic edifices can exist alone or in groups of adjoining crescents known as barchanic chains. These may be clearly defined or rather irregular when they are known as transversal chains or transverse dunes (Fig. 54). The latter are characteristic of several sand deserts, the most prominent example being the sand sea of the Taklamakan desert in Xinjiang, China. Barchans are infrequent in the Sahara except on the semi-arid northern and southern boundaries owing to the scarcity of appropriate sediments. As a result, the dunes are mainly longitudinal (Mainguet 1984b).

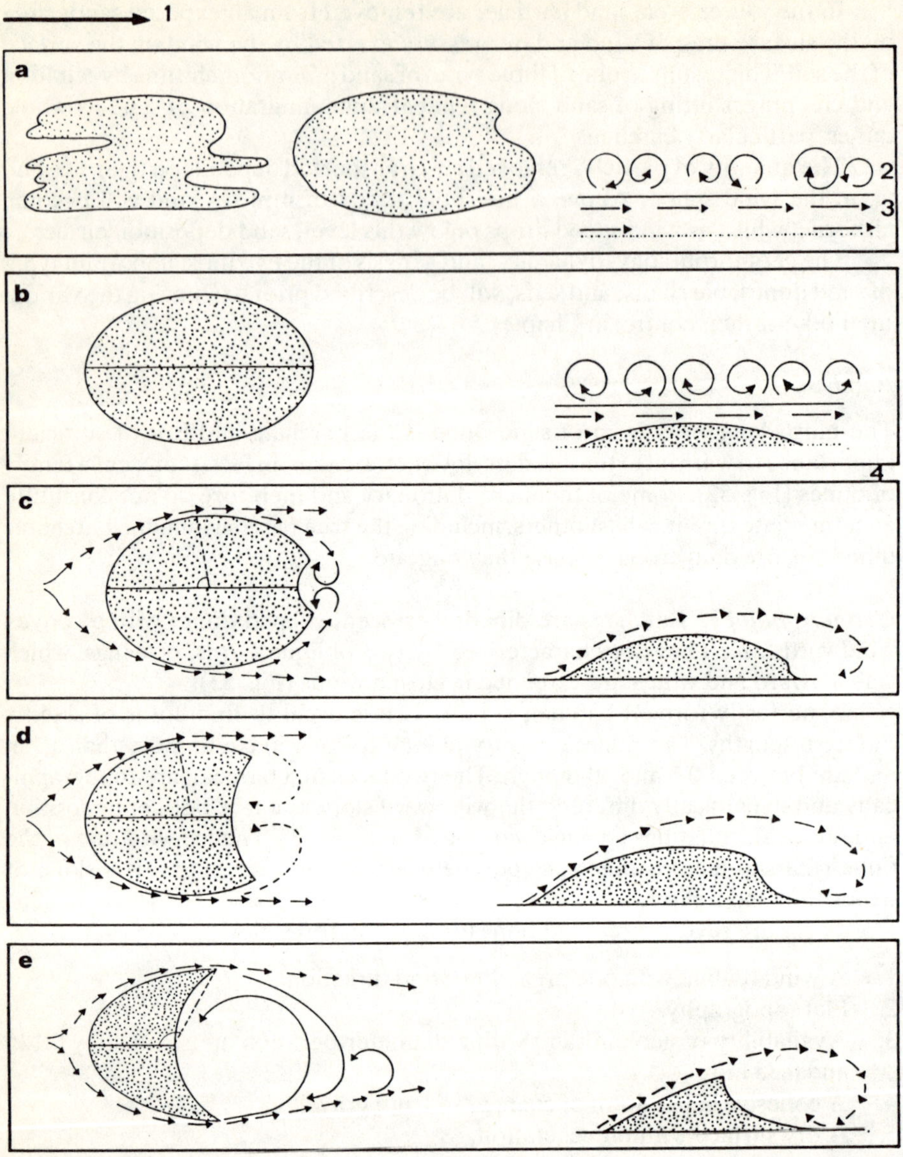

Fig. 52. The family of crescent dunes. **a** Irregular aeolian sand veneer. **b** Aeolian sand dome. **c** Barchanic shelter. **d** Barchanic dihedron. **e** Barchan. *1* Wind direction; *2* turbulent flow; *3* laminar flow; *4* threshold of the barchanic edifices

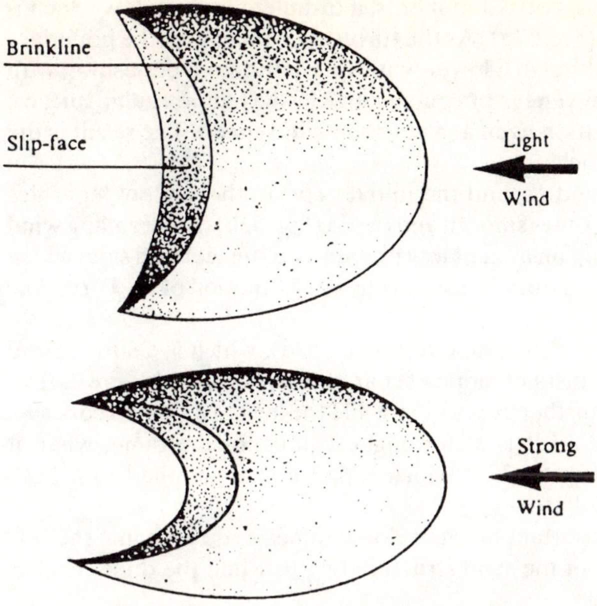

Fig. 53. The crescent shape of a barchan in light and strong winds. (After Wippermann and Gross 1986)

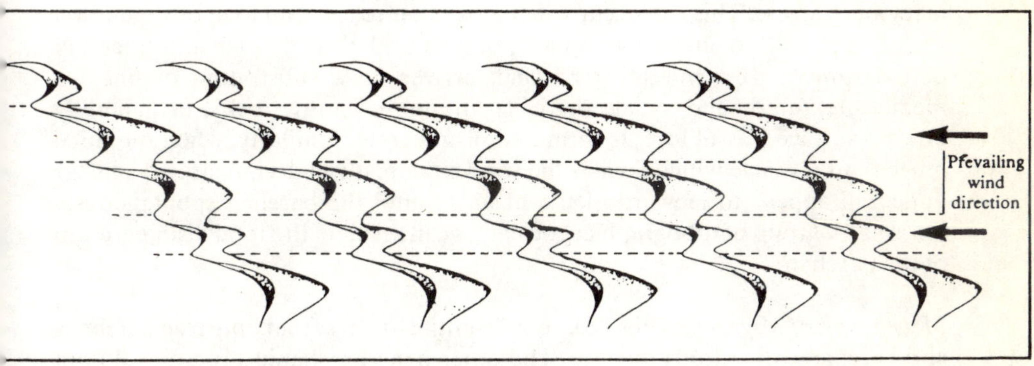

Fig. 54. Transverse dunes

Barchan Genesis and the Formation of Pene-Barchanic Structures. Prior to the formation of a true barchan, sand accumulation goes through a number of successive stages. Oldham (1903) showed that crescent dunes result from the transformation of a domal ovoid shelter form. Hogbom (1923) suggested that these dome dunes are the predecessors of the barchanic shelter dunes.

The origin of dome dunes is difficult to explain. They form without any topographical obstacle as a result of the horizontal (or less commonly, vertical) diffluence of sand-bearing winds. The flow field at the soil-air interface consists of a boundary flow which has two components, a turbulent layer below which is a more stable laminar flow (Fig. 52a). As the turbulence increases, the boundary layer becomes thicker, resulting in a lower wind speed and sand deposition with the heavier grains first. The veneer of coarse particles has an irregular surface, which increases sand deposition even at a constant wind speed. The result is the appearance of a sandy obstacle.

In the flow field over and around the initial deposit, the airflow separates laterally and the deposit becomes smooth and ovoid (Fig. 52b). A prevailing wind over this sand dome results in an area of low pressure on the leeward side, which causes a reverse wind flow and turbulence causing the formation of the barchanic shelter dome (Fig. 52c).

The next stage is the barchanic dihedron (Fig. 52d), which is characterized by the appearance of two distinct slopes separated by an angular crest. The windward face gives rise to further deposition while the leeward slope is concave. The crest moves upwind until it is at the highest point of the dune, when it becomes a true barchan (Fig. 52e), at which stage sand is supplied to it both horizontally and perpendicularly.

These processes are important because dome dunes and barchanic shelters do not migrate. Movement of the sand structure begins when the dune reaches the barchanic dihedron stage.

Avoidance of Encroaching Dunes. Barchans disappear or "melt" when they meet an obstacle. This can occur when a barchan reaches an area of vegetation such as a palm tree plantation or oasis, or a natural obstacle such as a steep hill or escarpment. If a migrating barchan arrives on a substratum of fine- or medium-textured sand, or alternatively an area of stony, rocky, or generally uneven surface, it will lose its form and disintegrate. Similarly, when the dune reaches an obstacle which divides the wind stream, it will disappear and a linear dune will appear to leeward of the obstacle until the barchan spontaneously reforms. A group of topographical obstacles can result in the total disintegration of the barchan.

Mechanism of Migration (Fig. 55). It is essential to stress that only true barchans and barchanic dihedrons migrate. The other pene-barchanic structures do not migrate and do not represent an immediate danger. It is therefore not a priority to "fix" these stationary dunes, although they may be fixed to avoid their becoming mobile structures.

The dynamics of the leeward face is more complex. It can receive sand deposited from the wind coming over the top of the dune or it can be fed from the reverse flow which transports sand back up the leeward face. The steep angle is maintained by vortices of air which keeps the area of recirculation free of sand (Fig. 56).

The Venturi principle is also the mechanism for the migration of dunes. The dune maintains almost constant shape and direction although as its size increases, the speed of migration decreases. The streamlines around a migrating barchan

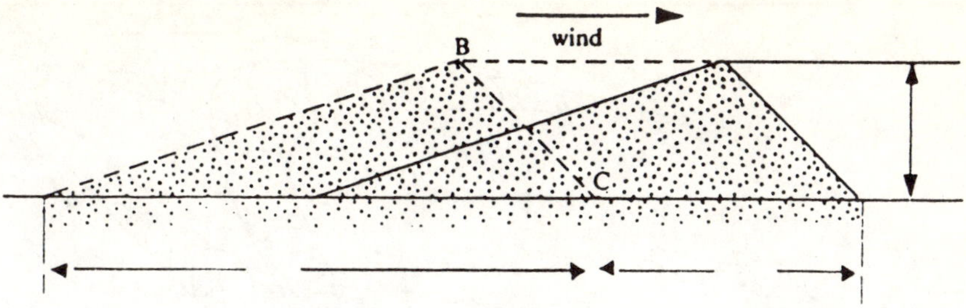

Fig. 55. Migration of a barchan (After Zakirov 1981). *B-C* Front of the dune, -initial location- - -

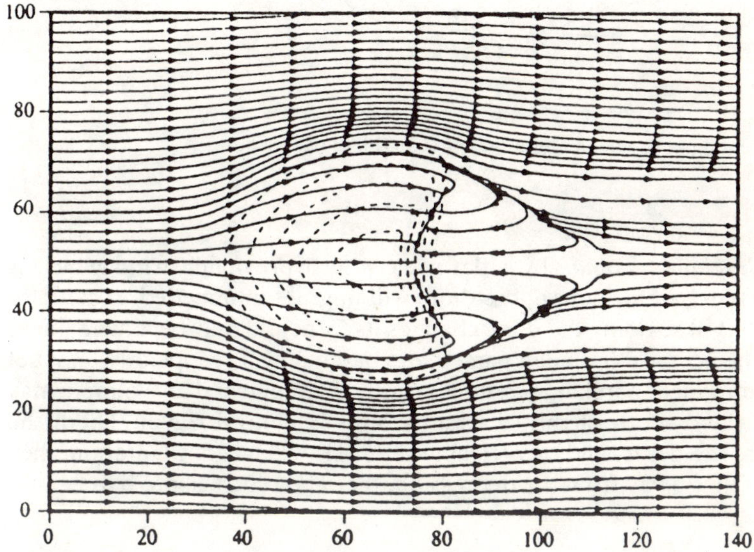

Fig. 56. The flow over and around a barchan represented by streamlines at a height of 1 m. (After Wippermann and Gross 1986)

are shown in Fig. 56. A rate of migration of 20 m a^{-1} has been measured in Faya Largeau in Chad.

When the speed exceeds a critical value, sand is eroded at the windward side and transported by saltation or creeping in accordance with the Venturi principle already discussed with reference to wind action. The sand dune acts like a Venturi tube: as the wind converges on the dune, wind speed increases and air pressure falls. Its passage over the center of the dune is equivalent to the neck of the tube, where pressure is lowest but velocity highest. On the leeward side of the dune, wind speed falls and air pressure rises, resulting in the deposition of airborne sand on the leeside (Fig. 57). This principle will be referred to again in connection with dune control and the sand-sweeping effect on roads.

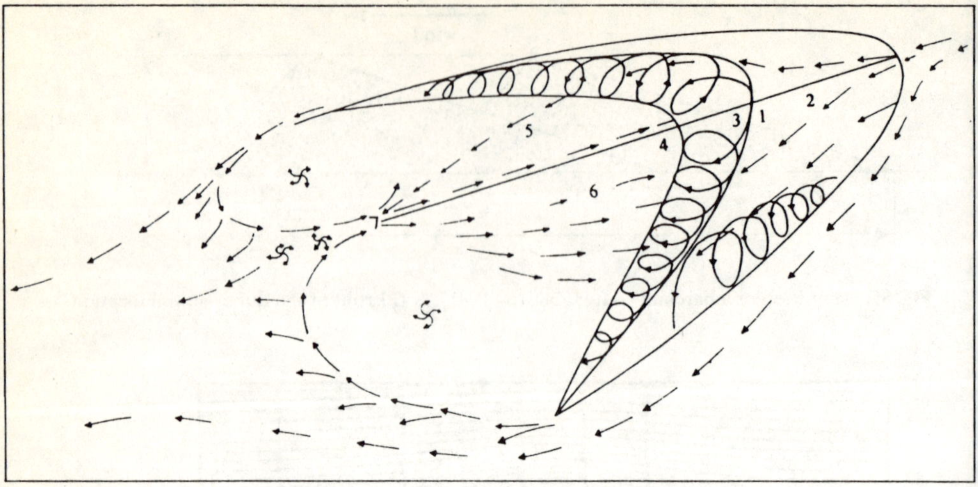

Fig. 57. The streamlines around a migrating barchan. Field observations

Permanent Barchanic Trend. In a flat area with a prevailing wind which is conducive to barchan formation, the dune will migrate unimpeded. It can also climb slopes but in so doing the dune changes its form and may become a linear dune. After the obstacle has been passed, the barchan will reappear (Mainguet 1972). This permanent barchanic trend is an important one and an example may be seen north and south of the El Kharga oasis in central Egypt. There the barchans cross the desert and on arrival at the oasis, village, or plantation, they break down. Once away from the depression, the sand reforms into barchans.

Seif Dunes

Seifs, or linear dunes, are the second most dangerous type of active dunes but they are even more difficult to understand and to stabilize than barchans. While barchans move as a whole, thereby threatening to bury human settlements, linear dunes are dangerous because as they grow, they elongate as much as 100 m a^{-1}, cutting roads and railways or covering villages, gardens, wells, and agricultural lands.

Geomorphology. Seif dunes are narrow, sinuous, and with structures which vary in length from some tens of metres to many kilometres. In the Grand Erg de Bilma, in Niger, a large seif is 40 km long, 50–100 m wide and 30 m high. In the initial stage they can become a straight "sand arrow", but as they grow, they appear as a line of open crescents linked together in a sinusoidal manner. In the desert of western China, this sinuosity is so pronounced that they are known as fishhook or honeycomb linear dunes.

These dunes are usually asymmetrically triangular in transverse section with two slopes divided by an angular crest. The windward flank is straight or slightly convex, while the leeward flank can be slightly concave and has a steeper slope. In some areas, such as in the Great Eastern Desert of Egypt, the asymmetry of the linear dunes reverses seasonally, probably because of the seasonal changes of the prevailing wind.

Seif dunes exist in three forms:

1. Symmetrical seifs or sand arrows. These are sharply pointed on the leeward end and have two faces on either side of the crestline, along which they are bilaterally symmetrical.
2. Asymmetrical seifs. These have a more developed windward slope and a slip face leeward and they tend to be sinuous.
3. Seif dunes with alternating slip faces. These occur where there are seasonally changing winds.

Location. The siting of seifs is a complex aspect of these dune types. They can be anchored on the leeside of an obstacle such as a hill or an escarpment and a good example of this is to be found in the oasis of Kaouar in Niger: the dune is formed at the foot of a sandstone escarpment, thereby limiting its extent upwind. An interesting example can also be seen in the Navajo Indian Reservation in northern Arizona, where a linear dune 2 to 5 km long runs towards the north east from the escarpment which separates Ward Terrace and Moenkopi Plateau (35°55 N, 111°15 W).

Alternatively, seifs can appear on coarse sand-sheets without any visible topographic obstacle. South of Sebha (Libya) the sandy substrate is the residue after winnowing of the fine particles of the sand-sheet.

They may also be juxtaposed or superimposed on other dunes (Fig. 58). In the Simpson Desert in Australia, for example, the seifs meander along the top of a substratum of longitudinal dunes and they show characteristic junctions which are open upwind. In the Grand Erg de Bilma in Niger, the parallel or confluent linear dunes lie on a sand ridge system with a different orientation (Fig. 59).

Northeast of this erg (or sandy desert), sand ridges 2 km wide separated by corridors 500 m wide and oriented NE-SW are crossed by seifs with an east-west direction. The difference of orientation is an angle of 34° and the entire Tenere consists of linear dunes superimposed on sand ridges of variable angles. The seif dunes are located on the northwestern slope of the sand ridges and their southeastern slopes have a surface with giant ripple marks. In the Great Sand Sea in Egypt, the angular difference varies from 9° to 35°.

Formation of Seif Dunes. Bagnold (1941) suggested that seif dunes could be formed by the asymmetrical extension of a barchan. When an aeolian wind flow meets an obstacle, an eddy flow forms in the obstacle's lee and sand is deposited in dunes at a tangent to the eddy area. In the northern hemisphere, these track dunes are usually oriented to the right, and in the southern hemisphere, usually

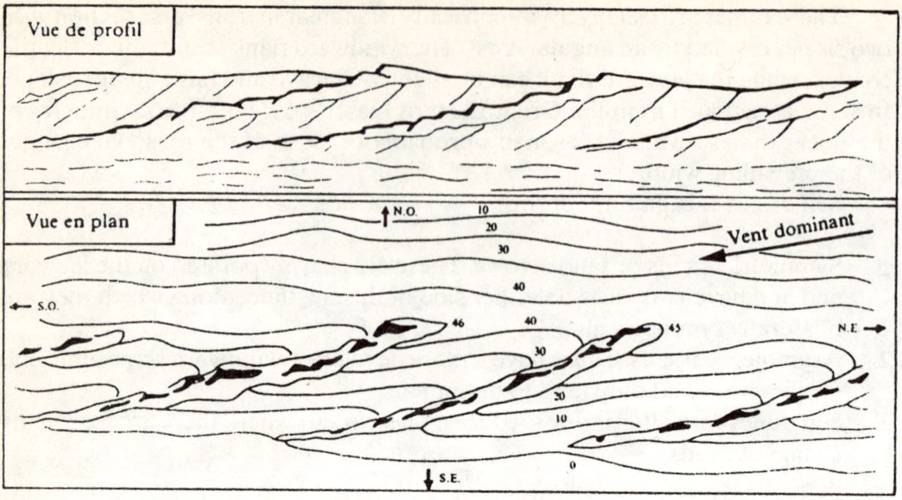

Fig. 58. Seif dunes superimposed on sand ridges in the Ijafene sand sea (Sahara). *Above* Cross-section; *below* plan view

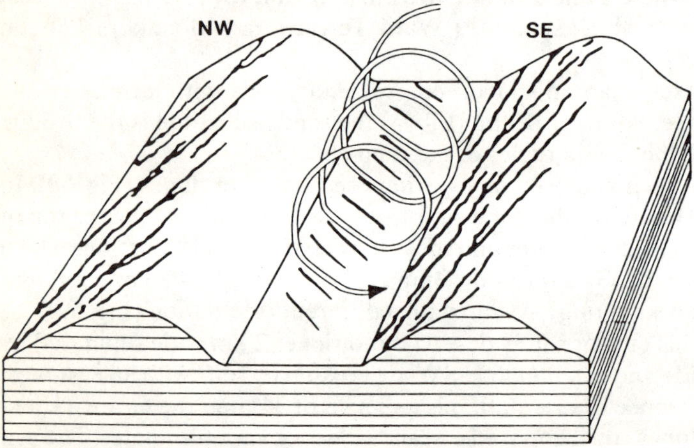

Fig. 59. Seif dunes on the northwestern slope of sand ridges in the Grand Erg de Bilma and Tenerc in Niger. The spiral flow in the interdunal corridor between two sand ridges may explain the genesis of the linear dunes by winnowing of the sandy material

elongated to the left. However, local topographic features can influence the orientation so that both orientations may occur in the same locality. In the active erg of Fachi Bilma in northern Chad, the seif dunes in the lee of the Tibesti mountains and the windward side of Air Mountain are elongated to the left, while in the erg itself the orientation is to the right.

The formation of those seif dunes that do not originate from barchans or obstacles has yet to be explained. Field observations and meteorological data show that such linear dunes occur in areas with two seasonal prevailing winds coming from different directions (Lancaster 1982; Tsoar 1989). This leads to the formation of a seif dune and not a barchan, which requires a regime with one predominant wind direction.

The means of elongation must be understood in order to control seif dunes. The wind crosses the windward flank obliquely while on the leeward flank the wind circulation is complex, causing sand deposition and avalanching, which explains the steepness and the loosely packed sand on this flank. In addition, a reverse flow occurs on this leeward flank, generating a spiral action which smoothes the sand and transports it both obliquely and longitudinally along the dune, causing ripple marks perpendicular to the crest line. Tsoar (1983) observed that the flow is divided into two components, one of which is parallel to and the other perpendicular to the crest.

The evolution of seif dunes therefore takes place by three simultaneous processes:

1. Sand reaches the dune, where some of it is deposited on the windward face.
2. More sand crosses the crest to the leeward face, where it is caught in the eddy and brought back to the dune and deposited.
3. Sand is transported along the flank of the dune.

Sand which is fixed to the dune migrates parallel to the dune axis, causing progressive elongation.

Control of seif dunes is difficult because they elongate at an oblique angle to the prevailing wind.

At the boundary between the Grand Erg de Bilma and l'Erg du Tenere, which both represent a sand-sheet of about 600 km, are located the oases of Kaouar (Niger), 18 oases at the foot of a little sandstone escarpment drawing a line north-south perpendicular to the continental trade wind (Harmattan). The Kaouar oases are slowly becoming desertified. The process is merely aeolian: shifting sand encroaching barchans and, elongating seif dunes are the agents through which the gardens, the palm tree plantations, the wells, and the villages are gradually being invaded by sand (Mainguet and Chemin 1982). The government of Niger has asked the FAO to propose solutions to save these oases.

Communication lines, irrigation systems, and pipelines built in an area of shifting sands, migrating dunes, or blowouts are vulnerable to deep burial by sand or to being undermined by deflation. Maintenance is particularly expensive in these areas. The road from Kharga to Asyiut in the New Valley (Egypt), the road from Al Ain to Abu Dhabi (UAE), and the railroad out of Plaster City, California (Cooke et al. 1982) are seasonally blocked by sand accumulations:

"The new road north of Al Kharga was built on a dam about as high as the local barchan dunes, which led initially to sand storage behind the dam (and temperory relief to the road), but dunes soon began to climb over the drifts, advance on the road, and partially bury telephone lines. Control measures, such as the dam, commonly fail because of a lack of understanding of the processes involved. One

particularly common error is to assume that by flattening dunes the problems ensuing from their movement will be eliminated: this is rarely so, because the winds will soon recreate aerodynamically suitable surface forms similar to the original dunes.
"Pipelines and similar features pose especially difficult problems, because deep burial by sand makes maintenance and inspection difficult; and unsupported pipes on active dunes can be left high above the ground as dunes move on, putting them under torsion and possibly causing fracture. Such was the fate of part of the phosphate conveyor belt at El Ayun (Spanish Sahara) (De Benito 1974)." (in Cooke et al. 1982).

Zakirov (1982) provides additional examples of shifting sands on communication lines in the former USSR and China, giving research methodologies and applicable solutions. The efficiency of wind as a process of degradation is under discussion. I wish to emphasize the difference between local and regional levels of wind action:

1. Local actions such as shifting sand, encroaching crescent dunes, and elongating linear dunes have been studied in the oases of the wadis Draa, Ziz, and Rheriz (south of Morocco) (Mainguet and Chemin 1979; Chemin 1985), in the Kaouar (as previously described), in the coastal area of Somalia (villages of Shalambot and Brava, for example) (Mainguet 1984d), in the steppe of Algeria (Bou Saada). These phenomena are very localized and are dangerous because of the presence of human settlements which are obstacles on the wind paths and creators of hazards: without these obstacles nothing would happen, the wind would not accumulate its sand load.
2. The main danger, however, is the highly damaging wind erosion on a regional scale: it is that of diffuse, insidious, and general deflation, a phenomenon which is detectable only on air photographs and on the ground. This process is responsible for the reactivation of the top of previously fixed dunes by winnowing away the fine particles, and makes it difficult to revegetate them.

   The threshold for the start of deflation is a function of the density of the vegetative cover. For Wasson and Nanninga (1986), sand transport on vegetated surfaces begins when the cover becomes lower than 35%. Deflation is the main process of soil degradation in the whole of the Sahel.

4.4.4 The Effects of Wind on Vegetation

The detrimental effects of high winds on the physiological processes of plants are of great importance. Transpiration from plants and evaporation from soil are jointly known as evapotranspiration, and its rate is proportional to wind velocity when the other causative factors are constant. High wind speeds also lower the humidity at the soil surface in addition to increasing heat convection from the ground. The word *sukhovei* is used in Turkestan when the plants wither in hot and dry wind.

Laboratory experiences and field observations show the negative effect of wind alone and wind-carrying particles on plant growth. Plants of sideoats grama (*Bouteloua curtipendula*), cane bluestem (*Bothrichloa barbinodis*), green sprangeltop (*Leptochioa dubia*), and sand bluestem (*Andropogon hallii*), were subjected to wind at 10, 20, and 30 miles per hour, with and without fine sand, in a laboratory wind tunnel. Wind alone had little effect, but blowing sand killed the seedlings or retarded growth. The young plants were more tolerant at the six-leaf than at the two-leaf stage. Blown sand ruptured plant cells, dried out the exposed tissue, and led to damage by insects and disease. *L. dubia* was the least affected (Fryrear et al. 1973).

Wind accelerates heat and water loss by the plant. Mainly on the underside of leaves are stomata (or small pores) through which gaseous exchange takes place. Each stoma can be opened and closed by the action of guard cells. Stomata are the major avenues of water loss for a plant because the rest of the leaf surface is covered with an epidermis which secretes cutin. This waxy substance forms an impermeable layer, or cuticle, which can be very thick in some plants. The walls of plant cells are kept rigid by high turgor pressure (turgor pressure is the pressure that the contents of individual cells exert against their walls). In high winds and with a high radiation load, the rate of transpiration exceeds that of water replenishment from the root system, particularly when the soil is dry. The result is that turgor pressure is lost and wilting occurs. If this continues, the permanent wilting point is reached and the plant dies.

The temporary response of the plants to excessive water loss is the closure of the stomata. This is not in itself a very effective water conservation mechanism as it only comes into operation when a lot of water has already been lost, but it does have the effect of reducing the rate of gaseous exchange and hence the rate of photosynthesis and growth (Figs. 60 and 61).

On a microscopic scale, the abrasive effect of the sand-loaded wind creates microscratches on the waxy epicutical on the skin or epidermis of the leaves and leads to evapotranspiration. The abraded leaves then dry out faster than the healthier ones.

Pollination is also adversely affected by strong winds because both pollen and the stigma can become desiccated. Moreover, pollen and seeds can be blown into areas where they are a nuisance or where they can be of little value.

On flower buds the wind effect can prevent flowering by drying out the flower before they open. Fruits can also be damaged by saltating sand. With winds of more than 6–7 m s^{-1}, scarring occurs. Wind at speeds of 11–12 m s^{-1} can also shake off the fruits. At wind speeds higher than 12 m s^{-1} and turbulent winds, cereals can be flattened with a loss of production of up to 40%.

Two experiences show the deleterious effects of seedling exposure to wind:

1. Yields of *Phaseolus vulgaris* plants were reduced 8% by exposure to winds of 15.5 m s^{-1} for 20 min during the seedling stage and by 14% when flowering plants were similarly treated. Pea yields were reduced 16%, but losses were not significantly affected by such treatment during the growth stage. Bean losses from plots treated during early growth were attributed to mechanical damage to the plant caused by the abrasive action of suspended sand particles and to wind dessication. Losses from plants treated at later stages were attributed to a loss of buds and blossoms. Physical damage was not as apparent in peas as in beans. Yields of beans and of peas were inversely related to soil loss when plants were exposed to wind at the seedling stage (Bubenzer and Weis 1974).

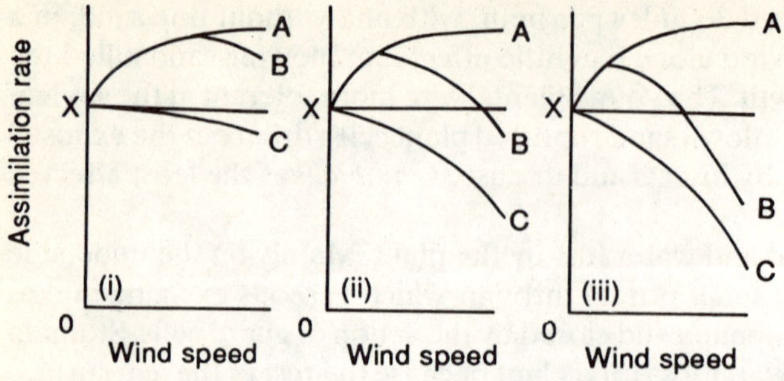

Fig. 60. Diagram to illustrate the predicted effect of moderate wind on plant growth rate. (Wadsworth 1964). **i** Small drying effect of wind; high humidity of the air and sufficient soil moisture. **ii** Medium drying effect. **iii** Large drying effect; low humidity and dry soil. Curve *A* suggests the potential growth in wind, curve *C* suggests the decrease in growth produced by the decrease in leaf water content caused by wind, and curve *B* represents the actual growth in wind; it is the result of curves *A* and *C*. *X* is the growth in the calm air

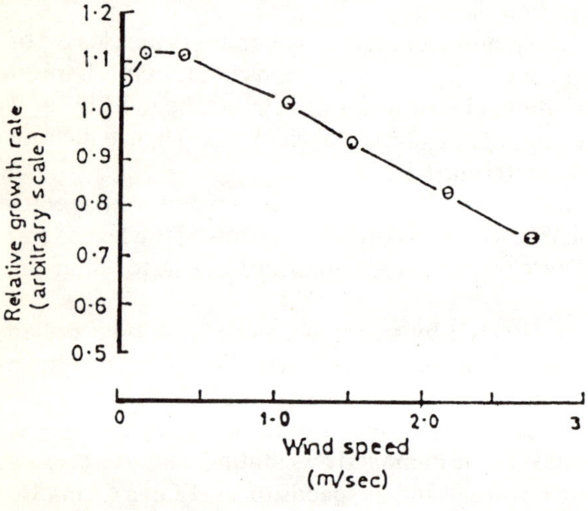

Fig. 61. Effect of air speed on the relative growth rate of leaf area. (After Morse and Evans 1962, in Ebomo 1983)

2. Estimated annual yield reductions due to wind erosion were given for wheat under two crop rotations in the Great Plains: grain sorghum in a wheat/sorghum/fallow rotation in the Great Plains and in Iowa with the addition of maize following autumn ploughing. Total yield reductions on sandy surface soils in the Oklahoma Panhandle, based on long-term yields, were estimated to be 106000 bushels (37352 hl) a^{-1} for wheat and 66100 bushels (23292 hl) a^{-1} for grain sorghum (Lyles 1977).

It is difficult to assess the detrimental effects of the wind on plant temperatures, soil evaporation, and plant transpiration and the ecophysiological and morphological corollaries. An increase in wind velocity causes an increase in soil surface evaporation and plant transpiration. This results in changes in the humidity conditions of the soil and of the lowest layers of air, in the heat transmission of the surface (which will be increased), and in the temperature of the ground and air.

Wind speed affects the boundary layer near the surface of the leaf and increases the convection of heat and the loss of heat by the plant. Loss of leaf turgor pressure results in stomata closure, which reduces plant transpiration. The most significant effect of stomata closure is that it reduces the carbon dioxide exchange between the plant and the air. Carbon dioxide is necessary for photosynthesis and growth in all plants.

4.4.5 Aeolian Soil Erosion

Field workers, technical assistants, specialists, and farmers alike should be able to recognize the various indicators of aeolian soil erosion, whether at the site of erosion, transport, or deposit. The principal indicators of soil erosion include the following:

1. Changes in soil texture and fertility resulting from deflation and winnowing.
2. Disappearance of the topsoil and the outcropping of the B horizon.
3. Scalding of the soil surface.
4. Exposure of the root system of the trees and shrubs.
5. Soil residues: nebkas and veneers of coarse sand.

Changes in Soil Texture and Fertility Resulting from Deflation and Winnowing. Initial evidence of winnowing comes in the form of ripplemarks, which are composed of residual particles of coarse sand. It is a common mistake to assume that ripplemarks are indicative of deposition: in fact their presence is a sign that sand has been removed. Nebkas, or sand arrows, are also good indicators of deflation and sand removal. They are formed at the leeward side of small obstacles such as rocks or plants, and they can reach a height of 50 cm. The obstacle causes a decrease in wind speed and deposition of the heavier airborne particles. Exposure of the roots of trees and bushes allows the thickness of the exported layer to be measured. Sand trapped by hedges is another indicator of deflation. For example, north of Maradi in the Hausa sand-sheet in Niger, the fields are often bounded by hedges of *Andropogon gayanus*. The surface of the fields have a concave profile and the majority of the topsoil particles are coarser than 400 μm. The hedges have trapped medium-grained sand (150 to 300 μm) blown off the fields, the finer particles being carried off in the wind.

An important factor in the relationship between wind, soil, and vegetation is that a soil surface without vegetation is a soil without organic matter. Organic matter in turn provides nutrients to crops, trees, and plants, increases the water-holding capacity of the soil, and improves soil aggregation and stability. Sandy soils are most vulnerable to wind erosion because one of their characteristics is an absence of organic matter. A lack of organic matter is also a characteristic of desertification in all types of soils.

The clearing of land for agriculture is one cause for the loss of topsoil by wind deflation. Areas seasonally or permanently vegetation-free or nearly vegetation-free in arid and semi-arid regions are potential source areas for dust, silt, and sand.

The Red Dune System found in the Sahel from Senegal in the west across to Mali, Niger, Chad, and the Sudan is a good example: this system of previously vegetated dunes has recently been turned into a source area through denudation caused by overgrazing (Mainguet 1985). Wind erosion, which removes fine particles from agricultural soils, reduces simultaneously the productivity of the eroded soil by eliminating essential nutrients normally attached to these particles. The ability of the soil to retain soil moisture in the root zone is also impaired with the loss of fine particles (Fryrear 1981).

Wind can remove substantial amounts of soil: the annual removal of 1 mm at the soil surface is equivalent to a loss of 10–15 t ha^{-1}, a rate far exceeding soil formation in the semi-arid and subhumid ecozones. At wind speeds of more than 4 m s^{-1} (15 km h^{-1}), soil particles can be removed and carried in the air at different heights according to their size, a phenomenon known as deflation. Deflation causes winnowing or the removal of the finer particles, the loss of which reduces both soil moisture and the quantity of plant nutrients. Winnowing can reach in the Sahel such a degree that all organic matter and particles less than 400 μm in size are exported. In this situation, the topsoil of coarse sand is practically sterile.

Wind erosion does not occur if the surface is wet. Bare soil which is wet is partially protected by the high surface tension binding the soil particles. However, it is important to stress that vegetative cover is the best form of protection against deflation and that without it, all areas become a potential source of dust, silt, and sand removal.

As shown in Fig. 62, wind and water erosion depend on rainfall. The shape of the curves C and D of wind erosion depend of the rainfall regime: with rains in the summer months the curve D falls more steeply, and more gently if the rains are in the winter season. According to this figure, the maximum water erosion corresponds to a mean annual rainfall of 400 mm and wind erosion to mean annual rainfall of 250 to 300 mm.

In the Sahel, the winds with the greatest particle-moving potential are those which precede the first precipitations of the rainy season, when the soils are denuded, and not the regular harmattan winds. Similarly, Brookfield (1970) pointed out that in Central Australia the winds with the highest erosion potential are the cyclones of the dry summer and not the anticyclonic winds.

The overriding influence of vegetative cover is shown in Fig. 63. The development of these parameters and their incorporation in management strategies involve the consideration of the state of the soil and of meteorological variables (Fig. 64).

Decrease of fertility is another proven negative effect of wind: wind erosion decreases humus and the total nitrogen content of calcareous chernozems. In the humus of eroded soils the proportion of fulvic acids increases and among the humic acids the fraction linked to calcium predominates. Due to erosion, the C/N ratio and the nitrification capacity of the soils decrease (Petrova 1982). Total

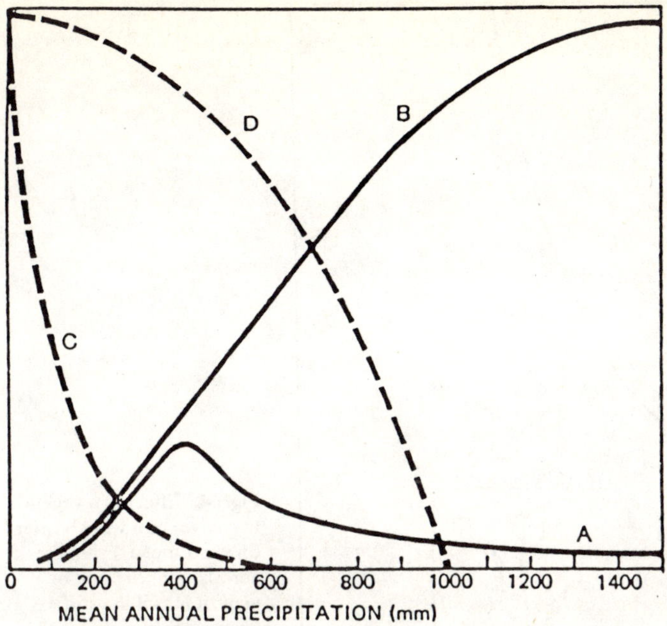

Fig. 62. Dependence of wind and water erosion on rainfall and vegetative cover (WMO no. 591 1983a). *A* and *B* Relative sediment yield (water erosion); *C* and *D* relative wind erosion

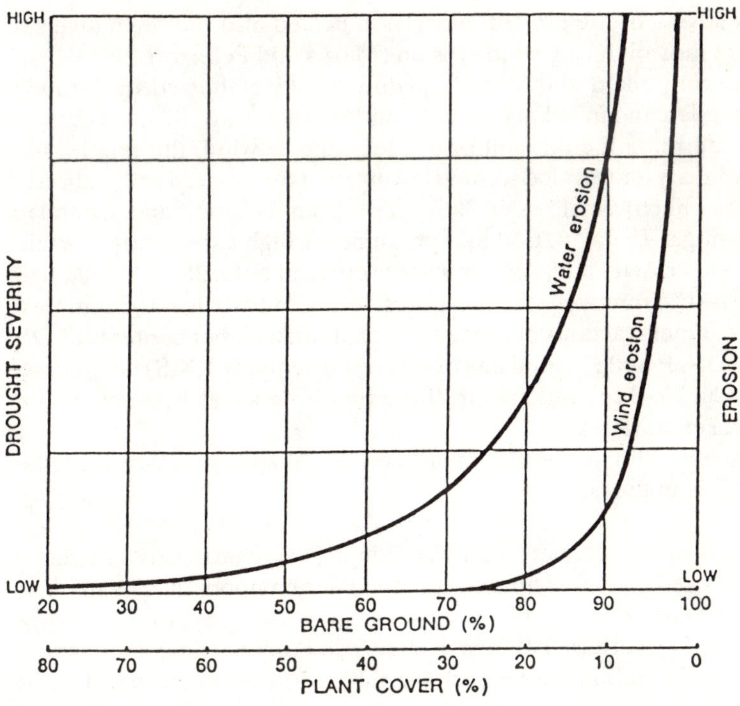

Fig. 63. Interrelationships between drought, plant cover, and soil erosion. (WMO no. 591 1983a)

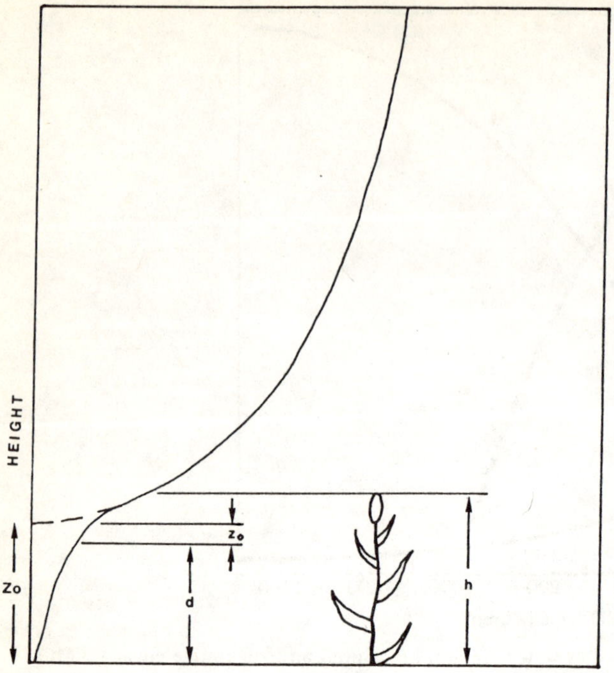

Fig. 64. Effect of a vegetative cover or a wind barrier on wind speed. From an aerodynamical point of view, it is as if the surface were more elevated at the height of 20 cm

phosphorous analyses of the eroded material indicated also that high levels of phosphorous are moved during wind erosion (Merva and Peterson 1983).

On a bare sandy plain and a sandy plain partially stabilized by *Lasiurus scindicus* at Chandan and in Bikaner (Rajastan India), during 1978, a study was conducted to determine the amount of soil loss due to wind, the relationship between wind velocity and eroded soil and the nutrients associated with it showed a soil loss of as much as 615 and 325 t ha^{-1} from Bikaner and Chandan, respectively, during a 75-day period in April-June. A high exponential correlation between wind velocity and soil loss was observed at both sites. Particle sizes in the range 0.10–0.25 mm were the most erosive, while particles 0.05 mm were the least erosive. Finer fractions of eroded soil contained higher contents of OM (organic matter), N, P, and K. Total loss of macronutrients (N,P,K,S) and of trace elements (Zn, Cu, Fe, Mn) was two to three times higher at Bikaner than at Chandan (Gupta et al. 1981).

The influence of deflation on chemical content of soils is demonstrated in the following three analyses:

1. Atmospheric dust collected at Ludhiana, Punjab, was calcareous and mainly composed of silt-sized particles; quartz was the major component. The mineralogy and physicochemical properties were very similar to those of Thar desert surficial materials. The surface 0–0.5 cm of 20 soils in Punjab, Himachal Pradesh, and Haryana, high in carbonates and soluble salts, was compared to soil below

1.5 cm in depth, during the pre-monsoon summer months when aeolian activity is maximal. The rate of accretion of aeolian carbonates decreased away from the Thar desert towards the Himalayas with a decrease in carbonate content (Sidhu 1977).

2. Wind-blown soil aerosol, generated during three dust storms in March 1974, was collected at 1.5 m above the surface in Plains (Texas) during periods of different wind velocities. The mass distribution of TOC (Total Organic Carbon) in the aerosol decreased with increasing particle size at all three of the wind velocities studied (7.3, 9.3, and 11.4 m s^{-1}). The percentage of TOC in each size fraction decreased with increasing wind velocity. Moreover, the aerosol was enriched in TOC compared with the bulk parent soil, the amount of enrichment being greatest for the smallest particle size. The enrichment factor, integrated over all sizes, varying between 1 and 10, increased with rising wind velocity. The mass distribution of the aerosol exhibited the expected shift towards smaller sizes and the percent of organic carbon increased in all size ranges, although particularly in the 10–100 μm size range. The content of organic C in the aerosol was 20 times greater than in the soil itself. Microscopic examination of the aerosol fractions showed the presence of vegetative residual material, typically with elongated dimensions and of low density (Delany and Zenchelsky 1976).

3. Exposure of soybean seedlings grown in sand cultures and exposed to wind at 13.4 m s^{-1}, with or without blown sand, for up to 40 min decreased nitrate reductase activity and increased nitrate content immediately after exposure, compared with untreated controls. At 24 h after the end of the exposure period the nitrate reductase activity in treated plants was higher than in the controls and this remained so for up to 40 days (Armbrust and Paulsen 1973).

In the process of deflation, the topsoil is gradually removed, leaving a layer of coarse sand of ill-defined or nonexistent structure. This process occurs especially at the end of the dry season. The first indicator of wind erosion on a sand-sheet or on a dune surface, therefore, is the concentration of coarser particles. In Hausaland in Niger all the particles of less than 400 μm have been removed and the coarser residue forms an infertile layer resulting in lower yields of millet. Fine-textured sand sheets are particularly vulnerable and the entire topsoil can be removed. The situation is aggravated in hot, windy climates when the vegetative cover is removed for cultivation or as a result of overgrazing. Similarly, when crop residues are removed and animal manure is not returned to the soil, the organic matter content is rapidly depleted since this component of the soil is particularly vulnerable to wind erosion since the soil particles are both light and very small. Without organic matter, the soil structure is broken down and soil fertility deteriorates.

When the topsoil disappears, the more compact B horizon is revealed and the wind then acts upon this. Grooves may appear as if it were rock and the soil can be scalded, (polished by saltating particles, rendering it impermeable to water) until it becomes sealed, facilitating rill wash and the development of water erosion. This is precisely the phenomenon called Wala Wala by the Bambara in Mali (Mainguet 1984c) or scalding, whereby the under lying "B" horizon is exposed. The vegetation suffers from both a moisture and an oxygen deficit, and seeds are unable to penetrate the indurated layer and therefore are either desiccated or blown away.

4.5 General Conclusion, Chapter 4

Air pollution, water pollution, fauna and flora deterioration and soil erosion are the major environmental concerns of today. It is certainly not possible to classify them according to their consequences, or to separate them: for example, accelerated water erosion depletes the capacity of production of the soil and produces pollution and eutrophication of natural waters. Pollution of the air is responsible for the acid rains which are responsible for the degradation of European forest. The various relationships between water and wind erosion can also be recalled: for example, water-eroded silts deposited on flood plains or in seasonal river channels can be the source of wind-blown silt that creates dust storms; sandy material washed down fields by rill erosion when dry can be deflated by wind and become a source of dune material. Wind and water erosion can aggravate the effects of each other.

Two schools of thought have developed on soil erosion. One believes that excessive soil erosion is a problem but that its effects are exaggerated. However, the real magnitude of soil erosion is almost completely unknown in the tropics. No standardized methodology of assessment exists. Little is known about the quantitative rates and assessment of the impact of soil erosion in economic terms. Only in highly developed countries like the USA, have technological advances greatly counteracted losses caused by erosion.

The second school believes that accelerated soil erosion and land degradation and consequent desertification must be considered as one of the most critical environmental hazards. Understanding the causative parameters, manifestations, and effects of the erosion processes will provide conservationists with the technical means to estimate soil losses and to find the best available methods to minimize their ill effects.

Available tools, however, do not assure implementation and successful soil conservation. Socio-economic considerations, regulatory and policy-making provisions, land tenure systems, watch-dog agencies, increasing population densities, and increasing needs for food, fibre, energy, use of marginal land for cultivation and grazing, size of landholdings, pattern of farming systems, prevailing political trends, cultural and religious taboos can restrict or nullify the implementation of available technology.

The final of this book, Chapter 5, will discuss possible solutions for remedying land degradation and for dealing with "desertification" when land degradation has become irreversible on the human scale.

Chapter 5 Prevention and Remedies

"Science tells us what we can know, but what we know is little, and if we forget how much we cannot know we become insensitive to many things of very great importance." Russel (1947)

"And he gave it for his opinion that who so ever could make two ears of corn and two blades of grass to grow upon a spot of ground where only one grew before would deserve better on mankind and do more essential to his country than the whole race of politicians put together." Swift (*Gullivers travels*)

5.1 Introduction

It will be noted that, in this book, I have chosen to use the expression land degradation rather than desertification. Nevertheless it is useful to keep the word desertification with its double meaning: (a) of a desert-like landscape replacing productive land, and (b) of irreversible degradation, without following the widespread trend which includes under desertification all types of degraded resources even in humid ecosystems. Even if restricted to these meanings: conversion to a desert and irreversibility, the word desertification does not eliminate ambiguity. The question is not just a theoretical one; there is also a practical problem, which is to consider the realities as they appear to our common sense and to draw conclusions only after analysis of existing cases and components of the problem:

1. Under our present terms of reference desert-like landscapes replacing productive lands can be observed only in seasonally or occasionally dry areas: e.g., the drylands of the American West, Australia, China, and the former USSR.

Land degradation occurs in all ecosystems: between 1961 and 1976 Canada lost more than 3.5 million ha of farmland (Standing Senate Committee on Agriculture, Fisheries and Forestry 1986). But this land did not become a desert and does not look like one.

The degradation of land resources in the Himalayas (India) has been a matter of great concern in recent years: overgrazing, deforestation, soil loss, siltation, flooding, and droughts occur. Nevertheless, the landscape has an aspect very different from the mountainous slopes of the Sahara massifs.

2. The desert margins can seem to have been turned into desert but the oscillations can be natural and even seasonal without irreversibility. Analysis of degradation in semi-arid and sub-humid ecosystems south of the Sahara leads to

the observation that the most severe degree of destruction of biological potential is in the domain of 800 mm annual rainfall. In this subhumid ecosystem the land is severely degraded, denuded patches exist with increased wind and water erosion, but still the landscape has not become a desert, but only degraded land.

Is degradation irreversible? Nobody knows. We have seen in 1988 how good rainfall in a semi-arid ecosystem (the Saharo-Sahelian) brought a good recovery of the grassland and high crop yields; but this area of the desert is sparsely populated. In contrast, the land in the more humid Sahelian and Sahelo-Sudanian ecosystems is intensively populated and used and it is there that we have to concentrate our attention and to try to apply solutions.

3. Risks of land degradation or desertification are greater in drylands than in humid ecosystems because of high climatic variability (rainfall, temperature, higher wind and water erosion potentialities, larger extension of poor soils, poor water reserves). All planning has to adjust to these natural risks. A consequence of these high risks is slow rate of recovery. Drylands recover more slowly from degradation than humid ones.

Sustainability and resilience are now the two leading concepts for the utilization and conservation of drylands. Sustainability of dryland agriculture means an agricultural system which functions indefinitely without degradation of the ecosystem. Resilience of dryland soil means the capacity for the maintenance and natural restoration of the soil. They remain only theoretical words if they are not linked to the physical and human local potential, the physical characteristics of the soil, the land use system, and the density of the human and animal population.

The Californian economist Douglass (1984) points out that "sustainability" has different meanings for different schools of thought:

- The food-sufficiency or productivity viewpoint, which "thinks of sustainability as supplying enough food to meet everyone's demand";
- The stewardship school, which "regards sustainability primarily as an ecological phenomenon," with a concern for maintaining an "average level of output over an indefinitely long period... without depleting the renewable resources on which it depends"; and
- The community perspective, which "pays most attention to the effects of different agricultural systems on the vitality, social organization, and culture of rural life." (Dover and Talbot 1987).

Because drylands are high-risk ecosystems, they cannot stand high human and animal pressure. It is obvious that the scale of damage increases with the increasing densities of populations. Traditionally, conservation was less necessary when there was lower density of populations and greater relative demographic stability. More and more with demographic increase marginal lands have to be looked after carefully. Moreover, in time of drought, the better-armed and organized pastoral people often moved on to conquer areas with higher rainfall.

Drylands differ in terms of climate, physical potential, and culture. The set of solutions has to be different from one area to another. I shall try in Chapter 5 to analyze first the possible ways of minimizing climatic risks, secondly to discuss ways and means of avoiding the threat to vegetative cover, with which are associated the solutions that avoid threat to soils, and finally express these solutions in social and economic terms.

5.2 Possibilities for Overcoming Climatic Risks in Drylands

5.2.1 Cloud Seeding and Weather Modifications

We are still not able to predict future climatic conditions, to identify the causes of droughts or to establish trends of climatic changes. Sufficient reliable and continuous data to permit the analysis and warning of droughts are not available.

Several countries have devoted much research and financial effort to cloud seeding. Cloud seeding requires water-bearing clouds and a rather moist atmosphere, which are most of the time absent in drylands. Positive results have been very limited and only in special situations. According to meteorologists, the rain would have fallen anyway but somewhere else. That means that the distribution of rain can be influenced in space but not in quantity. Reevaporation of the rain before it reaches the soil surface is one of the most acute difficulties of artificial rains in drylands. Regional climatic modification, for the time being, cannot be envisaged.

Since almost all the water in the atmosphere comes from the oceans, the effect of vegetation on the overall water balance of the earth is negligible. Green belts were considered as rain producers. The creation of a Saharan internal sea was envisaged and abandoned as physically difficult and even impossible because of the lack of slope between the Mediterranean sea and the internal "Saharan" sea and because of high evaporation rates of about 3 $m a^{-1}$, which would lead after two or three decades to a desiccated saltflat.

5.2.2 Strengthening of Meteorological Services

Downing (1986), with regard to climatic change and variability in eastern Africa, writes: "There is, at present, no adequate explanation of the causes of the potential climatic changes; it is wise to assume that climatic variability and drought are normal characteristics of the natural resources of eastern Africa."

National decision makers and development practitioners have to pay more attention to drought and its adverse effects and not forget it as soon it is finished. Drought is a recurrent phenomenon.

Hulme (1990) notes that "The changes in annual rainfall which have occurred within a single generation of the present century (between 1921–50 and 1956–85) represent about 25% of the fluctuations which occurred over the length of the Holocene. . . one should be cautious in regarding them as necessarily extraordinary."

National and regional meteorological services should be developed. Some of them have already become more active in providing decision makers with meteorological information useful to agricultural development. AGRHYMET in Niamey (Niger) is one of these. The World Meteorological Organization

(WMO) in an evaluation of climate and water resources for development of agriculture in the Sudano-Sahelian Zone of West Africa summarizes the points necessary for strengthening meteorological and hydrological services

"Development is required in the following sectors:
a) training of personnel and the establishment of staff structures which recognize the importance of their contribution to socio-economic development;
b) strengthening of observing station networks; using modern telecommunications technology;
c) rapid collection, processing, and interpretation of data in parameters and terms applicable to agriculture;
d) provision and prompt issue of bulletins, interpretations, and forecasts;
e) surveillance of rainfall by radar to provide better assessment of areal rainfall;
f) satellite imagery and aerial observation and photography for surveillance of weather development, state of pastures and cultivated crops, as well as the extent and conditions of flooded areas."

5.3 Avoiding the Threat to Vegetation

5.3.1 Vulnerability of Dryland Vegetation

Vegetation in drylands is resistant. It stands high temperature and long hours of insolation thanks to "xeromorphic foliage structure and function, biological controls over transpiration and metabolism, the development of moisture and nutrient storage organs, spiny or otherwise harshly textured external features, and the ephemeral nature of many annuals." (Armitage 1985). The vegetation has particularly deep roots (30 m for *Acacia radiana* in the Draa Valley, Morocco) and often a two-level root system. Vegetation is drought-resistant but sparse and patchy except in restricted favorable location (depressions, valleys). The dryland vegetation system has high natural resilience. We have seen how on different continents grazing may alter many parameters of natural pastures – biomass, species composition, productivity, and resilience – when it is not adjusted to carrying capacity. Trampling is damaging in sandy soil; the cutting hoofs of goats spoil more than their bites.

Suliman (1988) has analyzed on sand dune soil (called Qoz in the Sudan) changes in plant composition, plant cover, litter, and bare soil at three sites in the Northern Kordofan Region (Sudan), where the annual rainfall varies between 280 to 400 mm a^{-1} (At En Nahud and at El Mazrub) and 375 mm a^{-1} at El Khuwie. The most northerly site at El Mazrub, with a poor water supply, was least affected by desertification. "This suggests that aridity was not the major cause of desertification at the three sites during the study period and that the degradation was largely anthropogenic. High stocking rates and subsequent overgrazing as a result of the abundant water for livestock at En Nahud and El Khuwie sites may be the major cause." In the majority of cases it is easier to attribute degradation of grassland to overgrazing, as shown in the Sudan by Suliman (1988).

The impact of humans on dryland vegetative cover is known. Le Houérou and Gillet (1985) describe the principle stages:

1. Woody species recede first as a consequence of wood cutting, uncontrolled fires, overbrowsing, and sealing of the top soil due to trampling and

subsequent deflation. Under pristine Sahelian conditions the canopy cover varies from 5 to 30%, with an average cover of 15% (500 to 1000 trees and shrubs per hectare). The regression of the vegetation is greatest on the higher parts of the topography: tops of fixed dunes, hardpans, ferrugineous soils, gravel plains, and pediments. An observation made from air photographs and satellite images in December 1987 confirmed that in Mali an increase of runoff in depleted areas results in an increase in the density of the vegetative cover in the depressions and valleys of the Sahel.

The reaction of the vegetation is not only a reduction of trees and shrubs, but also a change in species composition: mesophytic soft, broadleaved species such as the Combretaceae are replaced by more xeromorphic spiny species such as *Acacia* spp. and *Balanites*.

- The thinning of the cover of trees and shrubs decreases the length of the growing season and the height of the grasses which have lost their shade.
- Perennial forage grasses are eliminated first: *Aristida longiflora* and *Andropogon gayanus*.

2. The mesophytic annual grasses of good fodder value disappear: *Brachiaria* sp., *Setaria* spp., *Panicum* sp., *Urochloa* spp., *Pennisetum* spp., *Digitaria* spp., and *Sorghum* spp. They are replaced by more xeromorphic, less palatable species of poor feed value: *Aristida, Stipagrostis, Eragrostis, Cenchrus,* and *Chloris*.
3. Less palatable bushes and shrubs invade the pasture land: *Cassia* spp., *Borreria* spp., *Limeum* spp., *Gisekia* spp., *Mollugo* spp., *Tribulus* sp., and *Blepharis* spp.
4. Denudation appears around settlements and in agricultural land. In grassland vegetation arcs or tiger bush of different patterns appear in the savanna areas with an increasing proportion of denuded interspaces. Air photographs and Spot satellite images show these patterns of vegetation perfectly in South Mali: in 1956, the vegetation covered 49.2% of the surface at 14°10'N. In 1987, at the same latitude, only 35.2%, revealing a loss of 14% in 31 years (Report I.G.N., ORSTOM, L.G.P.Z., for UNEP 1989), but this was before the good rainy season of 1988.

5.3.2 How to Obtain Biological Recovery?

Le Houérou (1975) calls biological recovery the reverse processes to degradation of ecosystems. He distinguishes between natural recovery and artificial recovery. By recovery he means "increase in biological activities, in biomass, in plant cover, in organic matter, in soil micro- and macroorganisms, higher fertility, better water intake and turnover, lower evaporation, and runoff".

5.3.2.1 Natural Recovery

Natural recovery may be obtained by exclusion of human influence: neither people nor cattle can penetrate the fenced area. The response is a regrowth of vegetation which depends on stability and resilience, i.e., ability to recover after deterioration. Recovery will be slower when climate is more arid, soil more shallow, and vegetation and soil more degraded, i.e., entropy to a system higher. Le Houérou gives several examples of natural recovery in Southern Tunisia, in Libya, and in Iran in the great Kevir, demonstrating that natural recovery can work in poor soils, coarse sandy soils, saline soils, even with rainfall lower than 80 mm a^{-1}. Sandy soil is preferable to shallow clay soil. Such exclusion can be either total or partial, but must be watched. The recovery of the Sahel after the good rainy season of 1988 is a proof that seed availability, or ability of species to propagate without seed, soil permeability, and water storage potentialities still existed even after a long period of drought. Another example: the Negev-Sinaì border after enclosure in 1949 revegetated naturally on the Israelian side, demonstrating that nomadism over millenia had not destroyed the ability of semi-arid grazing land to recover and had not lessened the resilience of the environment (Warren 1984).

5.3.2.2 *Artificial Recovery*

Artificial recovery involves first by intervention on topography and soils: contour terracing, scarification, plowing, water-spreading techniques, and fertilization. The second type of intervention is seeding of native (autochtonous) vegetation or suitable foreign species (exotic).

1. A first category of plants is fodder grasses, shrubs, and trees: Le Houérou (1975) gives a useful group (Table 24).

Seeding the existing grass without plowing the soil was the method used in the former USSR, Turkmenistan, and Uzbekistan (Babaev 1986b). This method was developed for use in areas of sparsely vegetated rangelands and on loose sand around wells. Seeds are placed on the surface and a flock of sheep is driven across the area. Before sowing, the seeds were immersed in a wet mixture of sand and clay. The clay particles sticking on the seeds make them heavier and ensure a supply of nutrients during the first days after germination.

"On the basis of experimental data, and practical experience, obtained in different localities of the desert zones of the Turkmen and Uzbek SSR, the following seed sowing rates were established: *Haloxylon aphyllum* (*Minkw.*) Jijin – 5 to 8.9 kg/ha, sagebrush – 0.5 kg/ha, *K. prostrata* – 3 kg/ha, *Salsola orientalis* – 6 kg/ha, Russian thistle – 10 kg/ha, *Calligonum* – 15 kg/ha, *Salsola Richteri Kar.* – 10 kg/ha, annual sow thistle – 5 to 10 kg/ha (Nechaeva et al. 1978 in Babaev 1986a).

"In the first year, the amount of aboveground biomass is insignificant, there being no grasses because of plowing. However, in the third and fourth years, a grass layer develops from the seed stock in the soil. As a result the total phytomass, (above and below ground) consisting of shrubs, semi-shrubs and grasses, reach 20 000 kg ha^{-1}. This is 6 to 30 times the natural productivity, depending on the location.

Man-made grasslands can be used for livestock pasturage as soon as they are 2–3 years old. Such grasslands have a long life and need little or no extra care: for early-maturing species with a short life cycle, the rangelands can last from 8 to 15 years; for species with a long life cycle, the ranges can last 14 to 30 years. The establishment of such grasslands allows one to change the seasonal limitations of natural ranges, by introducing autumn, winter, and year-round grazing. It also raised yields three to eight times." (Nechaeva et al. 1978 in Babaev 1986).

Table 24. Fodder shrub and trees. (After Le Houérou 1975)

Saltbushes	
Atriplex nummularia	*Atriplex semibaccata*
Atriplex halimus	*Atriplex glauca*
Atriplex canescens	*Atriplex vesicaria*
Atriplex lentiformis	
Acacias	
Acacia albida (W. Africa, Senegal)	
Acacia aneura (N. Africa, Sudan)	
Acacia ligulata (N. Africa)	
Acacia cyanophylla (N. Africa, Near East)	
Acacia salicina (N. Africa)	
Acacia victoriae (N. Africa, Near East)	
Acacia nilotica ssp. India (Pakistan, India, Somalia)	
Mesquites	
Prosopis juliflora (N. Africa, Mauritania, Senegal)	
Prosopis chilensis (N. Africa)	
Prosopis africana (Sahel)	
Prosopis tamarugo (chili)	
Prosopis cineraria (Pakistan, India)	
Cacti	
Opuntia ficus indica var. *inermis* (N. Africa, Brazil, S. Africa, Madagascar, Mexico, Texas)	
Opuntia inermis (Brazil, S. Africa, USA)	
Opuntia fusicaulis (S. Africa)	
Opuntia robusta (S. Africa Mexico)	
Opuntia streptacantha (Mexico)	
Nopalea cochenillifera (Brazil, Mozambique, Mexico)	
Saxaouls	
Haloxylon aphyllum (USSR, Iran)	
Haloxylon persicum (USSR, Iran)	
Tree salsola	
Salsola richteri (USSR)	
Salsola paletskiana (USSR)	
Calligonum	
Calligonum arich (N. Africa)	
Calligonum azel (N. Africa)	
Calligonum comosum (N. Africa)	
Calligonum arborescens (USSR)	
Calligonum caput medusae (USSR)	
Calligonum polygonoides (Iran)	

The pastures of East Africa, with high cattle population, are overgrazed. With the vegetative cover disappearing, these lands are eroding and losing their top soil. Cattle farmers need to learn to control grazing, to protect grazing land, and to plant fodder grasses and shrubs. Farmers who cultivate on slopes need to learn about vegetative methods for checking erosion and restoring eroded soils and gullies with useful plants. Wolstad and Botts (1984) propose a multipurpose group of three plants for rehabilitation of degraded land, combating water erosion, improving the soil structure and moisture retention, and providing fodder: two grasses *Chloris gayana* (Rhodes) and *Pennisetum purpureum* (Napier), and one shrub legume *Cajanus cajan* (pigeon pea).

2. The programs of biological recovery of the vegetative cover must also answer the crucial problem of firewood. In drylands each human settlement (city, village) in developing countries is nowadays surrounded by circles of woodless land, the radius of which can exceed 100 km (Bamako). Tree plantation of fast-growing trees for fuel in adequate sites is the answer. Le Houérou (1975) gives the list of the main species usable for firewood (Tables 25 and 26).

Rural afforestation and tree plantation for fuel must have more than these immediate objectives. It can also provide at the same time windbreaks and shelterbelts against wind erosion and high rates of evapotranspiration, around villages, cultivated areas, irrigated parameters and water points. This requires that the plantation be specifically designed in interrupted rows, as we shall see in our chapter devoted to control of wind erosion.

Combined effect

The strategy of reafforestation in drylands must simultaneously be a strategy of protection.

3. Beneficial outputs of well-controlled and moderate grazing were highlighted by Hyder (1978):

a) loosening of the surface layer during the dry season avoiding scalding,
b) production of litter in grassland promoting faster humus production,
c) conversion of plant nutrients into easily assimilated forms and soil enrichment,
d) maintenance of optimum leaf surface area in shrubs,
e) trampling of seeds into the ground,
f) reduction of the density of die-off plants likely to produce physical and chemical conditions unfavorable for the growth of young plants.

During the drought of 1968–1973 large numbers of animals died. I observed in 1974 in the grass steppe of Niger how the litter had built up excessively, disturbing the growth of young plants and retarding the production of vegetation. It is well known that organic acids and other compounds from the litter can hinder germination and plant growth.

g) Accumulation of dry vegetation favors the spread of fires, insects and rodents. Moderate grazing avoids this accumulation.

Burning in some oases is certainly more detrimental to vegetation than grazing. On air photographs, a fire can be detected several years after it has occurred by

Burning

Table 25. The main plant species usable for firewood north of Sahara. (Le Houérou 1975)

Tamarisk	
Tamarix aphylla	
Tamarix stricta	
Tamarix nilotica	
Gumtrees	
Eucalyptus microtheca	*Eucalyptus occidentalis*
Eucalyptus oleosa	*Eucalyptus salmonophloia*
Eucalyptus brockwayi	*Eucalyptus stricklandi*
Eucalyptus torquata, Eucalyptus flocktoniae	
Eucalyptus salubris, Eucalyptus lesoueffei	
Eucalyptus dundasii	*Eucalyptus gomphocephala*
Eucalyptus sargentii	*Eucalyptus eudesmioides*
Eucalyptus gongylocarpa	*Eucalyptus kingsmillii*
Eucalyptus woodwardi	*Eucalyptus gillii*
Eucalyptus intertexta	*Eucalyptus gracilis,*
Eucalyptus sideroxylon, etc.	
Acacias	
Acacia farnesiana	*Acacia horrida*
Acacia senegal	*Acacia raddiana*
Acacia victoriae	*Acacia tortilis*
Acacia peuce	*Acacia aneura*
Acacia ligulata	*Acacia sowdeni*
Acacia cyanophylla	*Acacia salicina*
	Acacia cyclops, etc.
Argan tree *Argania sideroxylon*	
Pine trees *Pinus halepensis*	
Cypress *Cupressus sempervirens*	
Cypress *Cupressus arizonica*	
Casuarinas	
Casuarina equisetifolia	*Casuarina cunninghamiana*
Casuarina tenuissima	*Casuarina stricta*
Russian olive *Eleagnus angustifolia*	
Schinus *Schinus molle*	
Schinus *Schinus terebinthifolia*	
Parkinsonia *Parkinsonia aculeata*	

a higher reflectance than in the neighboring grazed steppe. Burning in itself is not detrimental, it is the timing of the operation which is all important. Early burning means a cool burn, which is patchy, safe, and self-limiting. It also destroys litter, which if left to accumulate will produce a hot burn. A hot burn at the end of the dry season when the litter is at its dryest is capable of destroying seeds and seedlings, and of producing flash fires where the flames sweep through the

Table 26. The main plant species usable for firewood south of Sahara. (Le Houérou 1975)

Tamarisk	
Tamarix aphylla *Tamarix nilotica* *Tamarix stricta*	
Gumtrees	
Eucalyptus microtheca *Eucalyptus crebra* *Eucalyptus rudis*	*Eucalyptus camaldulensis* *Eucalyptus tereticornis*
Acacias	
Acacia nilotica ssp. *indica*	*Acacia albida* *Acacia senegal*
Mesquite	
Prosopis juliflora *Prosopis africana* The Neem *Azadirachta indica*	*Prosopis chilensis* *Prosopis cineraria*
Various	
Dalbergia sissoo *Khaya senegalensis* *Albizia lebek*	*Cassia siamea* *Anogeissus leiocarpus* *Parkia clappertoniana*

canopy destroying not only litter but life, trees, and bushes. Burning the vegetation destroys much biomass but results in young grasses of high grazing quality, destroys pests and diseases, and helps some seeds which require fire for regeneration (particularly Australian species).

Fire generally promotes grass and destroys trees. The fires are hottest 1–4 m above the ground. The growth points of grasses (where new leaves are formed) are usually low enough to survive the fire, while young trees are killed. Full-grown trees have a better chance of survival, especially as in their surroundings less grass will grow because of the shade. Young trees growing around full-grown trees have a better chance to survive than those amidst the grass. Therefore, the groups of trees can gradually expand into the grass areas.

In the analysis of land degradation in the savannas I have deliberately excluded the Sudano-Guinean and Guinean ecozones because the rainy season is longer than 6 months per year. Nevertheless, in these wet tropical savannas land degradation exists also, but the links between overgrazing, grass fires, and regrowth of trees are specific to these ecozones. One example of these links worth mentioning is the work of Hurault (1975) and Hurault and Lummaux (1979) in the north Cameroon.

Hurault (1975) shows the effects of overgrazing and grass fires on the physical environment in western Adamawa (Cameroon-Nigeria) between 6°30 and 7°30 north. In this area the annual rainfall varies from 1300 (Tignere) and 1850 mm (Banyo). The whole country: highlands of Banyo and Tibati (900 to 1300 m) and the Mambila plateau (Nigeria, 1500 to 1800 m) had no livestock, but only a dense population of farmers. Since the invasion by the pastoral Fulani at the beginning of the 19th century, the country has been overrun by herds and has suffered overgrazing to an increasing extent. Overgrazing has brought modifications in the vegetative cover: during the rainy season, on slopes and valleys, through the effect of grazing the herbaceous layer (mainly *Hyparrhenia*) degrades into pedestals and tufts (*Jabbe*); on the crest of the convex hills, where livestock circulates without supervision, the grass disappears, denudation of the soil *(karal)* becomes the rule; in a second phase the arboreal layer develops rapidly, evolving towards a continuous woodland which replaces the grass savanna. Hurault (1975) demonstrates that a dense grass cover scatters runoff. Grass savanna protects the soil perfectly because sheet-wash can appear, and this type of runoff has only a minor erosive power. In contrast, with overgrazing, the grass cover becomes discontinuous, runoff concentrates in rills, runways, and even gullies, and the disaster of soil erosion starts.

Hurault shows also that if man has not created the savanna, he has, thanks to fires, permitted it to subsist under ecological conditions that would normally have led to the re-establishment of forest. In the areas which are overgrazed, the propagation of fires is bad and the multiplication of trees is explosive (Hurault 1975). Contrary to the general opinion, Hurault demonstrates the beneficial effect of fires on the savanna: fires are favorable to grass regeneration, which prevents soil erosion and prevents the multiplication of trees.

In this case the afforestation of the savanna is dangerous for man and livestock because it is favorable to the installation and multiplication of tse-tse flies (*Glossina*), which are the vector of trypanosomiasis. *Glossina* produces one egg at a time. The egg is developed inside the body of the female until the mature larval stage, then it is laid in a moist shady place: the shade of a tree. The soil under the tree must be loose enough for the larva to burrow. After 1 month under the earth the larva emerges as an adult fly. Trees are necessary in the habitat of *Glossina* (Itard 1986), so that regrowth of trees, which in the savannas can be considered as an improvement of the vegetative cover, represents in fact a real danger for livestock and man.

A historical example of pastoralists running away from tse-tse flies is given by Hurault (pers. commun. 1989): in the large Sudanian plain which extends from Kouncha (Cameroon) to Gashaba (Nigeria), cattle have been grazing since the Muslim conquest (1820–1830). At the beginning of the 20th century, they were expelled from the area by a plague of tse-tse flies connected with the regrowth of trees in the overgrazed savannas. Nowadays, livestock can graze there only during the dry season and even then glossinas are numerous. The herders and their livestock stay only on the village lands, which are systematically cleared so as to eliminate the danger of tse-tse flies. To go there, the herders cross the infested areas during the night (Hurault and Lummaux 1979).

5.3.3 Solutions to Avoid Grazing Effects

Overgrazing is not specific to developing countries. It occurs in all drylands, as described above. One of the most alarming consequences of overgrazing is the alteration of the species composition and the genetic shifts of the pastures (Hacker 1987). The American rangeland scientists distinguish the species which increase after overgrazing: increasers or invaders which often are unpalatable and those which disappear, called decreasers. According to Ahlcrona (1988), in the Sudan *Acacia nubica* and *Callotropis procera* are invaders, and *Acacia senegal* and *Cenchrus biflorus* are decreasers.

Overgrazing is a very complex phenomenon which depends on both what is in the pasture and on what kinds of animals are grazing. Each animal selects its own different species of plants, which vary throughout the year according to the need for proteins, salt, minerals, and carbohydrates. Overgrazing can simply mean the overgrazing of one species and can still be controversial. The requirements of goats have been discussed for decades, but the goat still has friends as well as enemies.

5.3.3.1 Carrying Capacity

The basic rule for the optimum use of grazing land is respect for the carrying capacity which can be defined as the maximum allowable animal population. This definition is theoretical; the concept of carrying capacity varies in both space and time. Witness the difficulties faced by agronomists and range specialists as they try to include all the variables, which are:

1. The effect of the environment on fodder production levels.
2. The degree of stability environment.
3. Soil resistance and resilience to degradation.
4. Technologies which are used or can be applied: replacement of nutrients, organic matter, water, and reseeding of beneficial plants.

Energy inputs must be used principally to maintain the system rather than to add new material.

To avoid overgrazing and land degradation in semi-arid ecosystems and even perhaps to improve management of pastureland, the main objectives might be:

1. Reduction of animal numbers to suit the carrying capacity of the land and discouragement of indiscriminate well-digging for animal drinking purposes, mainly because it results in overstocking. The livestock herders in drylands lack the means to purchase food during and immediately after the dry season and during the drought periods. To avoid the drastic effects of herds being decimated, they tend to expand their livestock numbers, which results in overgrazing. Stocking rates must be adapted to carrying capacity.
2. To rationalize grazing periods to correlate with rainfall characteristics. This is a policy which may conflict with reservoir construction and well-digging.

Water availability should not override the corresponding carrying capacity in the dry season.

3. To avoid excessive grazing in the dry season of crop remnants. This is more and more difficult with the extension of farming to marginal lands. Instead the range should be used on a rotational pattern using different grazing techniques. Around Marsabit (North Kenya) the Rendiles nomads distinguish dry and rainy season grazing areas. The rainy season areas are those where surface water is scarce in the dry season. The dry season grazing land is nearer to the water points.

Orev (1986) explains how to improve the carrying capacity of a range composed mostly of annuals. It can be increased three- to fourfold by the addition of perennial shrubs. Even if the perennial species do not produce more forage, they are available for a longer time that annuals. "After the introduction of perennial shrubs the next stage is introduction of perennial grasses. Shrubs are rich in minerals and proteins but deficient in energy; perennial grasses are deficient in proteins and minerals in certain seasons, but rich in energy, so these two supplement each other."

For a winter rainfall regime of 70–200 mm a^{-1} (Mediterranean areas), he proposes the suitable combination of a shrub *Atriplex halimus* (saltbush) and a grass *Hyperrhaennia hirta*. Both are only moderately palatable, but for Orev we are dealing with countries of recurrent drought and the qualities to look for are: drought resistance, regeneration after grazing, ease of establishment, and then lastly, palatability.

Below an annual rainfall of 70 mm, the grass should be replaced by *Panicum turgidum*, which can be established on only 30 mm. *Atriplex halimus* does not grow well in sand and dunes. *Haloxylon persicum* (black saxaul of Central Asia) is more suitable.

"To correct the lack of energy of the saltbrush and to cover the ground between the shrubs in order to better utilize the rainfall." Orev proposes adding perennial grasses to the vegetation. In summer rainfall areas "the rains are only about 60% as efficient as winter rainfall for vegetative growth." The pastoral zones of summer rainfall areas extend to the 350-mm isohyet:

— from 100 to 200 mm the dominant perennial grass is *Panicum turgidum*
— from 200 to 350 mm the dominant perennial grass is *Andropogon gayanus*.

This grass is a prolific seeder and after 2 or 3 years of good rains or exclusion from grazing can revegetate even a badly eroded area.

Proper rotations or sequence of grazing must be worked out by long-term grazing trials.

5.3.3.2 Mobile Resources for Nomadic Pastoralism

In the semi-arid regions of the world Moore (1987) demonstrates that nomadic pastoralism must be considered as the least traumatizing grazing system: "the nomadic way of life does indeed enhance the efficiency of grazing systems and increase the likelihood of survival in times of drought." In fact, mobility, providing opportunity for special and temporal variation of diet, is the best way to exploit the semi-arid pasture land:

The vegetation of such areas in their different patches "contains a range of species that vary in their drought tolerance and in their palatibility to herbivores." These plants are perennial grasses, tuberous and bulbous perennials, annual grasses, herbs and shrubs, and even trees. The annuals are available only during the wet season. The dietary preferences of the animals varies according to the species: cattle prefer to eat grass (96% of their diet), camels mainly browse on shrubs and trees (95%), sheep, goats, and donkeys enjoy a more mixed diet, both browsing and grazing (Moore 1987).

The nutritional value of forage, varying seasonally, becomes minimum in the dry season. Moore points out the advantage in maintaining mobility and diversity of grazing animals as in the traditional pastoral system. This leads to the conclusion that settlement of nomads and even transhumance must be avoided in rainfall areas under 300 mm. Nomadism and transhumance are imposed by the ecological conditions in arid and semi-arid zones. Plant growth and production are too irregular in time and space to allow year-round feeding of livestock at the same place. Migration in search of pasture and water is necessary. Nomadic peoples turn to other activities (irrigated farming, trade, handy crafts, tourism, industry, army, civil service) only in an emergency situation.

5.3.3.3 *Game Ranching Instead of Cattle*

As well as introducing mobility in dryland grazing, wild animals instead of cattle can produce less damage. For the past 16 years a rancher in Kenya has been conducting grazing experiments with cattle and gazelles (Thomson's gazelle, Grant's gazelle, Coke's hartebeest, and wildebeest), in Athi-Kapiti plains 40 km south of Nairobi. On the ranch, 300 acres were fenced and divided into two plots, one for the gazelles, the other for cattle. After 3 years the gazelle enclosure had 32% more grass cover and twice as many species of vegetation (Amindo 1986). The difference arises because while cattle need to trek to the water point every day, gazelles require little or no water. Degradation around waterholes does not exist in the wild animal enclosure.

5.3.3.4 *Fodder Harvesting*

Making hay from temporary fodder surpluses is another method for saving grazing land ecosystems. In plains, lowland depressions and in general in all natural water spreading areas during years of excess rainfall, fodder production is greater than the needs of livestock. Traditionally, pastoralists are not accustomed to collect fodder and to keep reserves for the dry season or for dry years. This would help even if some of the fodder were otherwise wasted by trampling, insect consumption, and decrease in nutritional value. Trials of fodder collection in Tunisia, Senegal, and Niger were successful.

5.4 Some Solutions to Improve Dryland Agriculture

5.4.1 Drought-Adapted Plants

Dryland degradation and the apparently intractable corollary of famine make it clear that Africa will have to develop better ways to feed its increasing population. Part of the solution lies in cultivating plants adapted to aridity or droughts.

1. The first step would be to gather and evaluate all the crops available in the arid and semi-arid regions in the world.
2. The second step would be to create a resource base of such drought-tolerant useful plants on a world scale. In this respect, local knowledge is all-important. The bushman of the Kalahari, the aboriginees of Australia, and the Indians of the southwest United States for example, could be good advisers.
3. The third step could be the analysis of the multipurpose families of the plants such as cactus in the Saharo-Sahelian ecozones, beans in the Sahelian ecozones, and oil palm trees in subhumid ecozones and in the Sudano-Sahelian bushland. For example:

a) The family of Opuntia, native to American deserts, produces fruits which can be eaten raw or cooked (preserves), forage for animals (in Tunisia), goats are fed with the water-filled leaves and its gum and fibers are used for binding. In Tunisia the plants also provide living fences which make good windbreaks. In Somalia they were used successfully in the program of dune stabilization at Shalambot. In the whole of the Maghreb and in Mexico the fruits are sold as a dessert.

A spineless variety also exists. It is an excellent food for animals in the dry season when the pastures have dried up and there is no grass left. With this variety the drudgery of burning off thorns does not exist (FAO 1984).

A systematic collection of the existing literature and an update of the economic potential of available plants would be interesting and useful.

b) In the family of beans Vietmeyer (1987) describes: *Phaseolus acutifolius* (tepary bean) which growths in the southwest United States (Arizona Painted Desert) and northwest Mexico. It is an "ephemeral" which needs only one good rain to set its flowers and mature its seeds.

Bauhinia esculenta (marama bean) from the Kalahari. Its edible seeds have a higher amount of proteins than peanuts and twice the oil content of soybeans. It also produces an edible tuberous root.

Andean peoples, like other Latin Americans, depend also on beans, which at high altitudes need to be cooked at lower temperatures; for example *Phaseolus vulgaris* (nuñas).

Besides roots and beans, unusual grains are also found, e.g., *Chenopodium quinoa*. This seed "containing 12 to 19% proteins, is one of the richest sources of protein among grain crops. . . and possesses an exceptionally attractive amino acid balance for human nutrition because of its good level of lysine and methionine." (Vietmayer 1986); it is a good complement to maize.

Vietmeyer (1987) suggests also setting up research programs on *Voandzeia subterranea* (bambara groundnut) grown in the Sahel, and on *Psophocarpus tetragonolobus*, a winged bean, which "can be eaten raw, steamed, boiled, or stir-fried. Its flowers, tendrils, pods, leaves, seeds, and tubers are all edible. The seeds are comparable to soybeans in composition and the tuberous roots have exceptional amounts of protein".

c) Shrubs are also overlooked by forestry because they are not trees and by grazers because they are too tall. Nevertheless, shrubs can provide many different resources: food for humans — *Cnidoscolus aconitifolus* and *C. chayamansa* (chayas) growing in central America require little maintenance and are a nutritious vegetable, taken from the foliage of attractive hedges (Mexico to Costa Rica). *Cordeauxia edulis* (Ye-eb) from the Horn of Africa survives in an annual rainfall of 150 to 200 mm. Its nuts, which are available in Somalia markets, are as tasty as macadamia and pistachio. Ye-eb is also a forage plant.

There are hundreds of shrubby legumes: *Sesbania bispinosa* (dhaincha) according to Vietmeyer (1987):

"Its seeds contain a water-soluble gum that produces a smooth, light-colored, coherent, and elastic film that is potentially useful for sizing textiles and paper, as well as for stabilizing the mud used in oil drilling. The plant can be grown as a rotation crop to fertilize and improve soils. It also provides windbreaks, hedges, erosion control, and shade for crops. It is reported to make good cattle fodder and appears easy to grow on a large scale with little care or investment. Moreover, dhaincha shows remarkable ability to survive on saline and wet soils. Amazingly, agronomists outside the Indian subcontinent have largely ignored it."

5.4.2 Introduction of Lesser-Known Plants

"The "back-to-nature" movement, as it is sometimes called, has been gradually infiltrating the agricultural society. Some of the buzz words of this genre are "native crops," "polyculture," "organically grown," and "natural foods." Associated with this movement is the realization that some of the ancient crops fit the modern needs of a sustainable agriculture. Recent examples are jojoba, tepary beans, amaranth, quinoa, buffalo gourd, and guayule. Part of the philosophy involved is the broadening of the crop diversity of our agricultural base. Presently, the sustenance of modern society hangs in the balance of three foods: wheat, corn, and rice." (Yensen 1988)

Mankind has used about 3000 plant species for food during its history. Progressively, the trend has been to concentrate on a smaller range of varieties, and today the diet of the world is produced by at best 30 species (Fig. 65).

Nevertheless, large numbers of edible plants capable of development or crops exist, mainly in the tropics. The tropical ecozones have the largest fraction

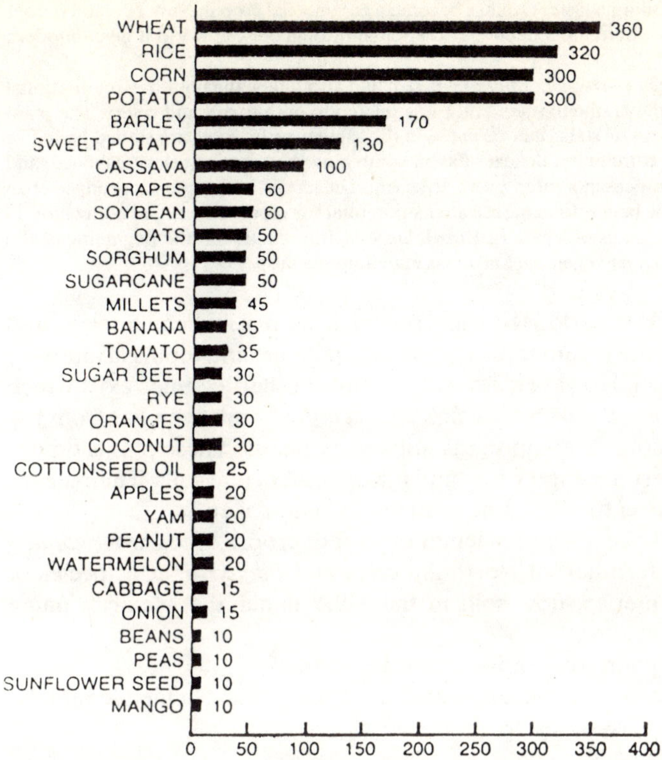

Fig. 65. Annual production (million metric tons) of the main edible plants

of the thousands of members of the underexploited plant kingdom for energy, pharmaceuticals, colorings, perfumes, industrial raw material, food, and other valuable products.

To answer the most pressing problems of food, economic botany and biotechnology should begin a systematic exploration of natural vegetative cover and sample its gifts before knowledge is lost through the disappearance of the natives of dry and humid tropical ecozones. This field of research already exists but has not yet been developed enough.

Vietmeyer (1986, 1987), senior program officer in Bostid and a specialist of forgotten crops, considers that the tropical world was neglected because research resources were concentrated in the temperate zones. When Francisco Pizarro discovered Peru in 1531, he "discovered" the potato, which was brought to Europe around 1550, ignoring several tens of other crops which were lost in the collapse of the Inca culture. Vietmayer quotes among them some remnant roots:

"Today, however, remnants of the Inca past remain, and among the "forgotten" crops is oca (*Oxalis tuberosa*). This exceptionally hardy plant is second in importance only to the potato as a root crop in the Andes. Its many varieties (not yet systematically collected) include some with a high sugar content

and others with a pleasant sourish taste. Oca has become a commercial crop in New Zealand (under the misnomer "yam") and would most likely be acceptable throughout the world if given modern agronomic attention.
Another Inca root, arracacha (*Arracacia xanthorrhiza*), could also follow the potato in international popularity. A member of the Umbelliferae family, it is related to both carrot and celery. The plant actually looks like celery, and its stalks can be eaten in the same way. Its roots are served boiled or fried. They have a crisp texture and a delicate flavor, combining the tastes of celery, cabbage, and roasted chestnuts. Andean peasants often grow arracacha because it is cheaper to produce than potatoes. The crop has now begun to demonstrate its potential for more widespread cultivation in Sào Paulo and other major cities of southern Brazil. Since its introduction at the beginning of the century, it has become a popular vegetable known as mandioquina salsa."

A great deal of work was done on grasses such as rice, wheat, maize, and sugarcane but among the legumes only soybeans, peanuts and lupins (Hamblin) have received attention. However, several dozens of other legumes exist which all have the same root nodules, by which bacteria convert nitrogen gas from the soil atmosphere into soluble compounds utilized by plants. These plants do not require nitrogenous fertilizer and withstand nitrogen-deficient soils. Their leaves act as a fertilizer because they become enriched with nitrogen.

Vietmeyer (1987) also proposes legumes as root crops: *Apios americana*, a groundnut, in the eastern half of North America and *Pachyrhizus* a species of central and South America, now sold in the USA under the Mexican name jicama.

The family of oil palm trees in Sudanian ecosystems:

In central America Vietmeyer describes other plants, which could be adapted to Sudanian ecosystems, for instance:

— *Guilielma gasipaes* (pejibaye palm) a chestnut-like fruit containing carbohydrates, protein, oil, minerals, and vitamins in almost ideal proportions is eaten in Central America but remains unknown in many parts of the chronically malnourished tropical world.
— *Jessenia polycarpa*, another palm of the Amazonian rainforest, gives fruits containing an oil similar in quality to olive oil, but is sold only in Colombia in spite of its richness. The African oil palm was unknown to commerce a century ago.

Brown (1986) describes another ancient oil plant: *Camellia oleifera*, cha you shu in Chinese, the tea oil tree (related to *Camellia sinensis* used to make the universally known tea) found in Hunan province (China) Assam (India), and North Vietnam. The oil is obtained by crushing the seeds. The Government has recognized the potential of tea oil production in China and incorporated its development into new forestry and agricultural programs. Primarily, the oil is produced from herbaceous crops (sunflower, sesame, groundnuts, rape seeds). With an improving standard of living, the demand for oil, which in China is annually 3 kg per head, will increase. Land shortages in China turned attention mainly to herbaceous crops, but there is now interest in oil-bearing trees.

Camellia olifera regenerates and coppices naturally, but its yields are low: 30 kg ha^{-1}. After selection, exceptional yields of over 1000 kg ha^{-1} have been obtained.

Balanites aegyptiaca (known in the vernacular as heglig, loba, shashoba, tau, saronga and g-og-g-ogat) is a very striking tree and heads the rapidly growing list of forest products. It lives for more than a century, producing annually an average of 125 kg of fruit. Its byproducts are animal fodder from the leaves, firewood from the branches, and pesticides and pharmaceuticals from the nuts. I have used the fruit successfully as a cure for stomach ache. It is rich in steroidal saponins which, upon hydrolysis, form diosgenin, an important source of steroid drugs such as corticosteroids, contraceptives, and sex hormones. The nut can also produce oil, crude proteins, vitamin C, and some minerals. The woody shells are highly combustible and can produce charcoal: "Every ton of whole fruit also yields a good half ton of hard woody shells." (Land 1986). Its main disadvantage is that it grows extremely slowly.

From its existing trees the Sudan could produce an estimated 13 600 tons of edible oil and 24 000 tons of kernel cake. Sudan alone could also produce 1200 tons of diosgenin a year, enough to satisfy half the world demand and earn an export income of $36 million (Ceres, FAO 1986, in Land 1986).

5.4.3 Reclaiming Salty Soils and Water

We have seen that, without proper agricultural and ecological methods, salinization occurs in all ecozones. Nevertheless, drylands are the most susceptible to salinization due to high temperature coupled with low rainfall, high evaporation rates, irrigation of crops, which is standard practice, and human activities of removing vegetation for firewood and field clearing resulting in a rise in the water table. We have already mentioned the historical example, in the Fertile Crescent of the Tigris-Euphrates rivers salted up 5 to 2 millennia ago and resulting in the collapse of the civilization. Similarly, the Peruvian Inca agriculture, located in the rich coastal valley, salted up the fields, and the farmers had to retreat to the Andean slopes.

Salinization is today a worldwide disease. Much of Latin America, Mexico, the United States, Canada, much of Africa, and especially Egypt, Arabia, India, the former USSR, and Australia have salt-affected lands. "There is already twice as much salty land as there is irrigated land." (Yensen 1988).

According to recent estimates, over 220 million ha of land are irrigated worldwide. Approximately 25 to 40% of that land is affected by salinization. The equilibrium level — above which conventional crops can become economically unproductive — between salt concentration by irrigation and loss of salt by drainage is approximately 3000 to 5000 mg l^{-1} (that of seawater is 35 000 mg l^{-1}). At this level economical agriculture is impossible. (Yensen 1988).

Salt-tolerant plants (halophytes) are one answer to this increasing salinity. The use of halophytes can, in appropriately designed systems, be a sustainable approach to the use of salt-affected land areas, in the 35 000 km of long coastal deserts and in drylands where brackish continental lake water exists. In all these areas halophytes are a better option than glycophytes, which are the conventional crops derived from plants adapted to soils of minimal salt content. Many regions owning vast saline aquifers are importing food at costs beyond their financial resources.

— How wild plants grow prosperously in very salty water is not well understood.
— How can a halophyte compete with the traditional glycophytes, wheat, maize, and rice, on the international market?
— Which among the bulk of halophytes will meet the needs of a particular region and can be grown economically?

Research in various countries such as the United States, CIS, Mexico, Egypt, Israel, India, and Australia to identify the potential of halophytic crops is ongoing. "There may be as many as 250 well-known potential halophytic crops" ... "To name a few: *Atriplex, Avicennia, Batis, Cressa, Distichlis, Juncus, Kosteletzkya, Limonium, Maireana, Rhizophora, Salicornia, Spergularia, Suaeda, Xylocarpus, Zostera.*" (Yensen 1988).

Yensen, noting that we have mainly chosen grasses among the glycophytes (wheat, maize, rice, barley oats, rye, pasture for meat and milk), proposes also a grass which is well adapted to saline drylands, *Distichlis*, on which he has worked for 15 years, and which produces a cereal grain. "*Distichlis* have the potential for reclaiming salt-ruined land, stabilizing soil from wind- and water erosion, providing pasture, and helping developing nations to feed themselves" writes Yensen, who has conducted selection and hybridization work resulting in a tenfold increase in grain yield, discovering that "the plant excretes the salt via salt glands, keeping the grain and tissue low in salt content. Thus, even the foliage can be grazed by cattle without ill effects." (Yensen 1988).

Nutrition studies revealed the low amount of the anti-nutritional factor of phytates that bind to the essential minerals, rendering them unavailable as nutrients, the amount of bran, which exceeds 40%, and a fiber content which is three times that of wheat. The taste test shows that it has good qualities as biscuits, "sautéed" on ice cream or in salads, or cooked as a rice. All these characteristics demonstrate the complexity of the studies which are necessary.

It is economically and environmentally sound to use the salinized farmland already in existence to cultivate salt-loving glycophytes or halophytes. There are probably cheaper than all possible trials of salt drainage. The relevant information already exists in the publication *HALOPH*.

Compiled by Aronson (1988), *HALOPH* is a database comprising 1560 species in 550 genera and 117 families of halophytic and salt-tolerant plants including life forms and, plant types of the primary habitat, their geographic distribution, maximum salinity tolerance, photosynthetic pathways, and economic uses.

Adapting crops to an environment rather than trying to change that environment can become a strategy for the rehabilitation and exploitation of salinized land.

"This has been achieved on thousands of hectares of salt land: in Tunisia using *Atriplex halimus*, *A. nummularia*, *A. vesicaria*, *A. glauca*; in Chile, *Atriplex atacamensis*, *A.* repanda; in Iran, *Atriplex canescens*, *A. lentiformis*; in Israel, *Atriplex halimus*, *A. nummularia*; in the USA, *A. canescens*; in

Australia, *Atriplex nummularia, A. vesicaria.* In Australia large areas of "scalds", i.e., eroded, saline, former agricultural land have been reclaimed by planting saltbushes and bluebushes, especially in New South Wales, which makes it possible to initiate the biological recovery processes. Other salt-tolerant forage species using drainage water have been utilized such as: *Hedysarum carnosum, Melilotus alba, Sporobolus airoides, Festica arundinacea, Trifolium fragiferum, Paspalum vaginatum* and *Puccinellia distans*. . . ." (Le Houérou 1968).

5.4.4 Plant Breeding

Twenty five years ago, the Asian Continent had a choice between chronic famine or economic dependency. In the 1960's the situation changed, when India and Pakistan introduced a high-yield and quick-growing wheat produced in Mexico: the green revolution had begun. Similarly, Mexico is nowadays developing a high protein maize in CIMMT, the Spanish acronym for the International Center of Maize and Wheat Improvement.

The maize problem lies in its deficiency of protein: only 9.5% and a low rate of amino acids (lysine and tryptophan). The improval of these low rates by CIMMT will be the solution if the transfer of these new technologies, new seeds, and new methods of cultivation is rapidly achieved (Hanley 1988).

Maize, originating in Latin America, now feeds 500 million people. It was brought to Europe from whence it spread to Africa and Asia.

Maize is a plant adapted to semi-arid ecozones. One quarter of the total production is consumed in east, south, and west Africa, the rest in Asia. It is also consumed indirectly through poultry food and oil.

In the Sahel, during the last decade, the production of maize can be considered as the best of all cereals. Based on several improved varieties, the maize project (Projet maïs) of the CMDT (Compagnie Malienne des Textiles), South of Mali, has contributed to increase the production from 1.5 t ha^{-1} to 3.5 t. The improvement results from good cooperation between farmers, extension technicians, and government; new techniques are proposed to the farmers (denser seedings, addition of fertilizers, new machines), and better financial help.

According to Kiabou (1989), the national production increased from 37 000 tons in 1976 to 211 000 tons in 1988.

Jones (1983), working on coconut tissue culture techniques, created clones of high-yielding top-quality oil palm and took plants through to their final stage ready for planting. The oil palm contains oil in the mesocarp and the kernel (endosperm). The objective of research was to produce progeny which bear fruit with reduced shell thickness and enhanced mesocarp yield with a consequent increase in palm oil content.

"Experience has shown. . . that **it takes an average of about 10 years to produce a new plant variety and another 5 to 10 years for it to gain widespread adoption.**" (Eicher 1986).

5.5 More Integrated Rural Activities for Agriculture Amelioration

5.5.1 Multicrop Agriculture (Fig. 66)

Multicrop agriculture, also called the intercropping system, or alley or strip cropping, is the simultaneous culture of two or more crops in the same plot. In the unpredictable climates of tropical ecozones this system gives the best defense against risks of crop failure. Peasant farmers have realized for centuries that intercropping is the best means to obtain stabilized yields, the best way for resource-use efficiency and for producing more food from less surface: "Aboriginal farming systems have probably always included trees." (Dover and Talbot 1987). "Temporal and spatial complementarity in resource use may enable intercrops to capture and convert more water and light (Natarajan and Willey 1986), as well as more nitrogen (Liebman and Robichaux 1988), than can be the case with single crops. This multipurpose character is an important quality for the Third World villager. Having the flexibility to harvest several products spreads the risk of failure and enhances the economic viability of his small land area. Modern crops are almost all single-purpose species, one of the reasons why they are sometimes not readily adopted in developing countries.

Cropping systems that require less water than do conventional monocultures of humid-derived domesticates are the solution proposed by Nabhan et al. (1988) to obtain "sustaining yields while reducing the inputs of water and nitrogen, the factors that most critically limit crop productivity throughout the world."

"Over the last 2 years, the University of Arizona's Ecology and Evolutionary Biology Department has collaborated with the DBG's research staff in experimentally evaluating a number of two-species intercrops. In 1987, both groups evaluated intercrops of tepary bean (*Phaseolus acutifolius*) and pearl millet (*Pennisetum americanum*) under different water and nitrogen regimes. In 1988, DBG shifted its designs to examine the effects of water and plant density on the growth and yield of sole crops of tepary bean and Sonoran panicgrass (*Panicum sonorum*) and their intercrops. Both of these latter species are exceptionally heat-and drought-tolerant, and were prehistorically cultivated in the Sonoran Desert. In this sense, the so-called new crops are really ancient domesticates being evaluated in a new agricultural context. The same is true for intercrops of pearl millet and an O'odham land race of cowpea (*Vigna unguiculata*) being analyzed for yield components at the University of Arizona's Campus agricultural center."

Three different types of agricultural exploitation exist in dry and subhumid zones without any link or interconnection: extensive pastoralism, rainfed agriculture, and irrigated farming.

These systems should develop different complementary activities to save vegetation. For example:

1. Extensive grazing serving as breeding areas while the irrigated farms act as fodder producer and fattening centers. The advantage of this method is to remove cattle from the natural pasture when the threshold of degradation is reached, which produces pressure on rangelands, and so allows for regeneration.

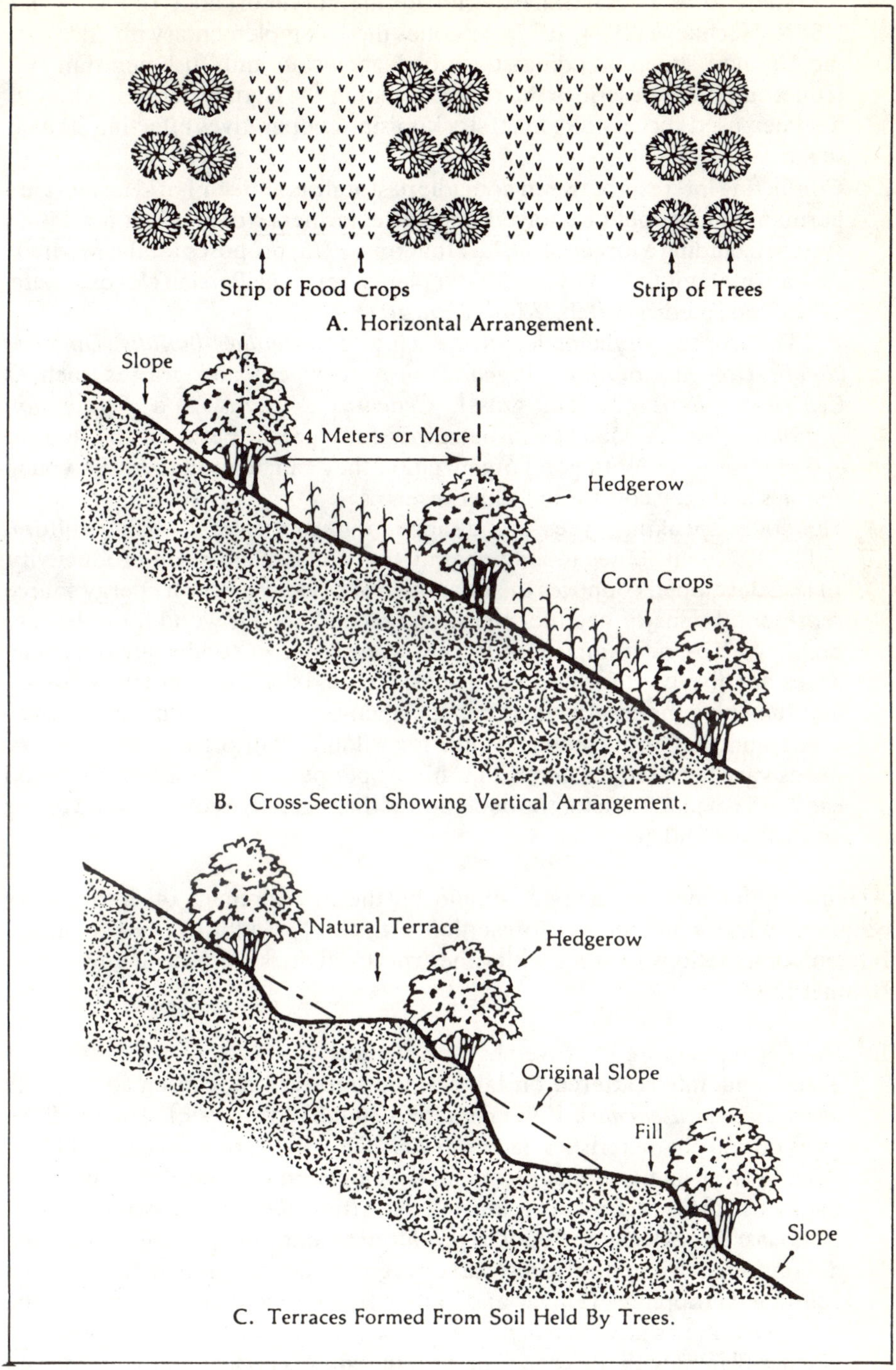

Fig. 66A-C. Multi-strip cropping with trees to form natural terraces. (Vergara 1982, in Dover and Talbot 1987)

There were several trial experiments in semi-desert areas of the former USSR. Nechaeva (1954, 1977) describes these complementary practices for the Karakul sheep-breeding states of Uzbekistan and Turkmenistan. Le Houérou (1975) describes the same system in the Syrian steppes, where it has been used successfully in 20 stock-raising cooperatives affecting 300000 sheep.

2. Conflicting interests between nomadic pastoralists and semi-arid farmers are permanent and could be solved only if the farmers would adopt a rotation system including a forage crop. Le Houérou (1975) proposes for the Mediterranean drylands the following fodder plants: berseem, Persian clover, alfalfa, subclovers, medics, vetch, Wimera rye grass, etc.

   The tropical drylands legumes such as *Stylosanthes humilis, Dolichos lablab, Dolichos biflorus, Vigna sinensis* (cowpea) and grasses such as *Cenchrus ciliaris* (buffalo grass), *Cenchrus setigerus,* and *Pennisetum typhoides* (fodder Millet) can be successfully grown between the 400- to 800-mm isohyets. With good organization they can be used to fatten young animals and they constitute fodder reserves.
3. Generally speaking, trees are disappearing, particularly on agricultural fields. This contributes to land degradation and declining soil productivity. In less-developed countries the quantities of wood used as an energy source represent the major need for trees. The second need is wood for poles and buildings. The next important is trees as a source of fodder for browsing. Trees are also used for iols, medicines, tannin, bark for fiber fruits, honey, and thatching material, and as beneficial environmental influences such as shade and shelter, including shelter for wildlife. World Bank estimates of needs vary between 0.5 and 1.2 $m^3\ h^{-1}\ a^{-1}$, per person. The search for wood can be the explanation for the depletion of the vegetative cover around towns and villages and along roads.

A third complementary activity should be the development of agroforestry programs which combine reafforestation with the agricultural need for trees. Fifteen combinations of trees with other rural activities were distinguished by Baumer (1987).

1. Rainfed agriculture improved by tree planting.
2. Trees randomly scattered on fallows; the trees can be chosen so as to fix nitrogen: e.g., *Leucaena* if the climate is suitable. Effects of *Acacia albida* trees on soil characteristics and crop yields in Africa are shown in Table 27;
3. "Taungya" is an agroforestry system developed in the late 19th century by the British in South East Asia, and in East Africa (Kenya, Malawi, Tanzania) in the last three decades. "Food is produced among newly planted trees, and discontinued when the forest canopy grows dense enough to preclude continued cropping." (Dover and Talbot 1987). In Sudan, for example, the system combines *Acacia senegal, Andropogon gayanus,* and *Vigna vexillata* (fodder plants) with the crop. The system requires noncompeting roots and the crop must tolerate shade:

Table 27. Effects of *Acacia albida* trees on soil characteristics and crop yields in Africa. (Dover and Talbot 1987)

A. Nutrients returned annually to the topsoil

Element	Amount (kg ha^{-1})
Nitrogen	186
Phosphorus	4
Potassium	76
Calcium	222
Magnesium	39

B. Increases in soil quality and yield under *Acacia* crowns

Item	Percent increase
Total soil nitrogen	33–110
Organic matter	40–269
Cation exchange capacity	50–120
Millet yields	37–104
Sorghum yield	105

4. Alley cropping is more suitable for subhumid ecozones. Crops are grown in between the spaces separating a row of trees. The trees are pruned frequently and clippings are laid on the ground where the mulch decomposes into organic matter;
5. Border planting to demarcate individual farm fields. Trees can be mixed with shrubs and grasses. They are used as property markers but also provide firewood, fodder and fruits, shade, and organic matter.
6. Multiple uses for small reserves of trees.
7. Shelterbelts and windbreaks planted to dissipate the force of the wind. In the Majjia valley (Niger) CARE's project is a successful example of shelterbelts of *Azadiracta indica* (neem trees) and lately also of *Acacia nilotica, Acacia senegal,* and *Acacia tortilis*. These shelterbelts in the harmattan wind system increase the crop production of the area by 25 to 30% which produced enough incentive among farmers to multiply the shelterbelts. Success was also obtained by controlling access of animals to the fields.
8. Trees in water conservation programs: drying out of swamps, reinforcing of river benches. *Eucalyptus* and *Populus* are particularly good "water pumps."
9. Trees in the gardens around the houses. They may be random or regularly planted.
10. Trees for industrial purposes: coffee, tea, cacao, arabic gum.
11. Trees for animal fodder.
12. Trees in dune-fixing programs.
13. Aquiforestry: plantation of *Mytragine inermis* or *Sesbiana* sp. on the border of semi-permanent water points, useful for animals. The roots of mangrove are good nests for shrimps and crabs.

14. Api-sylviculture: the trees are selected to produce nectar and pollen welcome to the bees.

Trees can play a multiple role in agroforestry. Another role which is less obvious is their action in soil rehabilitation. "They also minimize nutrient drain due to leaching and soil erosion and restore nutrient lost from the ecosystem." (Fig. 67) (Dover and Talbot 1987).

5.5.2 *Natural Bushland Exploitation*

Stiles (1988) proposes another approach to economic development of drylands by exploitation of the dryland bush of the Sudano-Sahelian ecozones.

The strategy proposed by Stiles is to determine what market potential exists. . . "how to make *Acacia-Commiphora* bushland sustainably productive and capable of supporting increasing population . . . and how to halt destruction of this land." (Table 28). The countries in Africa for which this study has relevance are (Stiles 1988): Burkina Faso, Cape Verde, Chad, Ethiopia, Gambia, Kenya, Mali, Mauritania, Mozambique, Namibia, Niger, Senegal, Somalia, Sudan, Tanzania, Uganda.

5.5.3 *Soil Degradation and Rehabilitation of an Agropastoral Land in Tanzania. Hopeful "De-Desertification": Case Study (Christiansson 1988)*

5.5.3.1 *Degradation*

The drylands (average annual rainfall below 800 mm) cover 40% of the country in the North and Central Tanzania.

"Throughout the drylands, symptoms of land degradation are evident. Hillsides have been cleared of natural vegetation and are intensely cultivated and overgrazed by domestic animals. This has resulted in excessive surface runoff during rains and soil erosion, which has removed considerable parts of the precious topsoil. The roots of trees and bushes are exposed on the surface of the ground. In some places alluvial fans and pediment slopes are dissected by gullies up to 10 m deep.
"Large amounts of water are lost through the increased runoff, and heavy loads of sediment in the streams rapidly silt up surface water reservoirs. These conditions have led to repeated food shortages and dwindling water supplies." (Christiansson 1988).

The annual rates of soil erosion on moderately grazed upper pediment average between 1 and 2 mm rising to 10 mm on overgrazed slopes. The sediment yields in cultivable lands an average of between 200 and 600 m^3 k^{-2} a^{-1}. This high production of sediments shortens the life of water reservoirs.

Inadequate rainfall for the crops, with ill-distributed and intense storms, leads to erosion and migration of people near villages or water-supplied areas.

The recent population increase around the site of Kisongo dam at Arusha, built in 1960 with clearance of the vegetation in the catchment area, explains the siltation. In the Mpwapma, east of Dodoma, 135 tons of soil ha^{-1} is lost annually.

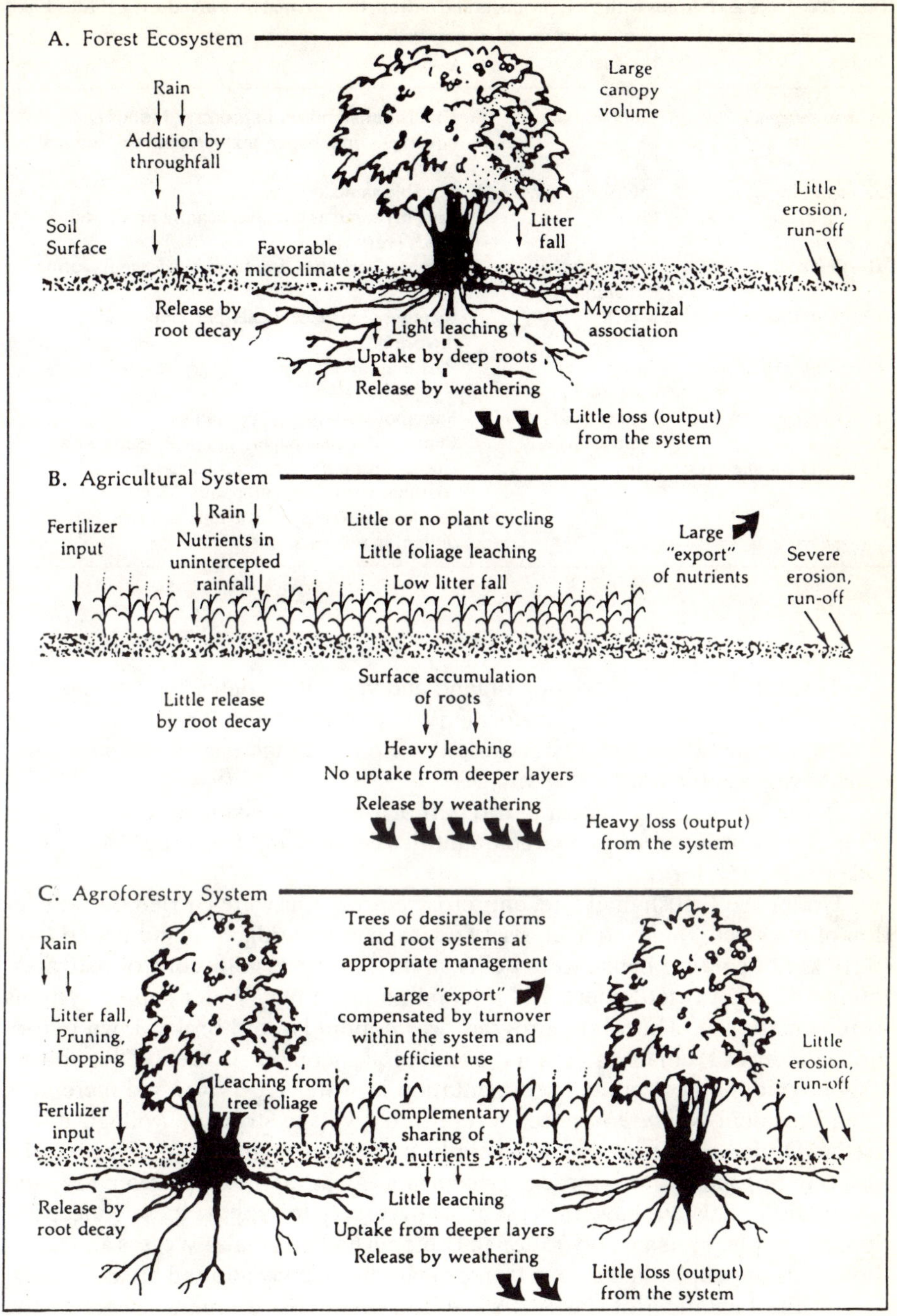

Fig. 67A-C. Schematic representation of nutrient relations and advantages of agroforestry systems (**C**) in comparison with forestry (**A**) and agricultural (**B**) systems. (Nair 1982, in Dover and Talbot 1987)

Table 28. Trees and shrubs with their products and current or potential uses in industry. (Stiles 1988)

Name	Product	Use
Acacia senegal	Gum arabic	Emulsifier/stabilizer in processed foods and beverages, ink, paper, textiles, pharamaceuticals, glues, etc.
A. mellifera	Inferior gum	Possibly same as above
Astragalus sp.	Gum	Possibly same as gum tragacanth, processed foods (very valuable)
Boswellia sp.	Frankincense, olibanum	Incense, perfumes, food and beverage flavoring
Commiphora myrrha	Myrrh	Incense, pharmaceuticals, perfumes, and flavoring
Commiphora sp.	Opopanax, opopanom	Perfumes and flavoring
Lawsonia sp.	Henna	Shampoos, soaps, and cosmetics
Sterculia sp.	Gum karaya	Pharmaceuticals, paper, and processed foods
Vernonia galamensis	Oil	Epoxies for adhesives, plasticizers, industrial coatings, varnishes, and paints. Meets EPA regulations to replace solvents as a reactive diluent in resin systems

The initiation of degradation of land and vegetation dates back at least 150 years, when woodland was transformed to open grass and farmlands, and the small and isolated settlements became threatened by increasing pressure from neighboring peoples and larger villages.

In the 17th century this semi-arid part of Tanzania became a caravan route, where the travelers required large quantities of food and firewood, and drastic soil erosion developed.

Under the British mandate, anti-erosive programs were carried out: reduction of livestock, range control, sisal hedges, contour ridging of cultivated land (3219 km of contour banks were built), animal manure, gully control reafforestation and even resettlement and antifamine programs. Water supply systems were constructed. The whole program was unpopular and broke down before independence. Human and animal population increased in the 1950's, and by 1964 degradation of soil and sedimentation in water reservoirs had increased. Then President Nyerere imposed a return to a village structure (villagization). In the Dodoma district 250 000 people were moved into 142 villages. The resulting impact of this imposed concentration was land degradation: the land settled did not always have the resources necessary to support these concentrations of people. The same operation in Ethiopia had no greater success and many thousands of people died during transport to their newly allotted areas.

At the beginning of the 1970's the Tanzanian Government, aware of the catastrophic state of the country and of the necessity to incorporate soil conservation measures into all land use plans and village master plans, decided in 1973 to initiate the HADO project, a soil conservation project for the Dodoma region with the financial help, field staff, and consultants of SIDA (Swedish International Development Authority). The objectives of the HADO project were as follows:

"To conserve soil and water and to recondition areas already affected.
To ensure that a sufficient number of trees are available to the inhabitants of Dodoma Region for their different needs.
To encourage the planting of trees which can provide shade and fruits and which can act as windbreaks.
To promote mass-planting of trees in the rural areas.
Apart from its emphasis on the development of arable agriculture. HADO rehabilitation activities have also involved the following strategies:
Construction of contour bunds to reduce the speed of surface runoff. To protect the bunds from breaching, they have been stabilized by means of vegetative techniques.
In gully reclamation, earthen and stone check dams have been constructed to reduce the intensity of discharge, trap sediment, increase percolation, and improve soil moisture conditions in order to facilitate establishment of plant cover in the gullies. Planting of grass on the bottom of gullies has also been found to be an efficient means of trapping sediment and reducing water speed.
In some gullies, peripheral live fences of sisal have been planted to check runoff from the sides and to prevent livestock from entering the protected gullies.
Quickly sprouting vegetation has also been used to trap sediment moving across alluvial fans as a result of torrential rain.
Reafforestation efforts have also been undertaken: these include establishment of nurseries, production and distribution of seedlings, and initiating tree-planting campaigns. These efforts also include demonstration of the principles of agroforestry, and educating the local population in the importance of protection of the vegetative cover and the detrimental effects of burning trees and undergrowth, particularly on steep slopes." (Christiansson 1988).

5.5.3.2 Results of the HADO Project: "De-Desertification" in Tanzania

In the Dodoma region, between 1973 and 1986 14 088 ha of eroded land were reclaimed, partly by construction of drains to divert excessive runoff from entering cultivated land. In the years 1983–1985 a total of 58 km was constructed.

Contour ridges with trees or grass for fodder on some ridges were built partly by the farmers themselves. By 1986 11 nurseries had been established with a production capacity of 100 000 to 500 000 seedlings. By 1986 the area covered with tree plantations was 3000 ha; unfortunately, 375 ha were destroyed by fire. An educational program was created which reached right to the village level.

In 1979 part of the Kondoa district was closed to livestock, with some 90 000 animals being driven out of the area. The heavily browsed shrubs and scattered trees were replaced by a grass cover with sprouting shrubs and trees, less surface runoff, decrease of sediment load into the rivers, a more permanent stream flow, and consolidated river banks usable for cultivation. There were, however, some negative aspects: invasion by birds that threaten millet and by monkeys and warthogs which attack the maize and sweet potatoes.

The disease conditions for livestock in the surrounding areas worsened, and cattle owners, having lost their rangelands, thus lost also some of their cattle; but the project appears to be mainly beneficial because of the improvement in agriculture.

The exclusion of livestock from the Kondoa region and Mvumi was later decisive in achieving success, but is not totally satisfactory because the zero-grazing program did not make use of the traditional activities, and increased the work load of the farmers. The economic and social effects must also be envisaged and understood, which has not been the case up to now.

5.6 Water Conservation Measures

5.6.1 Introduction

The advanced civilizations of the present and the past have been possible only where freshwater has been available in quantities sufficient to support the agriculture necessary to feed large urban centers. The first human settlements, and later nearly all the large cities are strategically located near a river or a lake.

The dry areas of the planet could be inhabited by larger populations if sufficient freshwater could be made available. Success and failure in drylands agriculture are determined by water availability. Low nitrogen and phosphorus content is the next limiting factor. Potassium, calcium, and magnesium are in general sufficient. This is why water conservation measures and methods for a more efficient use of water supply are of primary importance in drylands.

Water is produced by allochtonous alluvial valleys: in Africa, the rivers Senegal, Nile, Chari, and Niger, all of which have their springs in the humid tropical ecozone; in the Middle East, the Euphrates and the Tigris; in Asia, the Indus and the Ganges, or as in China rivers around the Taklamakan desert and the Huang Hé; and in the western part of North America the rivers which originate from the surrounding mountains.

The figures for rivers in the Sahel as recorded since 1903 for the Senegal, 1907 for the Niger and 1932 for the Chari have shown that these have received a severe decrease in runoff during the last two decades. The reduction in the scale of the annual floods has been catastrophic for the crops on the flood plains watered by overflow in the lower valley of Senegal and in the Debo delta (inland delta of the Niger 15°–17 °N).

In the western USA agriculture has made enormous use of its groundwater resources. Africa has large unused groundwater resources in its drylands, for example, the Saharan Nubian Basin in Sudan and the Chad Basin, in Chad. Others are known in Niger, Mali, Mauritania, and Ethiopia.

Ahmad (1988) thinks that the Sahel can no longer depend on rainfed agriculture and proposes the use of "extensive" groundwater "to combat the effects of drought", mainly in the Sahara Nubian Basin in Sudan where he has calculated that 1 million m^3 day^{-1} could be extracted and the Chad Basin resources, where 2 million m^3 day^{-1} could developed agriculture-based industry.

Nevertheless, there is a total consensus that deep groundwater in drylands is a nonrenewable resource. Even if the water level can be recharged, a plan for the control of inevitable runoff and a careful timing of use must be programmed. The amount of precipitation available for groundwater replenishment is low, so that aquifers contain almost entirely very old water fallen during more humid paleoclimates or are fed by rivers originating far away. In Libya, this excessive lack of groundwater is already for advanced, it is not expected to last beyond the middle of next century (Richards pers. commun. 1989).

5.6.2 Difficulties in Obtaining Water

5.6.2.1 High Technology Solutions: A Mitigated Success

Technology has proved more efficient in irrigation than in dryland farming.

> "Much of the settlement of dry lands by agriculturalists has been made possible only by advanced technology. In the USSR, North America, and Australia, deep wells could be drilled, prairies fenced, fertilizers and machinery imported, diseases controlled, cash crops taken to distant markets. These technologies opened the plains to many hundreds of thousands of farmers. In the Indian subcontinent, the huge canal irrigation schemes, made possible only by new engineering skills and materials, allowed millions to move into formerly barren plains. The same technology, on a smaller scale, allowed the settlement of river valleys in central Asia, Australia, North and South America.
> Retreat from these lands is now unthinkable. Pakistan is utterly dependent on irrigation, as are the northwestern provinces of India, like the Punjab and Rajasthan. The abandonment of the High Plains in Kansas may be conceivable, and the population has indeed fallen considerably since the 1930's, but there, too, there must be a commitment to maintaining settlement. Where retreat is not an option, people are forced to continue practices that may be reducing the resilience of the environment." (Warren and Agnew 1988b).

Much emphasis in combating desertification has been placed on technology, mainly in connection with large-scale irrigation projects.

Irrigation continues to be seen by many countries as a panacea for the food problem and to remove the threat of drought. It has inspired the most ambitious and also the most unrealistic management projects.

— The construction of a canal between the Mediterranean and the Republic of Niger is a French colonial idea of the first half of the present century. To fill the northern sebkhas of Libya and Tunisia with Mediterranean water is a more recent plan. Other possibilities include:

> "In Africa an interesting project envisages the transfer of Congo River water to drought-plagued regions of the Sahara, involving the construction of a 900 000 km^2 reservoir in the area of the old lacustrine water body of the Congo depression. From there, water would be taken by canal to Lake Chad basin from whence it would flow as a mighty river, greater than the Nile, through Sahara to Gabes Bay (Tunisia), allowing en route the irrigation of 60 M ha." (Zonn 1986).

- To carry the Siberian rivers towards the Central Asian deserts of the former USSR is a project that has been seriously analyzed in the last two decades, but is no longer considered, partly because of the possible impact on global climatic through an increase in the salinity of the Arctic Ocean and the consequent reduction in the Arctic ice sheet.
- To divert a part of the water of the Zambesi river towards the Kalahari (Botswana) is being seriously considered in the UNEP Zambesi action plan.

– To divert the river of northern Canada or the Mississippi to water the dry American West has been also envisaged.

"High technology has succeeded only in highly capitalized societies such as the United States, Canada, Libya, and Israel", as Warren and Agnew (1988b) found. ICIHI (1986), quoted by Warren and Agnew, assert that during the 1980's in the Sahel "for every new hectare. . . brought under irrigation, another irrigated hectare went out of cultivation because of bad design or bad management."

It is recognized that results achieved through high technology have fallen well below expectation; the causes of failure are multiple and different:

1. Heathcote (1986), analyzing dryland irrigation in Australia, concludes that unsuitable locations and soils had been selected, that a too-high proportion of the investments went to low-return pastures rather than to high-return crops, that the expenditure was focused on combating drought and not on the risks of economic failure.
2. Several big schemes have failed or run into severe difficulties because of human incompetence and conflict between the perceived needs of the administration and the farmers: lack of skill and knowledge, organization and management barriers, absence of motivation, inadequate guidance, insufficient capacity for maintenance, lack of capacity to switch once technological change is underway. For example, a large rice-growing irrigation scheme developed by a Chinese team near Bobo Dioulasso (Burkina Faso) was progressively abandoned by the farmers after the departure of the Chinese team.
3. The Jonglei scheme for the canalization of the White Nile through the Jonglei canal in the Sudan to provide more water for irrigation to the arid north and to Egypt promised to be a catastrophe for natural fauna and for the human population. In particular, it promised to reverse the seasonal flooding of the swamps of the Sudd.
4. Large-scale projects that alter the ecological situation over vast areas may bring temporary negative consequences. As we have seen in Chapter 2, in the course of the Karakum Main Canal construction, the development of a chain of new oases along a 1000-km strip resulted in tracts of trampled sand areas. Water seepage from the canal and its release on either side of the canal resulted in filtration lakes gradually occupied by solonchaks (Kharin and Kalenov 1978).
5. The area surrounding the Aral Sea is an example of "desertification" caused by big irrigation schemes. The steady diminution of the level of the Aral Sea caused by the diverted flow of the Syr-Darya and Amu-Darya rivers in order to increase the utilization of their waters for irrigation has resulted in the spreading of the desert over the vast delta areas of the Syr-Darya and Amu-Darya. The process is further aggravated by the disappearance of the delta's natural water supply. A sandy-salty desert is developing along the exposed dried-up sea bed. It increasingly affects the rate and nature of ecological changes in the Aral Basin. The dimension of this dry land is increasing, as the sea level drops. "A secondary (man-made) desert is being

developed, which may have far-reaching and as yet unpredictable consequences for the natural environment in the adjoining areas. The implication of such changes is frequently unpredictable." (Kuznetsov 1980).

"Despite the dilemmas associated with irrigation, irrigated farming in the arid zone remains the most efficient and effective approach to land management. Compared with dry land farming, it can return sixfold yield increases for cereals and four- to fivefold increases for industrial crops. It is not surprising, therefore, that the irrigated areas in developing countries continue to grow at 2.9% a^{-1} as opposed to 0.7% a^{-1} for dry land cropping. (UNCOD 1977)." (Zonn and Orlovsky 1986).

The value of supplementary irrigation emerged during the 10 years' analysis (1964–1973) of sorghum yields on a control plot in the experimentation stations of Tarna (Niger) (Table 29).

In the four deficit years of 1968–71, 1972, and 1973, the crops had been given supplementary irrigation. Sorghum in semi-arid areas is totally dependent on rainfall. A marked decrease in yields occurs with decrease in annual rainfall. With application of supplementary irrigation at critical periods there is a marked increase in yields.

The irrigation regime and water application rates, including bad drainage, have been at fault in many schemes. Among the major faults are leaking canals helping to raise the water table and groundlevel flood irrigation (borders and furrow irrigation) leading to salinity. Low level sprinklers, (better than canopy top ones) and subsurface and drip irrigation reduce this danger practically to nil.

Nevertheless, for developing countries the problem persists. Smith et al. (1983) assert that the cost per 1 million l water for high-tech sprinkler irrigation in Israel is US$ 82.00, while for the same quantity a low-tech method costs only US$ 0.55, but to this figure must be added water losses in the canals and in the proportion of water not used by the crop.

Large-scale irrigation is not only the most economically demanding technique. If poor cultivation techniques and use of marginal land are the causes of the most extensive degradation, we believe that salinization and waterlogging play the major role in intensive degradation of drylands.

Agricultural engineers have developed carefully calculated drainage systems that, when properly managed, will keep the salt out of the root level, although the slope must be sufficient without being excessive. Many nations in drylands do not have heavily subsidized agriculture. Creating a good drainage system in developing countries can be prohibitively expensive. After having put the system in place, it must be maintained and the salty drain water must be con-

Table 29. Yields of sorghum on control plots at Tarna, Niger, 1964 to 1973. (WMO 1978)

Year	Rainfall ± 50 mm	Irrigation mm	Yield kg ha^{-1}
1964, 1965	700	–	1756 and 2000
1967, 1969	600	–	614 and 1368
1966, 1970	500	–	600 and 448
1968, 1971	400	40 and 60	1412 and 1812
1972, 1973	300	320 and 200	2987 and 4725

trolled. Prevention is better than cure: drainage is expensive, but it is more costly to reclaim saline land.

5.6.2.2 *Low Technology Irrigation Systems and Water Conservation Measures*

Large-scale irrigation systems or small-scale ones? The dilemma still exists. Nevertheless in a rural economy low technology produces immediate results and community participation.

A variety of irrigation systems is offered to meet the range of imperative conditions which must be respected. They are reviewed by Armitage (1985):

1. Gravity systems which wet the whole surface: the methods are wild flooding, border check method, basin irrigation, furrow irrigation, and corrugation methods.

— Wild flooding is obtained from a graded supply ditch built along the upper side of the field and distributed freely down the slope.
— In the border check method flowing water is moved down slope by parallel earth ridges. Careful land preparation is necessary to avoid erosion. This is most suitable for medium texture permeable soils.
— Basin irrigation: the fields are divided in small units and filled with water which infiltrates. It is to be preferred for heavy soils.
— Furrow irrigation: channels carry water across or down the slope of the field, where it seeps into the bottom and sides of the furrow. This method requires high labor input, a high degree of skill and experience, and has the disadvantage of distorting the root system, which is aligned with the axis of the furrow.

The gravity systems require low investment costs, do not require water under pressure, but disturb the top soil and increase the vulnerability of the soil to erosion.

2. Corrugation irrigation combines furrow irrigation and overflow of water.
3. Sprinkler systems, which require water under pressure, can be used where moderate amounts of water are required and the topography is irregular, such as localized systems of trickle or drip irrigation.
4. Center-pivot irrigation which flourishes in the midwest of the United States and is very fashionable in dry areas (it can also be seen in the hyper-arid Libyan regions of Kufra and south of Sebha) is water-demanding. It is not recommended for areas where the groundwater is fossil.

5.6.2.3 *The Basic Requirements of Irrigation*

Irrigation must meet five basic requirements:

1. A sustained economic return: to satisfy demand for crop or tree production. The problem of economic feasibility is still difficult to solve because not

enough is known about the water requirements and responses of plants grown under irrigation, mainly in relation to different stages in their development, or about the different forms of plantations or the different optimal environmental conditions. To obtain a sustained economic return the other objectives are: a selection of the most appropriate species and plantation techniques suitable for the conditions of the area, to determine the profitability and economic feasibility of the plantation, to recommend the most suitable methods to the land users.

2. Conservation of water by minimal loss through evaporation control, minimal loss during conveyance and utilization, regular maintenance of water supply and distribution arrangements. Knowledge of groundwater movement and recharge rate is vital. It is also vital to ensure that extraction does not exceed recharge to avoid seawater intrusion and compaction of the aquifer.
3. Prevention of site degradation by avoiding waterlogging, salinization, and soil and groundwater pollution. Full attention must be given to slowing down salinization by better calculation of water requirements, in more efficient delivery of water to plants and in more adequate drainage systems.
4. Sociological and hygienic adequacy: proven need for irrigation, availability of land, adapted land tenure system, educational level, disease control, and esthetic achievement.
5. Appropriate organizational and management structures. Weak management and administration, poor training of staffs, and mediocre appreciation of social requests are often the mean reasons of failure.

Irrigation, after pastoralism and rainfed agriculture, is the third type of land exploitation in dryland. The great lakes formed after the ice age in temperate zones or those formed in endoreic depressions in deserts were areas of large concentrations of population because food was available there the whole year around. So very early settlements were possible. As these lakes dried out, some people tried to imitate nature and irrigation was born. Irrigation requires coordination and it also gives enough surplus production to support a small administrative and intellectual class. That is why the first great civilizations are based on irrigation at the edge of deserts.

Irrigation is very expensive and must be chosen only if other alternative designs or strategies do not exist.

In drylands, water must be considered as a nonrenewable resource. With modern pumps groundwater is used more rapidly than it can be replaced. This is the case in the Kalahari desert, which has little surface water. A hydrological meeting in 1987 in Gaborone concluded that, with the present agricultural and industrial development of the area, the groundwater will run out in less than 15 years.

Everywhere where groundwater is used the priorities of use must be discussed.

5.6.2.4 Deep Borewells

To increase the security of nomads and pastoralists, deep borewells that can supply livestock with water all the year round have been drilled. However, this is not totally satisfactory. The problem of lack of water is transformed into a technical problem of lack of pasture during the dry season and of land degradation around the wells afterwards. Traditionally, where the collective shallow wells are running dry, pastoralists move their herds on before the land becomes overgrazed and, before the rainy season, go to their private well where water is available.

"Concentration of rangeland sheep-farming around a few watering places in the Central Karakum encouraged the formation of areas of shifting sands near wells and along the cattle routes. The construction of new water wells and laying of aqueducts helped to scatter the stock, whereas the trampled sands are now held in place with vegetation." (Babaev 1986a).

Sinking wells has in many cases destroyed the traditional system and the refined balance between water and pasture. In Niger at the foot of the Aïr massif, wells were dug to give water security to the nomads. When the local authorities asked the nomads to move at the beginning of the dry season, they opposed this request with guns. Similarly, a program of well-drilling proposed in 1987 to the Rendiles people, West of Marsabit (Kenya), did not attract them because the management of communal grazing resources was not planned simultaneously.

The explosive multiplication of wells which results in the expansion of the grazing and cultivated areas can be more detrimental than advantageous in marginal, vulnerable dryland. "In Sudan almost 300 wells (boreholes) were opened during the period 1919–1982. . . . The great, drastic expansion of the areas under cultivation at the end of 1960's is a possible response by the farmer to declining crop productivity caused by drought." (Hellden 1988). The consequence here is not land degradation or decrease of productivity, but the shrinking of underground water resources.

South of Mali, different nations (France, Germany, Saudi Arabia) were competing in 1985 for digging wells without taking into account the declining underground water level (9 m) in the last three decades. The same decline in the underground water level is to be deplored in Cap Bon peninsula (Tunisia). The decline of 20 m in the last three decades obliges the farmers to spend almost all their income on improving the depth of their wells.

Wells are points of human and animal concentration and of overgrazing. To avoid this threat, an appropriate amount of such water supplies will lead to the dispersal of livestock over the whole area and so avoid overgrazing in the vicinity of widely spaced water supplies.

5.6.3 Water Harvesting

The main constraints for crop production in dry, semi-arid and semi-humid regions are firstly water shortage and secondly lack of soil fertility. Better management of water availability and improved conservation of soil fertility can increase yields. The techniques called water harvesting are traditional techniques which have kept their value and proved their use in sustaining the livelihood of farmers and nomads.

Rainwater harvesting is probably the oldest method, and has been used in China and the Middle East for centuries to direct infrequent runoff or dew to the rooting zone of trees or crops or to assist groundwater recharge in a depression. Nowadays, rainwater harvesting exists on all the continents: Asia (Afghanistan, India, Iran, Israel, Pakistan in the Thar Desert), Africa (Botswana, Burkina Faso, Niger, Sudan), Australia, and Mexico.

Water harvesting was a classic method used in the Nile Valley and Delta before the Mohammed Ali barrage and construction of the first Aswan dam, by yearly flooding of the fields. In this arid region agriculture was made possible. After the construction of the Aswan dam the advantages of greater water availability were counterbalanced by the difficulty in keeping the land fertile.

The water harvesting techniques can be modernized with one further restriction: nomads and farmers to whom proposals are made from outside to develop a water harvesting project for crop production do not readily accept the recommended techniques if these do not conform to their previous experience or culture.

A large part of Tunisia is in the semi-arid zone with a rainy winter and a dry summer season. The erratic rainfall being often insufficient for rainfed agriculture, runoff and flood water are captured and used for supplementary irrigation. The water harvesting methods were developed from the 13th to the 16th century by the introduction of Arabian-Persian and Spanish agricultural techniques (El Amami 1984).

Reijntjes (1986) describes the systems of water harvesting called *mescat* and *jessour* in Tunisia. The mescat system (Fig. 68) is developed between the 200 – 400 mm isohyet in a hilly landscape.

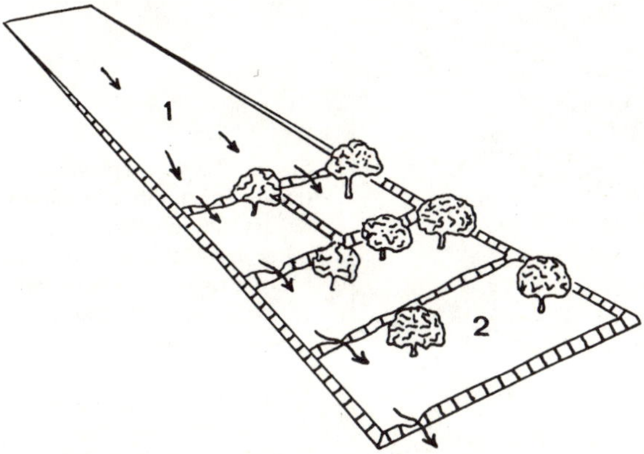

Fig. 68. The meskat runoff farming system. (Reijntjes 1986). *1* meskat: catchment area; *2* mankaa: terraces. Runoff from the catchment area (*1*) infiltrates in the cultivated terraces (*2*)

The mescat system is based on the complementarity between:

1. The upland which is the catchment area for mescat. This is often rocky with low infiltration rates and used for livestock production or forestry.
2. The slopes or lowest part or *mankaa* have a higher infiltration capacity. The water is guided and retained by construction of terraces and dams with 0.5-m-high contour bunds. Here the runoff infiltrates and makes it possible to grow crops. Nowadays, the system has deteriorated because of low prices for olive oil and the extension of olive plantation on to the mescat.

The Tunisian Government has started implementing a project to rehabilitate it since bad maintenance encourages slope erosion and flooding in the plains.

The *jessour system* (Fig. 69) is found in more arid (100–200 mm) and mountainous areas. It consists of little dams of earth and stones built across seasonal riverbelts. Soil accumulates upstream of them and terraces are built where water infiltrates. This water concentration allows fruit trees (olives, almonds, dates, figs), cereals, and legumes to be grown. The jessours require a high labor input and a large catchment area. They are now being abandoned.

In Tunisia the water harvesting systems were successful because of the high moisture retention capacity of the soils. The methods cannot be applied without improvement to the widespread red soils or alfisoils of semi-arid and semi-humid tropics, since in these soils the stability of aggregates is lower, the slaking of

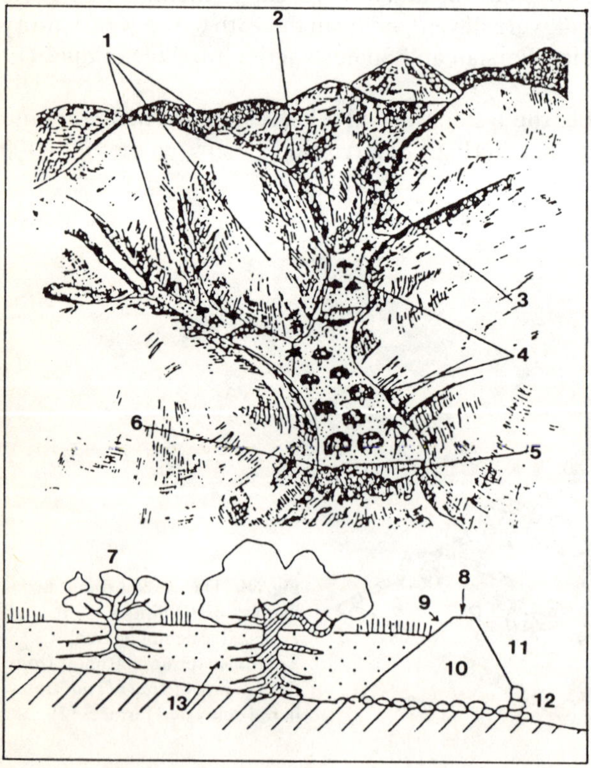

Fig. 69. View and profile of the jessour runoff farming system. (Reijntjes 1986). Runoff water and sediments from the hills are captured behind the dams (Tabia) in the valley bottom. *1* Chaabe; *2* Hernif or rills; *3* Katra; *4* Jessour; *5* Tabia; *6* Masrof; *7* Jesser or kliss; *8* Rass or thbaba; *9* Ouajah; *10* Tabia; *11* Gfa; *12* Sirra; *13* sediments

aggregates easier, and crusting more frequent, with poor infiltration resulting in increased runoff. Moreover these soils are very vulnerable to erosion.

Huibers (1985) demonstrated, according to a study of management of red tropical soils in India, that the water harvesting techniques developed in the dry Mediterranean regions are not adaptable to red tropical soils because of their low capacity for storage of water. He suggests different methods for red soil viz. graded bunds, which bring the runoff water to a reservoir located lower down the catchment. The water thus collected is then used for irrigation or supplementary irrigation of the rainfed crops during the dry season. The sediments collected are also used.

The introduction of a water harvesting system for reafforestation or other agricultural needs requires a high degree of social organization for the collective use and maintenance of the system. Water harvesting systems need high investment of land, labor, and capital, which must be justified by visible increase of yields. Water harvesting systems are of interest mainly in semi-arid areas. They must be based on the cooperation and skills of local farmers.

During the Nabatean and Israelitic periods in Palestine between 600 B.C. and 200 A.D., runoff farming was a prosperous method for the more efficient use of a scanty water supply. This technique, dating probably back to the Phoenician (pre-Roman) period, is still used in Jebel Nefousa (Libya) and in "Central Tunisia (Sousse, Matmata, Gafsa), where 10 million olive trees are cultivated with runoff farming techniques." (Le Houérou 1958).

Le Houérou (1975) quotes centuries-old olive groves in Beni Ulid area southwest Libya. The annual rainfall does not exceed 70 mm, but a large catchment basin concentrates the runoff. The system has functioned for almost two millenia. Le Houérou considers that this system has a potential equal to irrigation farming in terms of total area (in all probability 3 to 5% of the arid zone).

Runoff farming described by Orev (1988) (Figs. 70, 71, 72), is a technique which can be used in drylands to improve tree cultivation under low rainfall, even in a hyperarid ecozone (e.g., 25 to 80 mm) in the Kassala Province of the Sudan.

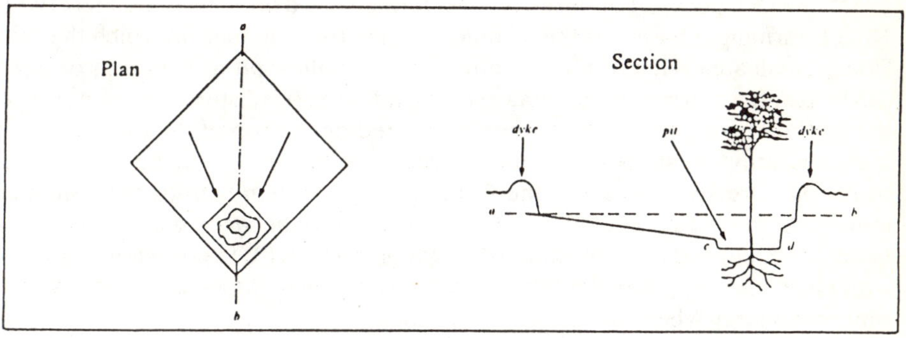

Fig. 70. A "*Nigarin*" system of microcatchment, showing the slope towards the pit (*arrow* on the plan) and the 40-cm-deep pit. Note that the base of the tree is drowned when the pit fills with runoff water. (After Orev 1988)

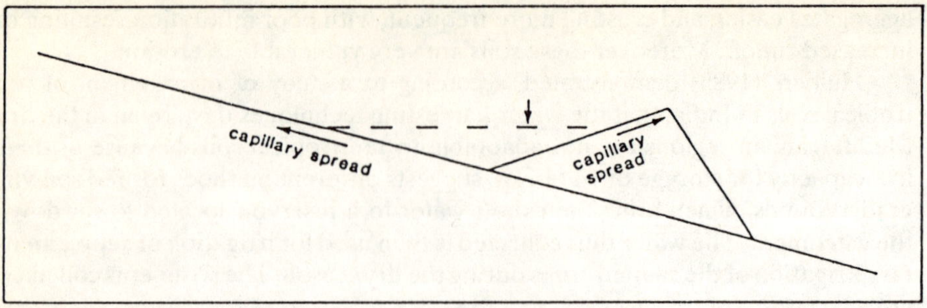

Fig. 71. A level dyke filled with runoff; the moisture spreads up the dyke and upslope by capillarity. (After Orev 1988)

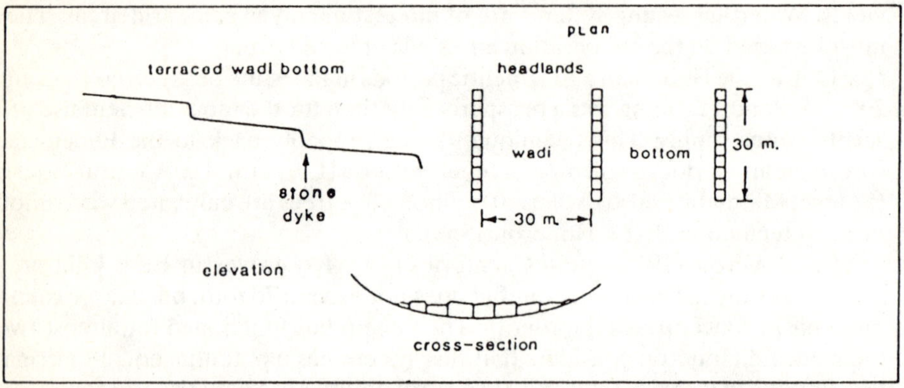

Fig. 72. Terraced wadi bottom; the top of the stone dyke must be level with the surface of the wadi bed: the slope between the stone dykes grows less steep and levels off with sediment settling. (After Orev 1988)

Rainfall which does not infiltrate is concentrated in a dam or in a small reservoir. Runoff farming is based on the technique of microcatchments in which the rain from a small area is harvested and directed to a shallow pit where crops or trees can be planted and so take advantage of this additional moisture. Orev improved the method by suggesting that trees be planted on the top of the dike to avoid lack of aeration of the roots and fungal diseases of the lower stem during flooding of the pit. Orev proposes also the Contour Dyke System with trees planted at either side of the water collection ditch. The distance between ditches and trees varies (1) to allow the use of animal draught or tractors, (2) to give enough runoff water to each tree, (3) to plant the trees close enough together so as not to waste time and energy when traveling between them.

In India more than 2 million ha of agricultural land have been equipped since 1948 with contour bunds. The farmers have not accepted the method without certain objections: to the division of the land, to prolonged water stagnation which can damage the crop, an increase in weed infestation, and a water deficit during dry spells.

The profile of the bunds can be either of basin or of ditch form (Fig. 73).

The Kambas of Kenya live in a very hilly area southeast of Nairobi, round Machakos. They have developed a system of bunds with ditches called Fanya juu which can store water on both sides of the bund. They are now introducing the use of trees planted on the lower side of the bund. The height and spacing of the bunds depend on the steepness of the slopes (Tables 30, 31).

In West Africa (Senegal, Mali, Burkina Faso) a similar system of water collection behind small dams has been introduced. The same system was developed in Kitui area (Kenya). The traditional techniques of water and soil (fertility) conservation used on the Mossi Plateau, Burkina Faso, are varied. They are summarized by Reijntjes (1986):

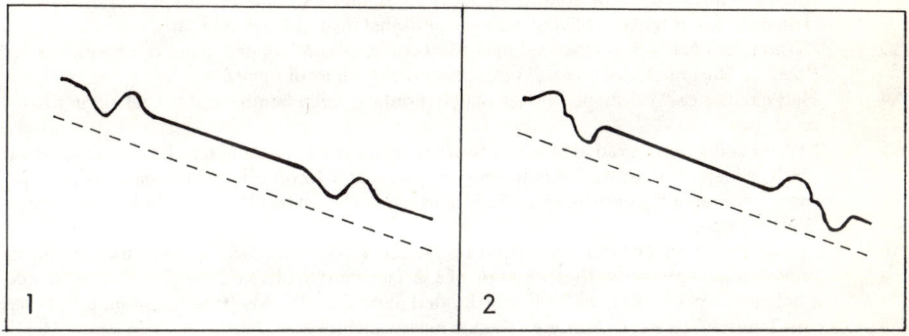

Fig. 73. Comparison between an usual type of contour bunds and Fanya Juu contour bunds. (After Ileia newsletter, no. 5 1986). *1* Contour bunds with basins; construction: 18 m per work-day; easy to cross by animals; overflow not allowed, basin has to be sufficiently large. Fanya juu bunds with ditch; construction: 16 m per work-day; cattle will be more afraid to cross the bund due to the ditch below; overflow over bund, but retained in the lower basin; the bund has to be protected by vegetation to prevent erosion

Table 30. Height (in cm) and spacing of the bunds (in m). (Reij 1983)

Height of the bunds (cm)	Percentage of slope 1	2	3	4	5
10	10	5	3.3	2.5	2
20	20	10	6.6	5	4
30	30	15	10	7.5	6
40	40	20	13.3	10	8
50	50	25	16.6	12.5	10

Table 31. Dimensions and volumes of the bunds ha^{-1}. (Reij 1983)

Height of the bund (m)	Spacing (m)	Length ha^{-1} (m)	Volume (m^3)
0.5	50	200	50
0.2	20	500	20
0.1	10	1000	10

N.B. The base of an earth bund generally is 2 times the height volume = 1/2 base × height × length.

"A — Systems with an external catchment area for collection of runoff water or flood water from small watersheds.

A1 — Agricultural use, without any special arrangement, of natural depressions where runoff or flood water is concentrated temporarily and water infiltration is relative high (traditional in, e.g., West and East Africa).

A2 — Simple techniques for water spreading and infiltration by means of low, permeable bunds (ridges) which follow the contour lines. These bunds are made of stamps, bundled sticks, crop residues or fences of living plants (traditional in, e.g., Burkina Faso).

A3 — "Water pockets" - holes for seeding, runoff collection and management of organic matter ("zai" in Burkina Faso, "covas" in Cape Verde, both traditional).

A4 — Half circular or "V"-shaped ridges mainly used for tree planting and rangeland improvement, new.

A5 — "Water collection", graded bunds or furrows are used for diverting runoff from agricultural fields, village, and waste lands to tanks situated at a lower level. This water is used for supplementary irrigation in dry periods or as full irrigation (traditional in India, new in, e.g., West Africa).

A6 — "Runoff farming", runoff water from a treated (e.g., by a chemical spray or the clearing of gravel-stones to improve the formation of a surface crust to increase the runoff) or untreated catchment area is diverted to lower-situated agricultural fields (traditional in, e.g., Israel and Tunisia: Meskat new in, e.g., Kenya: macrocatchment system).

A7 — "Runoff farming", runoff water and silt from small watersheds is captured by dams in seasonal streambeds or is diverted to agricultural fields. In front of these dams the silt builds terraces which are used for agriculture. The infiltrated water makes crop production possible (traditional in, e.g., Israel and Tunisia: jessour).

B — Systems for storage and agricultural use of flood water.

B1 — "Flood water farming": these traditional systems make use of the runoff concentrated by natural water sheds in seasonal or permanent river systems. The flood water is diverted from its natural channel by dams or barrages and led to the agricultural fields where the water is kept impounded by earthen dams around the fields. The infiltrated water is used for agriculture (traditional in, e.g., the Nile Delta before the Aswan dam; North India, "ahars and khadirs"; South Pakistan, "sailabas or kurkabas").

C — Systems with a "within-field" catchment area called "in-situ" water harvesting or "microcatchments".

C1 — "Negarin", runoff from a small plot (micro- or within-catchment) is captured at one side where it infiltrates the soil and directly contributes to the available moisture in the rooted profile of an individual productive tree or shrub (traditional in, e.g., Morocco, new in Israel and other countries).

C2 — "Contour ridges or bunds", the same system as C1 but instead of small plots, strips are used. Crops can be seeded in front of the bunds were water infiltration is concentrated (relative new in, e.g., India and Africa).

C3 — "Contourbeds", the same system as C2, but the beds are "W"-shaped, with alternating wide, and narrow ridges. The wide ridges serve as catchment zone, the narrow ridges as planting zone and the furrows can serve as drainage or irrigation channels; mechanization can be used," (new in, e.g., Brazil).

— Mulching with crop residues or dry grasses. This method of management of microclimate, soil humidity, and soil fertility is still used by some farmers.
— *Acacia albida* trees on the fields improve the microclimate. Also soil fertility is improved by the mulch of their leaves. These trees are, however, disappearing from the landscape.
— Stone bunds are used to harvest runoff water and to improve infiltration. This technique fell into disuse in the beginning of this century, but has been reintroduced by the farmers since the end of the 1970's.
— "Zai", pockets for water harvesting, seeding, and management of organic matter. This technique is still used by some farmers.
— Planting of shrubs (*Euphorbia* spp.) and perennial grasses (*Andropogon*) along the border of the fields decreases runoff considerably. This technique is still used by some farmers." (Reijntjes 1986).

Underground dams is another method for collecting water but the initial enthusiasm for them seems now to be declining: we cannot explain this. Climate is very important in assessing the potential for water harvesting. The water needs for rainfed agriculture vary. They are lower when the rain falls in the cold season (Tunisia, Israel) and only 100 to 200 mm are necessary to give a crop, and higher, because evapotranspiration is higher, when rain falls in the hot season (south of the Sahara), when 300 to 600 mm are necessary.

Water harvesting can be used in the semi-humid regions of the tropics for reafforestation and fodder grasses.

5.6.4 Desalinization Techniques in the Search for Freshwater

Among the four sources of freshwater: rainwater, surface water, underground water, and seawater, the two first are almost nonexistent or insufficient to sustain human populations in drylands. We have seen, by means of irrigation and wells, the solutions for use of underground water. Desalinization of inland brackish lake water (like the numerous lakes of North Africa called sebkhas) or of seawater is another way of providing a supply of freshwater.

Although surface and underground waters are the major sources of water, for many dry areas the distillation of seawater has been used increasingly in recent years. It is of high importance for the Middle East. (Table 32)

Table 32. Capacity of seawater desalinization plants in operation or under construction $Mm^3\ a^{-1}$. (U.N. Doc E/Conf on water resources 1977 in Zonn 1986)

Country	Power		
	Operating	In construction	Total
Bahrein	8.3	24.7	33.0
Kuwait	109.2	66.4	175.6
Oman	2.6	–	2.6
Qatar	10.4	18.6	29.0
Saudi Arabia	17.8	128.8	146.6
United Arab Emirates	2.0	–	2.0

Desalinated water supply exists in the United Arab Emirates (UAE). A total of 46 individual plants producing 25 Ml day^{-1} of freshwater were contracted for municipal use up to 1984 (Hornburn 1987). The first of these was a 0.1 Ml day^{-1} plant commissioned in 1964 for the town of Abu Dhabi. Nowadays, 16 plants are inoperative, about 30 units are producing 14 Ml day^{-1}, plus two units of 1.25 Ml day^{-1} and one of 0.5 Ml day^{-1}, which is solar-powered. All the existing techniques are used.

Many more desalination plants are programmed with a system of water distribution from shore-based seawater to inland areas lacking fresh or brackish water. At present financial limitation is not a problem in the United Arab Emirates.

The first major application for desalination was to provide freshwater to merchant and naval sailing vessels. It has been known since at least the 4th century B.C., when Greek sailors are reported to have used it.

The common desalination process is distillation by vaporizing the liquid content of seawater into steam that can be condensed and collected as pure water. Energy from a heat source is needed to boil and vaporize the water: the source can be derived from oil, coal, gas, wood, or other fuel. The least environmentally degrading sources available in arid land are solar energy and waste heat from geothermal and nuclear power plants.

Several methods of desalinization are described by Shearer (1986):

1. The use of the multistage flash technique developed in the 1950's avoids having to heat the water to 100 °C by the injection of hot water into a low-pressure vessel.

"Hot, salty water is introduced into a chamber at such a pressure that part of it will 'flash' into vapor which is then condensed as freshwater. The remainder of the salt water, now slightly cooler, then passes to a second chamber at a lower pressure so that more of it is evaporated. This is repeated in a number of stages – usually around 20 to 30 – until the salt water is quite cool." (Davies 1987).

2. In the compression distillation technique the vapor produced by the saline water is compressed — thereby increasing its temperature — and recycled through a heat exchanger to heat more seawater to steam. The energy needed is more for compression than for heating.
3. Freezing to -3 or -4 °C and washing out the salts is another method of extracting the water content. There the maximum energy needed is to freeze the saline water.
4. To extract the salt by using semi-permeable membranes is another technique based on forced filtering and reverse osmosis, i.e., the opposite of the natural tendency of freshwater to diffuse through a semi-permeable membrane into saline water. Reverse osmosis "depends on the ability of semi-permeable membranes to let water molecules through but to restrict the passage of salt molecules." (Davies 1987). This method is complicated: the filter must be fine enough to remove the salt, but strong enough to withstand the pressure

forcing the water through the filter. Reverse osmosis is an economic alternative to distillation.

5. Electrodialysis is a method which concentrates the ionic salt components in separate compartments leaving freshwater behind. The electric nature of salt molecules makes it feasible to extract the ions from the salt water by electrical forces using separate filters for the positively and negatively charged ions.

Ion exchange:

6. The saline water is passed through a solid cations exchanger, the organic resine exchanges its hydrogen ions H^+ for positive ions of the salt molecules in the water, the water becomes acidic and a second passage is necessary on another resin to exchange negative ions for hydroxyl ions OH^-. When the resins are exhausted they are washed, the first resin with an acid and the second with a base. This type of regeneration is prohibitively costly.

Desalinization techniques are now perfected and currently used to produce freshwater, but the cost is still too high for agricultural purposes because this would require very large plants which call for large amounts of energy, which only oil- or mineral-rich countries can afford. It is of little help for land-locked countries such as the Sahel. This source of water is too expensive for ordinary agricultural production but acceptable for drinking water or very high value crops, e.g., vegetables.

"During the Sudano-Sahelian drought, in the early 1970's, water supply to the Mauritanian capital, Nouakchott, was provided by seawater desalinization. When the water requirements, for economic reasons, exceeded one $Mm^3\ a^{-1}$ steps were taken to provide a supplemental water supply from groundwater sources. Presently, another seawater desalinization plant is being considered for construction for Noudhibou, Mauritania." (Zonn 1986).

North Africa has also developed some large-scale plants (Table 33).

Table 33. Desalinization plants in north Africa. (Zonn 1986)

Country	Location	Desalting method	Productivity, Ml day^{-1}
Libya	Benghazi	Electro-dialysis	19.3
Morocco	Ceuta	Multi-staged flushing	4.0

5.6.5 Conclusion

The section dedicated to water consumption and irrigation seems very pessimistic. It is more a critical analysis of the most prominent proposals and techniques than a selection of solutions.

The objective was to show that in many irrigation schemes physical and biological difficulties arise due to economic, social, and political factors.

Many irrigation projects have been abandoned because of (1) inadequate attention to the needs of either the individual growers or the responsible organizations at the regional or national level, (2) land degradation.

"The performance of irrigated farming in 25 years of independence is also sobering. In the Sahelian states of West Africa, the amount of new land being brought under irrigation each year (around 5000 ha) is roughly equal to the amount being abandoned each year because of neglect and lack of maintenance (Club du Sahel 1983, p. 34)." (Eicher 1986).

Proper attention to the sociological aspects of well-organized management and maintenance structures are necessary for sustainable success.

It was essential to give these warnings but it has also to be pointed out that well-managed irrigation systems are the solution to replacing unproductive dryland, at risk from climatic stresses, with productive plantations, high yields, and fast-growing crops.

Shmueli (1973) observed than less than the half of the 500 million ha of drylands potentially suitable for irrigation are actually used for this purpose. In 1800 10 million ha were used. It became 40 million ha by 1900, 160 by 1950, and 200 million by 1969.

5.7 Maintaining Soil Quality

5.7.1 Introduction

The soils of drylands are especially vulnerable and difficult to rehabilitate. The reasons are lack of weathering of rocks, and scarcity of organic matter, slow colonization of soil by plants, and almost nonexistent infiltration so that accumulated salts cannot be washed away. Thin and coarse, the soils hold little water and few nutrients. When sandy or silty, the water-holding capacity of soils is better but then they are vulnerable to wind erosion and the topsoil, which is poor in organic matter and can be easily eroded, when deeper horizons revealed. When the B horizon is a crust (salt, carbonate, iron, aluminum or silica) the soil becomes useless. Beckman and Coventry (1987) have shown the predominance of such soils in dry Australia. Deep plowing reveals them in Algeria and Israel (Noy-Meier and Seligman 1979).

The soil-forming mechanisms are reduced by aridity and increased with increasing humidity. This applies to the vegetative cover, the organic compounds associated with it, leaching, and hardpan formation.

In semi-arid and semi-humid ecozones the soils are more developed but hardpans formed under wetter conditions are frequent. If erosion occurs they are denuded. This phenomenon is the rule in the Sudanian ecozone (e.g., Guinea, Mali) where, during the last three decades, outcrop of iron crusts have become detectable on air photographs.

The overgrazing impacts on the soil are as diverse as on vegetation: soil shattering and loosening trigger deflation processes mainly in sandy soils, shifting sands soon follow and a hummocky active dunes pattern topography develops. Yet moderate soil loosening by livestock can have a beneficial effect by improving soil aeration and incorporating seeds: around the wells in the Ferlo (Senegal), the vegetative cover is denser after a good rainy season, the variety of species higher near the well; trampling and burying of animal dung containing seeds are the mechanism by which this is brought about.

Overgrazing can, in heavier soils, cause soil compaction and therefore decrease infiltration and increase runoff. As a result, soil fertility decreases and the vegetation suffers soil moisture shortages.

Soil erosion, like deflation, scalding, or soil compaction causes the productive capacity of rangelands to decline and the pasture areas to shrink as some sections become unsuitable for pastural use.

Overgrazing and overuse for agricultural purposes are not the only causes of soil degradation. Cutting of trees and shrubs for firewood leads to intensive deflation and the expansion of shifting sand areas around towns, oases, wells, and human settlements, where all combustible plants tend to be cut far beyond recovery.

Industrial and urban development in drylands can also produce adverse environmental effects. They involve creation of new settlements, construction of roads, gas oil fields, irrigation work. All these practices bring to dry areas heavy equipment and machinery with often uncontrolled traffic and roads, which take up extensive areas of productive grazing lands, break the soil surface, and set in motion sand deflation foci.

Dryland roads can become several kilometers in width, visible even on satellite images, each driver creating his own track. Near wells and human settlements such roads converge in a devastated star.

Dzhanpeisov and Dzhamalbekov (1978) summarize the changes in a range ecosystem in Kazakhstan caused by soil erosion (Table 34).

In Africa, from the desert boundary to the Sahelo-Sudano ecozone, the types of soil degradation differ.

Wind erosion occurs on sandy steppes or sandy savannas with winnowing effects of grains smaller than 400 μm (in Niger and Mali), removal of the topsoil with development of pavements of coarse particles or outcropping of various kinds of hardpans (calcareous gypsums are dominant in the northern side of the Sahara, ferrugineous or aluminous hardpans south of the Sahara) or the land is eroded down to hard geologic substrata. Such hardpans or bare rocks remain barren even under wetter climatic conditions.

Water erosion, sheet erosion, and rill and gully erosion interact with wind erosion in the Sahelian ecozone. The problem increases with steepness of slopes. Badlands appear in the most serious cases. Sealing of the bare surface and mainly of the B horizon by raindrop splash make the soil surface almost impervious. The infiltration decreases as the runoff increases, resulting in the death of perennial shrubs. The removal of perennial plants results in a sharp decrease in organic

Table 34. Water erosion and deflation impact on rangelands. (Dzhanpeisov and Dzhamalbekov 1978)

Elements of range ecosystem	Impact	
	Erosion	Deflation
Vegetative cover	Reduction of soil depth and loss of soil nutrients; xerophytization and degradation of vegetative cover; washoff and burial by runoff water	Reduction of growth basis and volume of nutrients; xerophytization and inhibition of all species; burial and cutting by flowing sands
Root system	Exposure; reduction of root mass	Exposure; reduction of root mass
Sod	Wash-out and landslides	Burial
Upper horizon	Sheet and rill erosion, greater amounts of stones	Blowout, coarsening of texture, aeolian forms of topography
Humus soil	Reduction of depth, organic-matter content, nitrogen, and other nutrients	Diminution of volume and reduction of fertility
Hydrophysical soil properties	Decrease of moisture-holding capacity, soil permeability, and strength of soil aggregates	Reduction of moisture content, destruction of soil aggregates
Underlying rocks	Denudation wash-out	Denudation
Groundwater	Lowered water table, reduction of ground flow	Decrease or increase of water table

matter which has many feedback effects on structure, chemistry, and biology of soil components, in particular in the ratio between humic and fulvic compounds, reduction in soil porosity, permeability, stability, water intake, storage capacity and oxygenation, and therefore in microfauna and flora.

"We believe that of all threats to sustainability, threats to soil fertility are the most serious. These are the threats of wind and water erosion and the salinization of irrigated land. Waterlogging is a major constraint on current production, but not a long-term threat to sustainability." (Warren and Agnew 1988a).

Salinization and its consequences were analyzed in the previous chapter. This chapter will concentrate on water and wind erosion. Water erosion is often considered as the main physical mechanism of soil degradation. I cannot agree with this point of view. In drylands, wind erosion must be taken seriously and can be a more dangerous threat than water erosion up to the 400-mm isohyet. In addition, water erosion is relatively easy to observe, quantify, and cure, which is not the case with wind erosion: the diagnosis is less easy and the remedy more difficult to apply; for these reasons, water erosion solutions will be treated briefly and wind erosion control at greater length.

5.7.2 Halting Water Erosion

Halting water erosion entails:

1. Counteracting sheet-wash and rillwash.
2. Gully reclamation.

5.7.2.1 Control of Sheet-Wash and Rillwash

Soil particles, as we have seen, can be moved by runoff on slopes without creating any rill in the case of sheet-wash. Ephemeral or perennial rills appear with rillwash.

Three types of remedy exist:

1. Contour strips (Fig. 74), which are the most effective method to control erosion.

   Trees and shrubs in combination with grasses are planted along contour bands. The method is applied on slopes where the erosion problems are not urgent. The width of the strips and the distance between them are determined by the amount of rainfall and the degree of steepness of the slope.

   Table 35 can be used to calculate the spacings.
2. Terracing (Fig. 75) is used when water erosion requires stronger measures to check it and when the main objective consists in catching precipitation

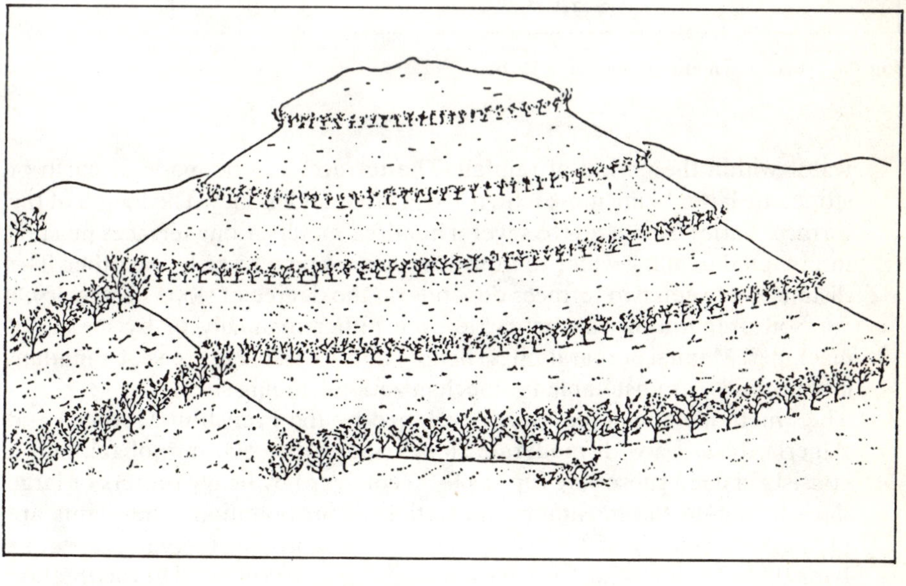

Fig. 74. Contour strips. (Anen-UNEP 1988)

Table 35. Width and distance of strips with respect to slope value. (Anen-UNEP 1988)

Slope	(W) Width of strips	(D) Distance between strips
0	2	50
5	4	47
10	5	43
20	8	38
30	10	33
40	13	28
50	17	24
60	20	20

Basis 0–600 mm mean annual rainfall.
In areas with rainfall between 600 and 1000 mm increase W by 20%, decrease D by 10%.
In areas with rainfall of over 1000 mm increase W by 50%, decrease D by 20%.

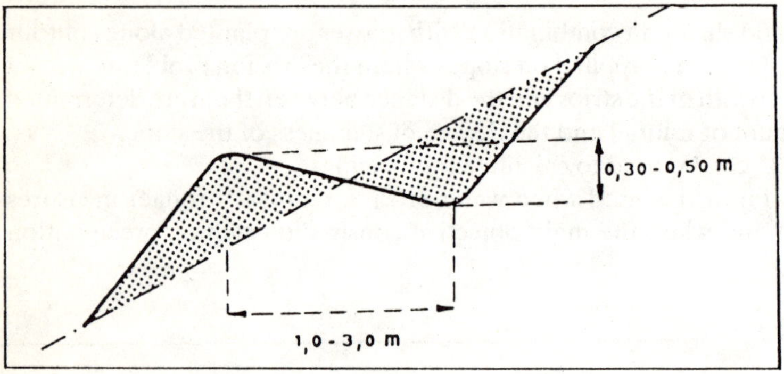

Fig. 75. Terracing methods. (Skoupy 1976)

water within the period of rainfall. The terraces can be made of earth or stone, the latter being more expensive in labor and money. The length of the terraces varies from a few to several hundred meters. Long terraces must be interrupted to allow water to flow down from one terrace to the other. The distance between two terraces depends on the degree of slope (Table 36).

Soil preparation can be carried out either manually or by means of machines. Manual preparation is cheaper and advisable in areas of abundant labor or in areas unsuitable for mechanization (sloping sites).

3.. The "méthode steppique" was developed by the French in Morocco and Algeria. In areas with moderate slopes and 300 to 500 mm of rainfall, it consists in deep subsoiling (up to 60–70 cm deep) by heavy rooters or large discs to obtain the maximum infiltration of precipitation. The plants are planted at a spacing of 3 × 3 or 4 × 4 m. Often subsoiling is combined with a bank 0.5 to 1 m high and 2 to 3 m wide. *Eucalyptus* was planted by this method in the Mamora region (Morocco) and *Pinus Alepensis* in eastern Morocco.

Table 36. Distance of terraces with respect to soil gradient. (Skoupy 1976)

Slope gradient %	Distance between terraces (m)	
	Horizontal	Vertical
3	67.0	2.0
6	42.0	2.5
10	30.0	3.0
15	23.0	3.4
25	16.0	4.0
35	13.0	4.5
50	10.0	5.0

5.7.2.2 Gully Reclamation

Two basic methods predominate in gully reclamation:

1. Trees and other vegetation can be planted in the gullies. The vegetation helps in the formation of a soil layer. Trees are not recommended at the bottom of the gully because they might narrow the channel and increase the speed of water and therefore the erosion.
2. Changing the topography of the bottom of the channel by creation of a staircase profile. In Somalia in the gullies cutting the fixed dunes, a trial of check structures with jute bags filled with sand was achieved. Another method consists of building check structures or small dams in gabions or piles of stones so that water will flow through them and thus slow down its speed. Trials with *Opuntia* were also made.

5.7.3 Control of Damaging Wind Effects

5.7.3.1 Methods for Preventing Progression of the Barchan

Four methods of halting barchan migration are feasible:

1. Fixing the sand in the source areas.
2. Fixing the barchan itself.
3. Creation of an artificial dune.
4. Destruction of the barchan.

1. *Fixing the Sand in the Source Areas.* A source area can be fixed if it is a small area, but this may be more difficult where the source of sand is large or in a region far upwind of the dunes, perhaps in another ecozone. The source may also be the surface along the path of the barchan. For example, the El Kharga oasis is threatened by the entire desert upwind of Kharga because the substratum itself

has produced loose sand by corrasion or deflation. This is common where there is a basement of sandstone or fluvial deposits.

2. *Fixing of the Barchan Itself.* All the methods used for fixing the dune itself are based on the same principles. They include fixing the windward face of the dune to prevent sand being carried from there to the leeward face. The technique used depends on the biological environment and the availability of inexpensive local products such as palm fronds or other dry vegetation. Plastic nets and chemical products are generally expensive and should be avoided.

The windbreaks must be erected as a consecutive series of parallel bands, on a grid pattern, or placed one on top of another as the sand accumulates and they must be sited at right angles to the wind direction (i.e., to the axis of symmetry of the barchan). Note, however, that a single set of low barriers such as brushwood placed perpendicular to the wind direction will be quite insufficient and will be submerged by the advancing dune.

Dune stabilization using revegetation is the second and best step method because it ensures effective long-term protection against wind, with the plants acting as a natural trap for airborne soil. However, plant introduction requires care, and plants adapted to local ecological conditions should be established in nurseries. Failures have occurred by sowing tree seeds directly on dunes: germination takes place while the sand is moist, but the abrasive effect of moving sand kills the young trees. The sand should be immobilized long enough for the establishment of vegetation, but once established, plants will improve soil conditions by lowering soil temperature, anchoring the soil, and improving soil structure. However, soil moisture will fall owing to evapotranspiration, although this will be offset to some extent by the increase in soil organic matter which tends to have a high moisture retention capacity. Fertilizers and irrigation may be necessary and it is important to exclude livestock during the whole establishment period.

3. *Creation of an Artificial Dune.* The creation of an artificial dune occurs when the arriving barchan is broken down by the use of temporary physical barriers erected in parallel, or by placing permanent palisades in a lattice work, as described above. It must again be emphasized that a single barrier constructed downwind but at right angles to the axis of symmetry of the barchan is useless, as it will be buried by the advancing dune. The resulting artificial dune grows and as it does so, the fences must be raised. This stage must be followed by biological dune fixation as explained above.

4. *Destruction of the Barchan.* The traditional and least expensive method is to scatter large piles of rocks on the windward side of the dune. An alternative method is to change the profile of the dune by causing a dispersing of the sand. This technique has been used in the area of Mhamid (Morocco).

5.7.3.2 Control of Seif Dunes

Control of seifs is difficult because the dunes elongate at an oblique angle to the prevailing winds. Three methods are feasible:

1. *Preventing the Sand from Arriving at the Dunes.* This is achieved by reafforestation of the area upwind of the dunes. If the source area is too large and if reafforestation or revegetation is not possible owing to lack of water, an alternative is to deflect the wind by placing shelterbelts perpendicular to the wind direction NOT perpendicular to the dune.

2. *Halting the Longitudinal Migration of Sand Using Barriers.* Barriers are placed in a herring-bone pattern perpendicular to the prevailing wind in the corridors between two dunes (Fig. 76). These intercept and deflect the sand. They can be constructed from local materials such as millet or grass stalks, twigs, brushwood, or palm fronds, but shelterbelts made from living vegetation are better. The distance between two barriers should be ten times the height of the fence. This method is being tried in Shalambot, south of Mogadishu in Somalia, where barriers of trees such as *Euphorbia biloculata* and *Commiphora* are being used.

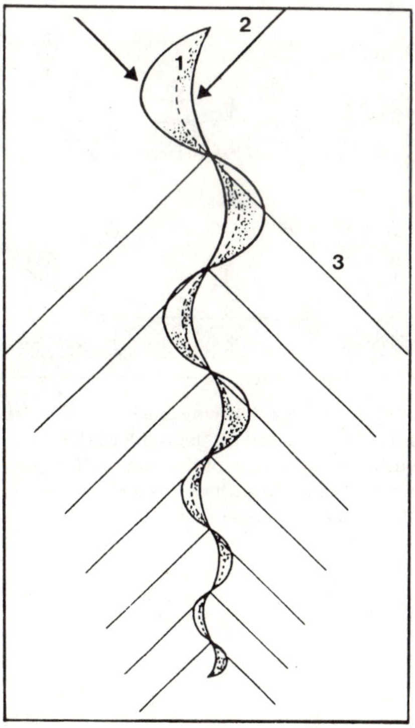

Fig. 76. Stabilization of a seif dune by a herring bone palisade. *1* Linear dune (note that the head of the dune often has no angular crest); *2* directions of prevailing winds; *3* wind barrier perpendicular to the prevailing winds

3. *Fixation of the Dune Itself.* This can be done independently or in addition to (1) and (2), since revegetation with cactus and euphorbia is primarily for the purpose of sand fixation and has little economic value: tree planting is preferable where practicable.

It consists of revegetation in three stages (Fig. 77).

a) In the first stage, living barriers are established to reduce the abrasive effect of moving sand on young trees.
b) Two or three years later, pioneer trees should be planted either in the corridor or in the flat land between the linear dunes.
c) The third stage entails planting the lower third or half of the slope leaving the upper part of the slope bare for several years to allow wind to flatten the top of the dune by deflation. When the dune is flattened, then trees should be planted on the surface.

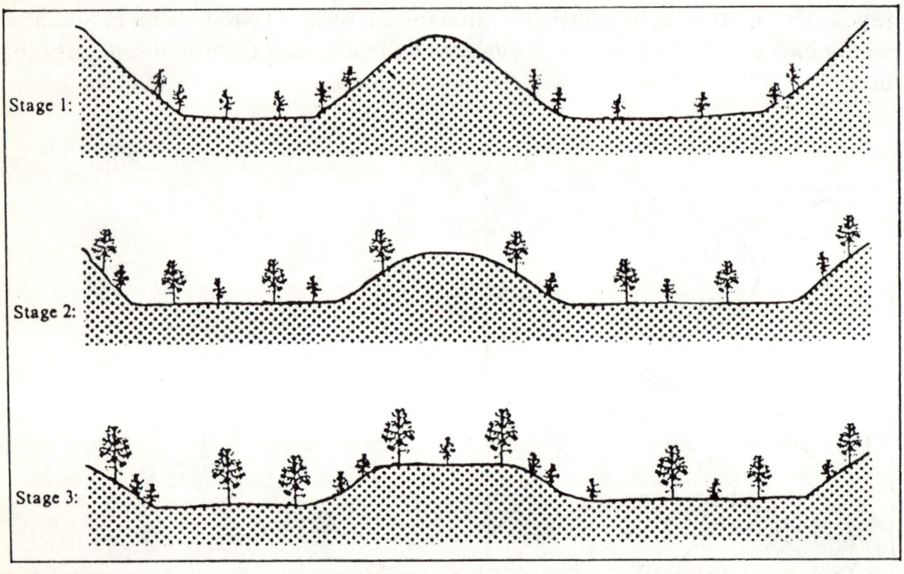

Fig. 77. Different stages of reforestation on linear dunes. *Stage 1* Living bush barriers are planted; *Stage 2* planting of the interdunal areas, with shrubs planted on the flank of the dune; *Stage 3* planting a third or a half of the flanks of the dune while the top of the dune is flattened by wind deflation. The flattened top is planted. Since revegetation with cactus and euphorbia is primarily for the purpose of sand fixation and has little economical value, tree planting is preferable.

5.7.3.3 *Strategies for the Control and Prevention of Sand Encroachment*

In this section, methods to keep roads and settlements free of sand are discussed. In addition, other physical and biological techniques are reviewed. All these methods for prevention of encroachment should be simple, cheap, and effective.

1. *Road Construction in Areas of Drifting Sand.* There are three elementary precautions that must be taken in the design of roads where encroaching sand is a threat.

a) Sections of the road (or railway) should not be sited at right angles to a strong prevailing wind. If this is not possible, the surface of the road should be inclined upwind to create a Venturi effect where the constriction of flow causes an increase in wind speed. Over a smooth, nonerodible surface, this constriction and slightly reduced friction result in a road-sweeping effect, keeping it free of sand.
b) Roads must be built on flat terrain to ensure that the surface and the sides of the road are smooth. This will increase surface wind speed and avoid sand deposition. Note, however, that two apparently contradictory methods can be used: one is designed to hasten the movement of sand over the road while the second prevents the sand from getting onto the road in the first place.
c) In dry, sandy areas, a 150-m-wide strip on each side of the road must be protected against grazing, wood-cutting and disturbance of the soil surface during the construction of the road. This procedure is routine on coastal roads in Denmark.

2. *Prevention of Sand Deposits on Existing Roads.* A sand deposit on a road can result from sand grains blowing by saltation, by the migration of barchans, or by the elongation of seif dunes created by human-made obstacles such as heaps of subbase material, wrecked cars, or public shelters.

3. *Halting Migrating Dunes.* A migrating dune must be stopped at a distance from the road which should be 10 to 15 times the height of the dune. There are four methods of halting the dune:

1. by placing loosely scattered durable objects on the surface of the dune. This will create vortices strong enough to scatter the sand and accelerate its dissemination;
2. erecting fences with a gap at the base. This will increase deflation by creating a turbulent flow with strong upward air currents;
3. creating a stable artificial dune that can be established in order to halt the moving sand upwind of human settlements. The creation of such dunes should precede biological fixation of natural sand dunes;
4. by application of the Venturi principle. This can be applied to the Somali coastal dune system as well as to the road-sweeping effect for clearing roads of sand. Both uses will be discussed in the next section on case studies.

4. *Creation of an artificial dune to stop encroaching sand.* Plantations, villages, and fields located in the transport area of wind-blown sand can be endangered by encroaching sand in the form of moving barchans or elongating linear dunes. In areas where strong winds (greater than 4 m s^{-1}) are responsible for deflation and sand flow, even human settlements located on a sand flow can become an obstacle initiating sand deposition, but an artificial dune (Fig. 78) can also be used to stop large quantities of sand from arriving. The object is to encourage sand accumulation upwind of settlements. This method requires building a palisade or wall at right angles to the main wind direction 200 m upwind of the area to be protected. The palisade forces the sand to accumulate and will eventually be buried by sand deposition. Then the first palisade can be lifted or another palisade put on the top of the first. Alternatively, a parallel palisade can be put up 100 m from the settlement (in front of the first palisade).

The optimum porosity for the palisades is about 50%, as this prevents reverse flow and sand accumulation behind the fence or "digging" in front of the fence.

The material for these should be cheap and locally available, e.g., palm branches, rice straw, small branches, or bushes. Fibro-cement barriers can be used if the area to be protected is small. Each section should be about 1 m × 0.9 m and have 150 holes each with a diameter of 3–4 cm. The sections should be lifted once or twice a year. When the fences are three-quarters buried by drifting sand, they can be raised several times or another fence can be added upwind. The distance between the fences must be 15–20 times their height.

The artificial dune is built from natural aeolian sand by first reducing wind speed so as to stop particle migration and halt the barchans. When no further sand accumulation takes place on the created dune, a second artificial dune must

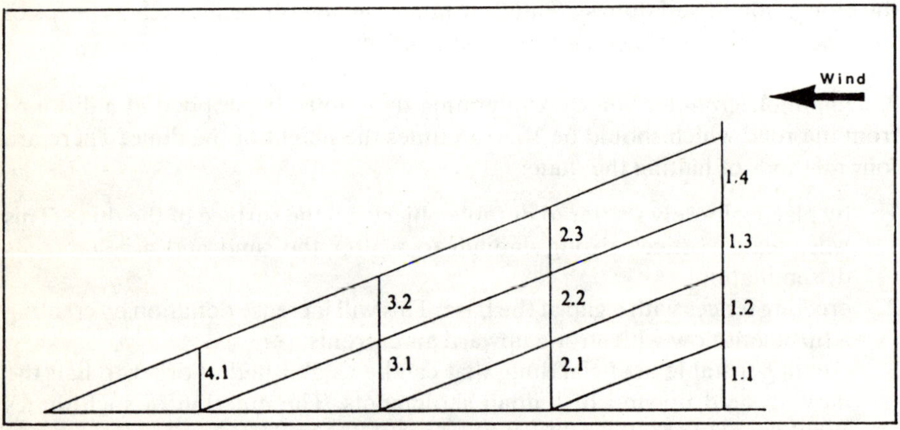

Fig. 78. Phased installation of windbreaks and creation of an artificial dune. Phase 1. Wind barrier *1.1*; Phase 2. Wind barriers *1.2* and *2.1*; Phase 3. Wind barriers *1.3*, *2.2*, and *3.1*; Phase 4. Wind barriers *1.4*, *2.3 3.2*, and *4.1*. The high point of the artificial dune gradually shifts upwind towards the center line. The wind barrier must be fixed from upwind to downwind

be created upwind of the first. This process will have to continue indefinitely as long as nothing is done about the sand in the source area. When the artificial dune no longer collects any more sand, revegetation becomes important because the spontaneous regeneration of plants is conducive to sand stabilization.

5. *Protection of Sandy Areas with a Multidirectional Wind Regime.* Areas with variable winds are rare but they do occur in semi-arid ecozones where the wind direction in the dry season is different from that in the rainy season. The method of controlling wind erosion in an area with such a regime is to cover the whole area with a lattice work of partially buried squares of fencing of 4 to 16 m^2 in size.

One hundred m^2 is needed per ha (UNEP-UNEPCOM 1986a). The fences should be 25–30 cm high and 4–5 cm thick, and can be made of straw or local grasses. The fences are perpendicular to sand transfer.

At Shapotou, in the Tenger desert of China, this method was used. Trees were planted together with shrubs such as *Robinia pseudoacacia, Eleagnus angustifolia, Populus cupidata, Hedyseoparium, Cargana korshinskii,* and *Salix cheilophila.* In another area of the same desert *Hadybarum sciparium, Calligonum mongolicum*, and *Salix flavida* were used.

These plantations in the Tenger have proved successful as a result of a number of factors such as the establishment of a research center, the use of irrigation water rich in silt (400 g l^{-1}), surface biocrusting, active soil formation, and the persistant efforts of the Chinese.

6. *Methods for Altering Wind Speed to Avoid Sand Deposition.* Roads, railways, and water channels are all obstacles to wind and sand transport whether they are above the surface or in a cutting, particularly if these structures are at right angles to the wind. In order to prevent sand accumulation on roads or railways, there are two strategies which can be adopted:

1. Saltating sand can be stopped before it reaches the road or railway by the construction of an artificial dune or barrier 25 m before the obstacle. This barrier can be built of any available material such as stones, branches, or even old vehicles.
2. Another strategy is to accelerate sand transport over the road or railway. This method is based on the principle that when the wind is blowing down a slope it decreases in speed because the streamlines are expanding. When the wind is blowing up a slope, the streamlines are compressed and the wind speed is increased.

If the road or railway is built on a gentle incline (no greater than 5°) in an upwind direction, and if the leeward side of the road is also slightly inclined, then this will prevent sand deposition because the wind will then sweep the road or railway.

In the case of preventing sand deposition in water channels the solutions are somewhat different. The water channel should be built as parallel to the wind as possible. Owing to the presence of water in the channel, sand will accumulate whether or not it is above or below the surface topography. The channel should therefore be as level as possible with the surface.

7. *Physical and Biological Techniques for Sand, Sand-Sheet and Dune Stabilization.* Stabilization of sand sheets requires fixing the sand for long enough to enable natural or planted vegetation to establish itself preferably without watering. The objectives are to stop the abrasive action of sand on recently germinated plants and to avoid their being buried by drifting sand.

Physical stabilization means using nonliving material such as branches, palm fronds, weeds, mats, sacks, fibro-cement, wooden planks, or corrugated metal for the construction of barriers. This method is not ideal because of their limited lifespan, so they are only used to give immediate protection around villages, communication lines, transportation routes, and arable land. They represent an interim measure until vegetation can be estalished.

8. *Construction of Physical Barriers (Table 37).* Long, low barriers 1 m high made from any available material should be erected. If the wind regime is mono-directional, the barriers must be erected at right angles to the wind direction and if bi-directional, cross-barriers must be erected, each at right angles to the main wind to form a grid pattern.

The barriers must be maintained until vegetation on and around the area which is being fixed has become established; natural vegetation can only colonize the sand if the sand is completely immobilized and only if the surrounding vegetation will produce seeds.

Table 37. Guidelines for the construction of palisades

Rainfall zone	Areas below the 500-mm isohyet
Objective	To decrease wind speed at the soil surface and to avoid sand arrival in gardens, plantations, and villages, and to promote sand deposition
Direction	Checkerboard structure perpendicular to the main wind direction(s)
Distance	Ten times the height of the barrier or 2–3 times the height on a windward slope or on a dune slope
Material	Any available organic or inorganic material

9. *Mulching.* Dense vegetation is the best protection against wind erosion, so if plant cover is depleted, the soil is vulnerable to erosion, especially if winds are strong and the soil is loosely packed and dry for most of the year, and if the terrain is flat, allowing deflation. Mulching involves evenly covering the sand or dune with materials in order to break up the smooth surface of a bare field into a rough surface, thereby slowing down the wind.

Organic mulches such as plant residues are best because they are often locally available and inexpensive. Leaves, twigs, branches, or grasses can be laid on top of the soil surface or dug into the soil. Branches can also be tied up with wire, rope, or strong grasses into small panels for covering the ground.

Man-made materials include sheets of plastic for small areas or acrylic fiber (50 kg ha^{-1}) or acrylic wire netting for larger areas. The latter comes in rolls of two widths and three mesh sizes (2.5, 3, or 5 mm). Oil byproducts such as asphalt or kerosene (paraffin) and various chemicals can be used. All these materials are often difficult to obtain and very expensive and they often have limited effectiveness. For instance, asphalt forms a protective layer on the sand surface, which is vulnerable to breakage by animals. Kerosene is the most widely used and it has proven to be more effective than asphalt because it can penetrate the topsoil to a depth of 10–20 cm, its effects last for 3–4 years, and it still allows vegetation growth while stabilizing the sand surface (see Table 38).

Chemical sand stabilizers have been used to fix slopes and coastal dunes in temperate zones, but have not been used in dry zones so they should be tested first to determine their capacity to form erosion-resistant aggregates and to increase water retention to stimulate plant growth. They should also be cost-effective, easy to apply, nontoxic, and biodegradable. Some commonly used chemicals are listed in Table 39.

Table 38. Guidelines for using kerosene mulch

Heat the kerosone to 50 °C and scatter on the sand, sand sheet, or dune surface		
Quantity	3–4 m^3 ha^{-1} with rainfall	< 300 mm a^{-1}
	2–3 m^3 ha^{-1} with rainfall	> 300 mm a^{-1}
	5 m^3 ha^{-1} with high dunes	

Table 39. Some chemicals used as sand stabilizers

Name	Form	Application	Effects
SAND STOP	Powder	Spray at 20–25 g m^{-2}	Absorbs moisture to become an homogenous coating
URESOL 156E	Liquid sprayed at concentration of 3.4–6.4% at 1.5 l m^{-2}		Penetrates down to 4–5 mm and forms aggregates
URESOL 310EA	Yellow liquid applied as above		As above
HULS 801	Dark brown liquid or white emulsion applied as spray. 10–40 g in 2 l water; 10–50 g m^{-2}		Hardens in 2 h
AGROFIX	White liquid sprayed at 30–50 g m^{-2}		
UNOSOL	Latex-based. Sprayed with fuel mixed with water at 150 g m^{-2}		
TEXAND D	Net of very fine fibers scattered on slope like a spider's web. 5 g m^{-2}		

10. *Permanent Stabilization by a Vegetative Cover.* Permanent stabilization of dunes can only be achieved by the establishment of vegetation cover by one of three methods:

— natural regeneration;
— semi-natural regeneration;
— replanting programs.

a) Natural regeneration. This is the least expensive method. Where human over-exploitation has destroyed the vegetation cover, further degradation can be prevented so the land can return to its natural state. The time it takes for regeneration depends not only on the effectiveness of efforts to protect the area (by such measures as fencing) but also on rainfall and soil parameters such as soil organic content and salinity. Experience with natural revegetation in Burkina Faso has shown that in areas with a rainfall of greater than 300 mm a^{-1}, 2 or 3 years of protection have sufficed. However, if annual rainfall is only 100–120 mm, as in the case of southern Tunisia, 10 to 12 years of protection is necessary for complete regeneration. Even below the 80-mm isohyet, natural revegetation is still possible because protected sand can retain moisture at a depth of 50–60 cm even in the height of the dry season.

b) Semi-natural regeneration. This method involves providing inputs to the degraded area in the fort of reseeding with xeromorphic plants or drought-resistant grasses to complement natural revegetation. Irrigation or fertilizer application is sometimes carried out. The area must be protected before regeneration can occur.

c) Replanting projects. In some situations the natural vegetation cover is completely destroyed and regeneration is impossible. Establishment of new vegetation cover on an active sand sheet or dune field may involve any one of a number of options. These include:

— seeding with or planting of appropriate drought-resistant species;
— irrigation in the early stages of growth;
— soil improvement, e.g., in intradunal corridors where alkalinity or salinity is a problem;
— physical stabilization, e.g., fences.

Seedlings must also be protected from animal trampling.

Successful regeneration can be brought about by breaking the soil crust by superficial tillage after rain and subsequent reseeding; the seeds will be trapped by the broken surface and by grass, bushes, stones, and other obstacles. This can be followed by mulching; germination and survival of the young seedlings will be enhanced because breaking the soil crust leads to a greater infiltration of rainfall and an increase in soil moisture (Tables 40, 41).

11. *Revegetation with Trees and Bushes.* The most effective biological techniques of sand, sand sheet, or dune stabilization involve revegetation, which

Table 40. Guidelines for grass cover rehabilitation on sandy soils

Soil	Sandy
Action	Superficial breakdown or loosening without crumbling of the soil and slight packing to firm the soil
Location	On sandy slopes according to the contour lines
Time	Onset of rain
Tool	Cultivator — roller type whose effect is to scarify with slight packing
Effect	Soil loosening without breaking and improvement of infiltration and regeneration of grass

Table 41. Guidelines for grass cover rehabilitation on sandy soils which have lost the A-horizon (topsoil)

Soil	Degraded sandy soil with outcropping of compacted and smoothened horizons
Action	Ploughing
Ridging	From 3 to 10 m, perpendicular to run-off directions. Dense ridging in the first year is not recommended and can aggravate erosion
Tool	Blade plough
Effect	Infiltration is poor because of compaction, and must be improved
Results	Runoff is collected in the ridges where it penetrates the soil and is conserved longer at the beginning of the dry season. Grasses regenerate spontaneously, grow better and stay green longer

provides both a protective canopy and plant residues for the soil which reduce soil loss by trapping sediment. A canopy intercepts turbulent pulses of wind and reduces the erosivity of rainfall at the soil surface. Soil loss is proportional to the kinetic energy which reaches the soil and surface residues.

Trees are particularly effective in protecting against water and wind erosion. Not only do they enhance rainfall infiltration but they increase soil organic matter through leaf fall. They also recycle nutrients which have been leached to lower soil horizons and they may also produce cloud formation and rainfall.

In semi-arid ecozones, especially on sandy soils, the use of trees when scattered in fields is particularly beneficial in reducing wind erosion. Empirical evidence has shown that a density of one tree per 15 m^2 increases yields of millet, reduces deflation, and increases soil moisture retention. Recommendations for the use of *Acacia albida* under such conditions are shown in Table 42.

Other uses for trees such as *Acacia albida* include (Table 42)

— establishment of living fences, a cheaper and better alternative to nonliving ones;
— provision of fencing materials;
— planting in rows along irrigation channels; and
— windbreaks.

Acacia albida needs either underground water or rainfall of 400 to 600 mm.

Table 42. Guidelines for using *Acacia albida*

Objective	To reduce windspeed and promote the re-establishment of a vegetation cover
Recommended Soils	Sandy soil or flood plains
Spacing	10–15 m between trees planted in lines
Density	40–60 trees per ha
Cultivation	After 4 months in a nursery, transplant. Optimum growth occurs where rainfall is less than 600 mm a^{-1}
Protection	Thorny branches or hedges to deter grazing cattle
Benefits	Pods for animal fodder; increase in crop yield owing to improvement of the soil conditions as a result of nitrogen fixation

12. *Plants for Effective Sand Stabilization.* The choice of species to be planted on sand, sand sheets, or dunes should be made after carefully considering a number of factors. These include:

— their adaptation to sandy soils;
— their tolerance to high temperatures, both ambient and surface;
— drought resistance and their development of good root systems capable of reaching lower layers in the soil;
— adaptation to low and seasonal rainfall, or prolonged rain;
— season and high humidity;
— tolerance of the abrasive action of drifting sand; inundation by sand and adaptation to active dune ecology.

In Morocco and other North African countries both sugarcane and *Tamarix articulata* have been successfully used for hedges. The sugarcane is planted in 15–20-m squares. *Ammophila arenaria* and *Artemisia monosperma* can also be established in dense clusters.

Ipomea pescaprea is a creeping plant and ideally adapted for fixing active dunes or sand sheets. This plant is not eaten by livestock. It is effective because it is an invading plant and its creeping roots compete well with existing vegetation cover.

If possible, indigenous vegetation is preferable. Table 43 gives some further examples of sand fixation methods using various types of vegetation for road and railway construction in China.

13. *Additional Benefits of Biological Stabilization.* Besides controlling wind erosion and desertification, there are additional environmental benefits to revegetation. These include:

— reduction of the wind velocity which reduces the damaging effects of the wind upon agricultural areas, grazing land and human settlements;
— changes in topography: increased surface roughness causes sand deposition and the reorganization of the sand surface;

Table 43. Sand fixation methods used in road and railway construction. (Babaev 1986b)

Landscape type	Sand fixation techniques
	Railways
Steppes with shifting sands and semi-fixed and fixed dunes	Planting of *Artemisia halodendron, Caragana microphylla* on dune ridges, of *Lespedeza dahurica* on leeward slopes, and of *Salix flavida* on windward slopes. This protective system is known as "a tree you need in a place it should be"
Deserts with shifting sand dunes of great length	Straw checkers 500 m wide as mechanical guards with plantings of *Hadysarum sciparium, Calligonum mongolicum,* and *Salix flavida*
Arid zones and Gobi, where sand transport presents serious threat	Forest belts of *Populus cupidata, Eleagnus angustifolia, Salix cheilophila* (often irrigated)
	Motorways
Deserts	Forest belts of *Artemisia ordosica* and *Salix cheilophila.* Rock debris of salt slabs covers. Mechanical guards in the form of checkers

— reduction in sand movement which decreases airborne sand and thus encourages soil development; and
— production of commercial products such as firewood, charcoal, building materials, and crops such as fruit and nuts.

Preference should be given to biological control measures. The mechanical methods are more difficult to maintain and generally more expensive. However, biological methods are more difficult when the rainfall is less than 300 mm a^{-1}. In recently revegetated areas or where sand has been reactivated, fencing is very important to prevent the access of animals and to allow natural regeneration.

In areas of continuous sand deposition which is undergoing geological changes the solutions are more difficult, more expensive, and in some cases may not exist.

5.7.3.4 Windbreaks, Shelterbelts, and Wind Barriers

Although wind barriers have been discussed in previous sections, they are so important in the control of wind erosion that they deserve greater elaboration. A wind barrier is a strip of trees, shrubs, and/or grass which slows windspeed, reduces evaporation, and prevents overheating of the soil. Windbreaks refer to two to four rows of vegetation while mini-windbreaks are wind barriers constructed of permanent grasses such as *Andropogon gayanus*, which is often used for thatched roofs. Other such grasses include *Sporobolus festivus, Setaria pallidefusca,* and *Oxytenanthera*. When the strip of vegetation is wider than four rows of trees, the term shelterbelt is used. The first shelterbelts were planted in 1955 in Cameroon, north of Mount Kapsiki, when rows of *Cassia siamea* were used.

A good wind barrier has a number of purposes, for example:

- reducing wind speed;
- preventing deflation;
- making airborne sand drop behind the barrier and trapping it so that rows and ridges gradually form. Plant growth then becomes possible in the protected areas behind the ridge;
- producing firewood, building materials, human or livestock food (such as nuts or fruit), and medical products.

1. *Construction and Effects of Wind Barriers.* There are five principal factors that must be considered before constructing wind barriers: wind, area of erosion, selection of species, and social factors.

a) Wind regime and prevailing wind direction. Winds which back and veer around a prevailing direction should be considered in addition to the main wind direction. Since shifting sand or dunes are influenced by a complex set of variables, their direction of movement may change seasonally. It is wise to observe sand movements for 1 year and one should seek out local knowledge with land users.

b) Assessment of the wind system. Except under extreme conditions, it is more effective to stabilize the area of origin of shifting sand rather than the area where sand is deposited. It is therefore very important to understand the area as an entire wind action system.

c) Area of wind erosion. Once the wind system and prevailing direction are known, the area most exposed to wind erosion must be selected. Dune crests or interdunal corridors are often the most vulnerable. It was suggested that the top of the dunes should be left unplanted while the bottom third of the slope was vegetated. For longitudinal dunes with corridors threatened by wind, revegetation must be started in the corridors where soil erosion is greatest.

d) Selection of suitable vegetation species. This is an important aspect of constructing suitable wind barriers. The species must be adapted to local soil and climatic conditions, especially to the quantity and variability of rainfall, as well as being compatible with indigenous species in the area.

e) Respect for local traditions. Wind barriers should not offend, alienate, or violate local traditions. Similarly, they should not disturb human and animal movements. They should not obstruct railways, water channels, roads, livestock, or village trails. However, gaps in the rows should be avoided as much as possible because wind is funneled through them, the wind stream being compressed and accelerated, resulting in further wind erosion. To avoid this, paths through wind barriers should be placed at an oblique angle and not at right angle to the rows of trees (Fig. 79).

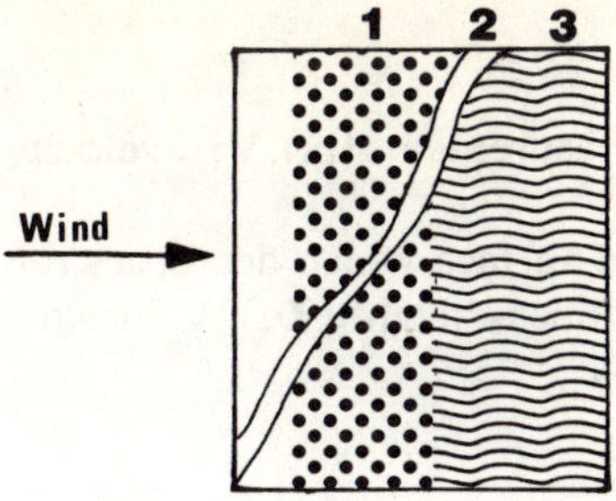

Fig. 79. Path through a wind barrier (After Weber and Stoney 1987). *1* Windbarrier; *2* path; *3* cropland or pastureland. Any path through the wind break should be at an oblique angle rather than at right angle to the tree rows. This will allow people and livestock to move through the wind break without opening a gap for the wind to roar through and to benefit from a shaded passsage way. Wind barriers and shelterbelts can be laid out to include roads, trails, or driveways for livestock

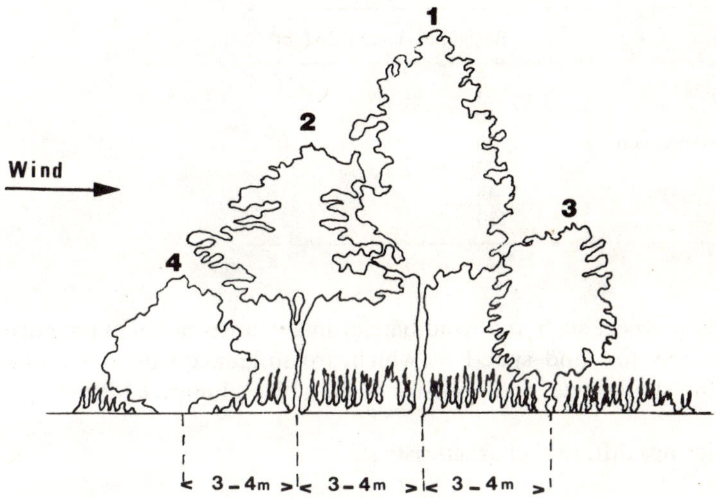

Fig. 80. Ideal structure of a wind barrier (After Weber and Stoney 1987). *1* Row of high trees; *2* row of shorter species of trees; *3,4* auxillary rows of bushier trees and shrubs. Wind barriers can furnish protection for downwind areas up to ten times the height of the trees. The wind break consists of at least two rows of plants of different heights. Large trees should be chosen for one row (*1*). Fast-growing species can be mixed with slower-growing, longer-lived trees, depending on local preference. Row (*2*) should be composed of shorter species, chosen if possible for their byproducts. and rows (*3*) and (*4*) are auxillary rows. These are planted with lower, bushier trees, shrubs, and grasses. A well-chosen vegetation mix for windbreak composition will not only provide protection from the wind, but will yield secondary products as well

The effectiveness of a wind barrier depends on its height, its permeability, and the texture of the vegetation used. Ideally, a wind barrier using trees must consist of at least three rows of plants. The center row should be planted with a tall, preferably fast-growing species, with the second row consisting of a shorter species, and the third row of short trees or bushes. Figure 80 illustrates this design.

The relationship between wind speed and height was expressed by Bagnold (1941) in the equation:

$$V_2 = V_1.Z.o,$$

where V_2 = wind speed in m s^{-1}, Z = height of the measurements (m), V_1 = velocity at a height of 1 m, o = 0.46 Karman constant.

Plant cover causes a roughness in the ground surface which decreases the wind speed up to a certain height, called the roughness height (Zo). Values for different types of vegetation are shown in Table 44.

Table 44. Roughness height of different vegetation types. (Caborn 1957)

Soil or plant cover	Roughness height Zo (cm)
Smooth surface of snow	3
Short grass	10
Medium grassland and bracken	20
High grassland	30
Turnip field	45
Wheat field	130

Any protruding object such as a wind barrier in a windstream will produce turbulent eddies, the size and speed of which are influenced by the initial windspeed, and the shape, size, and porosity of the object. Figure 81 illustrates this wind flow.

A wind barrier has different characteristics:

— In the disturbed flow the windward side has an air cushion with low wind speed. Upon reaching the obstacle, the air cushion stretches from the ground to the top of the barrier.
— The flow over the barrier is accelerated because the streamlines are forced to rise. In a permeable barrier, some streamlines filter through to produce a large windless cushion from the top of the shelterbelt to the ground on the leeward side. The wind gains its normal speed at a distance 10 to 15 times the height of the wind barrier.

If the wind barrier is not permeable (solid board fences, cement plates) eddies appear on the leeward side owing to the low wind pressure. A suction effect results and the air currents above the barrier are drawn downwards into an area of turbulence. The wind then resumes its normal speed a short distance from the wind barrier.

When the thickness of a wind barrier increases, its permeability, and hence its efficiency, decreases. A forest several hundred meters wide is less effective than a strip 20 m wide.

Figure 82 shows that the reduction of wind speed by a wind barrier is a function of the distance from the barrier (after Gloyne 1955).

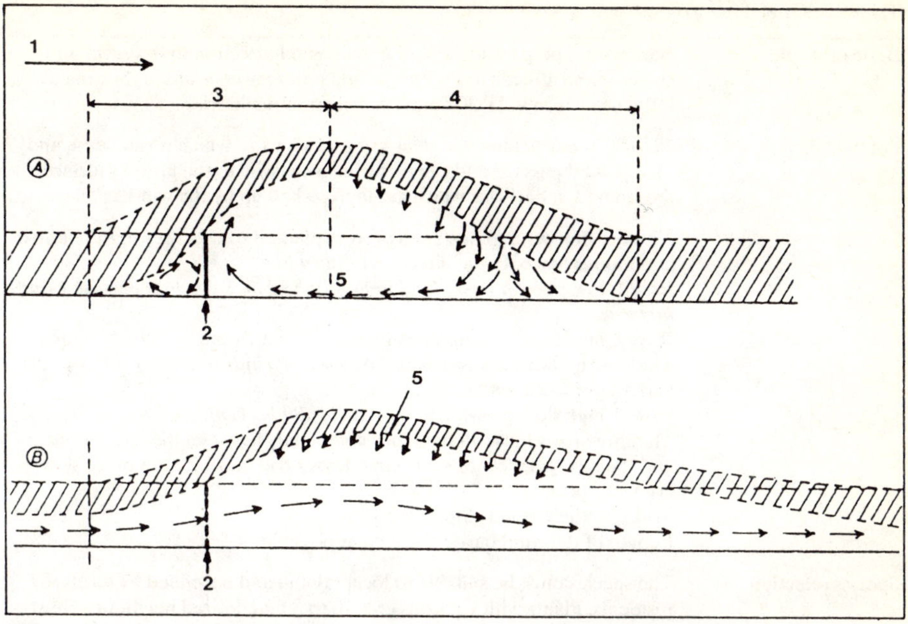

Fig. 81A,B. Effects of a wind barrier on the flow of wind. (Guyot, FAO 1987) **A** Impermeable shelterbelt. **B** Moderately permeable shelterbelt. *1* Wind direction; *2* wind barrier; *3* contraction of the streamlines; *4* expansion of the streamlines; *5* turbulent area

2. *Spacing and Design of Wind Barriers (Table 45).* To obtain the greatest decrease in wind speed, the ideal spacing of barriers should be five times the height of the wind barrier. However, this may not be compatible with agriculture or land tenure in the area, so a realistic choice of spacing should be opted for which does not compete with these factors. The recommended spacing might therefore be between 5 and 25 times the height of the wind barriers.

If the wind is not very turbulent, for example in areas of varying topography such as in valleys swept by local winds or in corridors cutting escarpments, the spacing must be reduced to five times the height of the barrier. The spacing also depends on the topography of the windward slope where the airflow is compressed and the wind speed is accelerated. This requires a denser spacing of wind barriers. On the contrary, on the leeward slope, where the windstreams expand and the wind speed is slower, the density of the wind barrier can be decreased.

The shape of a wind barrier must also be taken into account. The various shapes or designs of wind barriers depend on topography and the wind direction. The following designs of wind barriers have been described by Ebomo (1983) and they are shown in Fig. 83.

a) The straight-line type is the conventional shape. It is made up of rows of trees with its length greater than its width and usually runs at right angles to the direction of the prevailing wind. Establishment is easy. However, the com-

Table 45. Guidelines to constructing wind barriers

Basic principle	Barriers are perpendicular to the main wind direction to intercept air currents and affect wind velocities and temperatures within the area of influence which is 25–30 times the height of the barrier
Permeability	40 to 50% permeability is ideal to avoid a barrier which is too dense and blocks wind passage and causes turbulence close to the ground surface loosening soil particles which are then picked up by the wind
Layout	Number of rows: 2, 3 or 4 rows of trees planted so that they mature into a continuous fence. In the direction of the wind: Row 1: low trees or bush; *Bauhimia, Euphorbia, Salvadora persica, Combretacae.* Row 2: high fast-growing short-living trees which can be cut when the wind barrier becomes too dense; *Azadirachta indica* (neem), appropriate varieties of Eucalyptus. Row 3: high slow-growing trees, *Acacia albida, Tamarix, Parkia, Nere.* The three rows must fill the space completely between the soil surface and the top of the canopy. Spacing: 1.8 × 1.8 m and 1.2 × 1.8 m for slow-growing species. Width of the ground strip: 1.25 to 2 m Length of the wind barrier: as long as possible
Species selection	The species must be suitable to local rainfall and be valued by local residents. Plants with vigorous growth to attain desired height in a short time are best. They should be protected by law and preferably indigenous
Design of wind barrier	Basic rule: linear design perpendicular to wind. In cases of complex land ownership or for mobility, straight continuous rows of trees may not be feasible, so it is possible to replace them by staggering trees around villages or towns
Drawbacks	High cost of protection (fencing) against overgrazing
Benefits	Soil conservation, reduction of wind damage to crops at a distance 10–15 times the height of the barrier, increases in crop yields and production of fuelwood, poles, fruit, and medicine

promise between relief, wind direction, and land availability has led to the modification of the straight-line belt to some other forms like the cross, the checkerboard, the manx-leg, and the circular type.

b) Cross-type. This has a main belt of sufficient width to make the development of a dense barrier possible, and two arms join this main belt at right angles (Osborne 1975).
c) Manx-leg. This is used to give protection against winds from various directions (Osborne 1975). It consists mainly of a central group of trees and three short legs making it appear like irregular clumps.
d) Checkerboard. These are belts planted in square or rectangular blocks protected by the belts on all four corners. This pattern is used in regions where wind direction is indeterminate and where land is scarce. There are only a few rows of trees.
e) Circular. Circular belts are good for grazing areas on hilly terrain where winds come from all directions. The rows of trees are usually circular in arrangement and enclose the protected farm or rangeland.

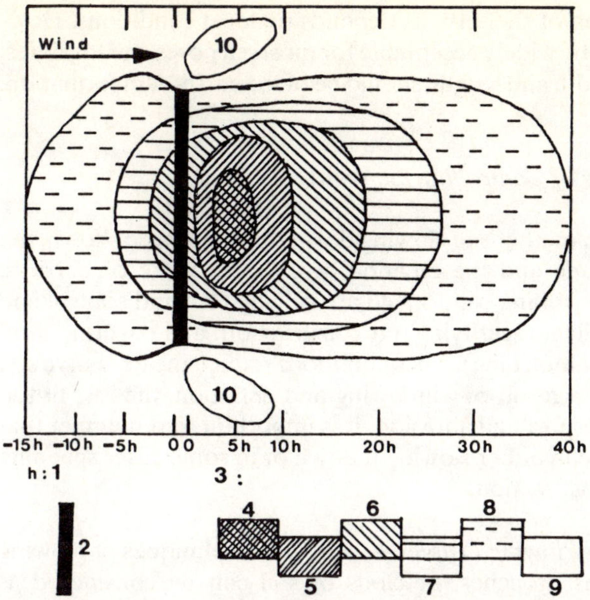

Fig. 82. Distance from the wind barrier and reduction of the wind speed. (After Gloyne 1955). *1* Height of the wind barrier; *2* wind barrier; *3* wind speed in percentage of the initial wind speed; *4* < 20%; *5* 20–40%; *6* 40–60%; *7* 60–80%; *8* 80–100%; 9 ≥ 100%

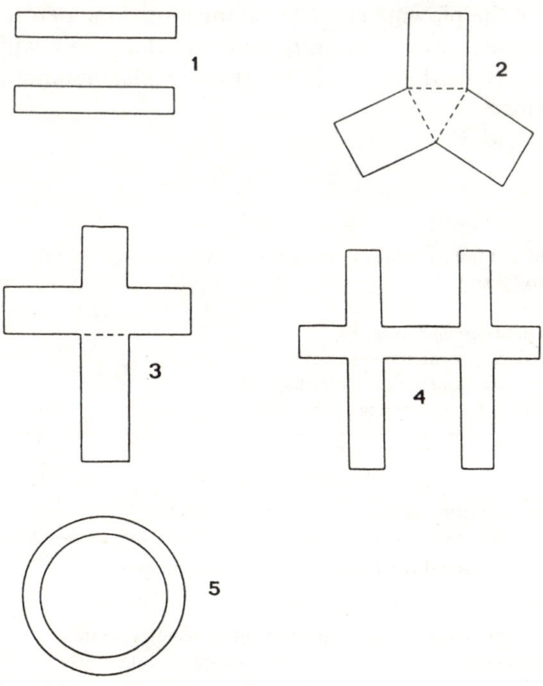

Fig. 83. Different shelterbelt designs. (After Ebomo 1983). *1* Linear type belt; *2* manx-leg belt; *3* cross-type belt; *4* checker-board-type belt; *5* circular belt

The design and construction of shelterbelts depends on local conditions. However, multiple belts seem to be widely acceptable for most purposes and localities; the profile, whatever the width and length, should be of the earth-dam formation.

5.7.3.5 Additional Methods of Controlling the Wind

In the previous section, two methods of slowing down the wind were shown, the creation of an artificial dune and the creation of a lattice work of barriers. Amongst other ways of controlling wind speed are ridging for wind control and deep plowing, minimum tillage, scarifying, pit construction, and covering sand surfaces with clay (a type of mulching). These methods reduce the excessive soil erosion that takes place as a result of winnowing and deflation, the loss of top soil and minimize sand transport and abrasion. It is important to remember that controlling the wind can mean either slowing it down or in some cases speeding it up, as was shown in the last section.

1. *Wind Ridging and Deep Plowing (Tables 46, 47).* The techniques of plowing or digging to create ridges, trenches, or clods of soil can be considered as protective techniques against wind erosion, but they must be used very carefully since deep plowing (10–15 cm in depth) is only beneficial when soil can be divided into clods.

Deep plowing is not recommended in dry conditions with sandy or loamy soils because they do not form clods and the existing wind erosion can worsen. It is more effective if the plowing is carried out immediately after the first rain, when the soil is wet to the base of the plowing layer, but it must be done before weed growth begins. The protection is only temporary, since wind and rain will eventually break down the clods. In medium- or fine-textured soils, minimum tillage or scarifying is recommended.

Table 46. Guidelines for wind ridging

Ridge height	30 cm above the natural surface (to provide 3 m of protection)
Depth of furrow	According to soil texture
Ridge material	Cloddy material
Spacing	3 m maximum between each ridge
Direction	Perpendicular to the prevailing wind direction
Extent	Whole of drifting area and area of potential drift
Starting area	Preferably upwind of the source of drift

Table 47. Guidelines for plowing for wind erosion control

In region affected by water erosion	Avoid ploughing up and down on hills, contour ploughing only
In flat region affected by wind erosion	Avoid planting row crops parallel to the prevailing winds; requires ploughing and planting rows perpendicular to the prevailing winds

Wind ridging and surface roughening will also remove grass roots and promote the establishment of woody shrubs.

If wind erosion is severe, these solutions will be insufficient. This is especially the case for soils low in moisture, organic matter, or aggregating ability, or if the crop does not produce effective cover (such as cotton, sorghum, or beans). In such cases tillage may have the following adverse effects:

1. Soil aggregates are broken down into sizes which make them susceptible to wind erosion.
2. Plant residues are buried.
3. Organic matter and soil moisture is lost more rapidly.

In Algeria's Hautes Plaines, for example, soil erosion started to become serious in the early 1950's when tractors were introduced; plowing was previously done by animal-drawn harrows. It was easier for tractors to plow up and down the slope, but there were serious consequences: water erosion developed in the form of sheet erosion, rills, and gullies. The speed of runoff increased and the infiltration rate fell. Wind erosion ensued owing to the loss of vegetation and to the drying out of the soil, the latter being exacerbated by the use of moldboard and disc plows, which plowed deeper, thereby exposing moist soil to desiccation and removal by wind.

2. *Minimum Tillage.* This is cultivation involving minimal disturbance to the soil during planting, harvesting, and the fallow period; it can also refer to the practice of simple scarifying. It is recommended for sandy soils. Combining minimum tillage with the mulching of crop residues is even more effective as a control measure against wind erosion.

3. *Scarifying.* Scarifying is useful where scalding has occurred in order to break up the hard polished horizon which forms when the topsoil has been eroded. It is also particularly important on certain overgrazed land where a hard algal cap develops. This is known as crusting, and the surface is both impermeable to water and prevents the successful establishment of germinating seeds. It is often so hard as to require mechanical breakage using heavy machinery, but cattle hooves can also work. Once the crust is removed, plant establishment is much easier.

Scarifying is the technique of cutting or scarring the soil surface in bands 1 or 2 m wide and at a spacing of 8 to 10 m. It can be done by using ordinary tools such as pickaxes, jembes (grub hoes), and dabas (common African cultivating tools), or a narrow harrow pulled by an ox or donkey can also be used. The bands must follow the contour of the slope. Scarifying can also be combined with seeding in areas where natural reseeding does not occur.

The purpose of scarifying is to increase water infiltration through the loose layer of topsoil which is exposed to greater air circulation by the scarring of the soil surface. Roots have a better opportunity to develop and seeds can penetrate the soil more easily as they are protected against heat and wind. In Botswana, scarifying is done by pulling branches along the ground.

4. *Pit Construction (Table 48).* Pit construction has been extensively used as a method of soil reclamation for sandy soils threatened by wind. Pits are dug in a staggered pattern and their primary purpose is to trap runoff water and increase infiltration. The method was first used in Australia in 1968 on soil damaged by scalding (Keetch 1981) and the pits were still effective 12 years later. Measurements after 8 mm of rain showed that infiltration reached 3–4 mm on a scalded area but 200–250 mm at the bottom of an adjacent pit.

Pitting can only be used on a slope of less than 5% because a greater slope will increase runoff and water erosion. On loose soils, pitting degenerates because of silting, raindrop splash, and cattle trampling.

5. *Staggered Furrows (Table 49).* These have proved successful on slopes susceptible to water erosion such as sheet-wash and rillwash, and where pits are ineffective after several heavy rains. Staggered furrows are better than continuous furrows in areas of wind erosion. For example, both pits and staggered furrows have been used successfully in Australia in the Alice Springs Dust Control Project (Keetch 1981). The reclamation techniques effectively reduced water and wind erosion, thereby increasing soil moisture and forage production.

Where winds are multidirectional, staggered furrows can be dug in the shape of a spiral covering 1 ha but only on flat surfaces. The coils of the spirals should be spaced 3–5 m apart.

These techniques often require tractors or specialized machinery, and unfortunately this is the main factor restricting their widespread adoption.

Table 48. Guidelines to pit construction in sandy areas

Dimensions	Length 1.8 m	Width 38 cm	Depth 30 cm at deepest point
Water-holding capacity	60–80 l		
Location	along contour lines		
Layout	As long and straight as possible		
Spacing	Double rows 5 m wide separated by a space of 10–12 m		

Table 49. Guidelines to staggered furrows

Location	Flat or sloping areas
Length	8–10 m
Spacing	3–4 m long 10–30 m wide
Layout	Furrows are arranged so that a gap of 3–4 m between two furrows is covered by a furrow in the adjoining run

6. *Sand Covering Surfaces with Clay.* This method is a form of mulching which involves the spreading of clay, gravel, or other locally available material over a sandy surface. The procedure must be done with care and rainfall should not exceed 300 mm a^{-1}. In the Shalambot desert of Somalia, clay was spread on previously fixed dunes, but it did not succeed because the rainfall was too high and water erosion appeared.

This method has also proved useful in Argentina, where 75% of the north of the country is arid or semi-arid, especially in the west. All this area is subject to wind erosion. The active dunes are called medanos and can reach a height of 20 m. The stabilization techniques involve two steps:

a) Leveling of the dunes by sand sprayers, graders, or with wooden beams pulled by animals so that the dune loses its crest and the leeward face becomes less steep.
b) Before the rains, the dune is covered by a thin mulch layer and in autumn (October-November) a grass mixture is seeded.

5.8 Training and Research

5.8.1 Training

5.8.1.1 What Is Not Working?

It is recognized that results achieved through conventional training have fallen well below expectations. Performance-oriented training seen as a part of a management system rather than as an end in itself should be a rule.

Education and training programs are still too often established according to systems developed in northern countries which are adapted to temperate ecosystems and to developed economies which are too theoretical.

A considerable amount of ecological knowledge has been built up on arid zones over the past 50 years. Generally, it is not possible to transfer techniques developed in a particular ecosystem with a particular culture during a long process of experience to another region with different practices without making adaptations and modifications.

Agriculture does not fit well into the nomadic strategy of survival, in which mobility plays the major role. Technically, rangeland can be improved, but the probability that nomads will accept the changes must be seen as very low. To introduce changes into a nomadic pastoral society is very risky because almost all aspects of their life are found to be traditional, adaptive strategies. Gans (1986) describes how the Turkana of northeastern Kenya, who live in a hot dry area (188 mm annual rainfall with extremes of 19 and 498 mm), have appropriate adaptive strategies: species diversification of livestock, herd mobility and dispersal, community rituals (birth, initiation, marriage, death, etc.), and offer an example of a harmonious relationship between man and nature.

5.8.1.2 What Can Tradition Bring?

"To incorporate in development planning indigenous knowledge is a courtesy to the people concerned, in an essential first step to successful development, emphasizes human needs and resources, rather than the material ones alone, makes possible the adaptation of technology to local needs; is the most efficient way of using western 'research and development' in developing countries; preserves local knowledge, encourages community self-diagnosis, and heightens awareness." (Brokensha et al. 1980).

Each traditional practice has an inherent wisdom which must be respected. They are traditional techniques and forms of social organization which have great value and prove their utility in sustaining the livelihood of nomads and farmers in dry areas.

Local populations have lived for many generations in these dry areas and have thereby accumulated a deep knowledge of local resources and discovered rules for survival. If they are neglected or overwhelmed, these rules become lost.

5.8.2 Research Results

Foreign research or results from a national experimental station do not necessarily bring about a miraculous solution: new techniques or strategies will be eyed distrustfully by the farmers, new methods will be more easily adopted once their efficiency has been demonstrated.

Moreover, the solutions are neither country-, region-, nor ecozone-specific, but most often village-specific. Donors' perceived priorities, farmers' perceived priorities, international organizations' perceived priorities must be discussed and reconciled. Scientist-inspired research should complement research inspired by farmers' needs. Nevertheless, the basic processes of land degradation, which are still imperfectly understood, should continue to be investigated through applied research: essentially these are the physical processes of soil degradation, the variation of their dynamics in the ecosystem, in the local environment, and through their impact on people's activities.

The golden rules of research are:

1. Careful study of the local situations with the land users. Livestock development strategies and agricultural innovations should integrate the research with the indigenous knowledge and special skills of the pastoralists and farmers.
2. This will involve careful sampling of the opinions of land users' wishes, requirements, needs, and suggestions.
3. There should therefore be involvement of traditional land users in setting up research priorities and testing new techniques, since successful agricultural innovations require the active participation of the land users.

Rehabilitation and development projects where conservation is the major concern require scientific monitoring and periodical assessment to evaluate the natural and human potential, to identify the rate of changes, the risks of degradation, and the status of all the parameters, and to identify and correct faults so as to take advantage of them immediately and in new programs.

Research of this kind has already worked in the USA for several decades, in the former USSR, in China, in India, in Israel, and in Africa. But there are too few ecologists, range managers, geomorphologists, and other specialists in developing dryland countries, where the administrative, legal, and medical professions enjoy a higher status.

New methods offering helpful information are being developed and improved every day. Of these, satellite images are certainly among the most fruitful.

The *Handbook of Ecological Monitoring* edited by Clarke (1986) is designed to provide information on the range of techniques now available for obtaining environmental data on people, economic and social facets, animals, plants, and the Earth itself. Interdisciplinary research and the efficient monitoring of the collection of information combine research on the ground, from the air (human observer flying in light aircraft and aerial photographs), and from space using information and satellite images such as Landsat, Meteosat, and Spot.

5.9 General Conclusion, Chapter 5

Remedies for land degradation are numerous and interdependent. In Chapter 5 of this book only some of them have been given and classified according to the five most important points

1. Climate: to circumvent unpredictable risks. In the last decades rainfall in the Sahel has dropped, but it has, in fact, always been extremely variable. It seems as if this variability has increased, but more information is needed before it is certain that rainfall is decreasing. Well-based programs must deal with the most adverse data.
2. Vegetation: to save the vegetative cover, to avoid damaging grazing methods, to introduce new species, and adapt more resistant varieties to water scarcity and salinity, to improve crops genetically.
3. Water: to make better use of water, to emphasize the modest water harvesting solutions and use salt-tolerant plants.
4. Soils: to avoid the threats to soil fertility and to combat water and wind erosion.
5. Education: to combat ignorance by training programs at all levels, to monitor and evaluate projects in relation to socio-economic data.

Technically, "desertification" can be halted and many degraded areas could be restored to production. However, the solutions are expensive. They require strong political will and power, and even more the agreement and participation of the rural population, since the rehabilitation programs interfere with their daily life and activities.

To combat land degradation requires a collective organization based on the rational use of natural resources and the management of environmental quality. Such organization must be based on a long-term integrated development of all social and economic sectors.

5.9.1 The Specific Solutions for China

China feeds one-fifth of the world's population on one-fifteenth of the world's agricultural land. About 11.5% of the country is arid, semi-arid, and semi-humid, as seen in Chapter 3 of this book. In 1988 drought and flooding on the Huang-He necessitated the provision of food aid for more than half a year for 20 million people.

In 1949, after the civil war of the 1940's when Chairman Mao Zedong chose to make China self-sufficient in grain and ordered the farmers to plant cereals on every available unit of land, he was not aware of a new difficulty which would arise: increasing soil erosion, which resulted in one of the worst famines and ecological disasters of Chinese history: 16 million persons died between 1959 and 1961. Then Mao Zedong organized the battle against erosion at the level of technological solutions large-scale cooperative efforts of terracing, dam and canal building, but not at the level of socio-economical restructuring. After the death of Mao in 1976, the Chinese changed their objectives and put more emphasis on research for the most efficient crops. They grew adapted crops on each type of land: the improved yields on the flatter lands freed the steep slopes for sylviculture and improved more nutritious fodder crops, which allowed the animals to be kept in stalls. Under Chairman Mao Zedong all the land was farmed under the commune system, the food collected and redistributed. This system has now been abandoned and replaced since 1980 by a responsibility system: each family receives a plot of land for a precise period, and pays tax on only a fixed part of the harvest. China, understanding that the prices paid by the government to the farmers were too low, raised the prices paid for their crops and allowed the farmers to sell the crops on the open market and not only through government control. Reward is offered proportionally to output, and soil erosion has come under control.

Zhu Zhenda (1984) describes the measures successfully used in China to combat environmental degradation.

1. In semi-arid lands the measures adopted are:
   a) Cutting down dry farming areas and enlarging the proportion of land used for forestry and animal husbandry; a combination of agriculture and forestry with shelterbelt networks as the framework.
   b) Reduction of livestock population on steppe land and the building of farms to produce forage only.
   c) The protection of natural vegetation and fencing off of degraded land to rehabilitate the vegetation.

   d) Tree planting in the interdunal depressions and plantations of sand-holding species on the slopes and tops of active shifting dunes.
   e) Birth control to give a birth rate lower than 1%.
2. The methods chosen for the rehabilitation of their arid lands are based on the rational use of water and land resources and the establishment of systems of shelterbelts.
   a) Planning of the drainage area of inland rivers, and of the proper distribution of water, so as to adjust the ecostructures of agriculture, forestry, and animal husbandry.
   b) Sand fixation by plantation and sand barriers where farmland is subject to shifting sand and encroaching dunes.
   c) In the oases: establishment of a shelterbelt network inside oases, plus windbreaks of tree-shrub-grass at oasis edges and fencing of sand dunes at the periphery.
   d) Birth control to give a birth rate lower than 1%.

In October 1990, a visit to Yanchi Research Station proved that China is successful in applying global integrated solutions. In the Yanchi County Research Station, Chinese scientists are gaining beneficial experience in combating desertification by establishing direct links between fundamental research and the grass root level. According to Song Bing Kui, Head of the station, the following results are observable:

In a semi-arid ecosystem with an annual rainfall of 288 mm, an annual average temperature of 7 °C, an annual evaporation of 2597 mm and sunshine of 2876.5 h, the research station is located on the southwestern edge of Maowusu Sandy Land (Ordos Plateau). The area belongs to the administrative village of Liuyangbu Town, which comes within the jurisdiction of five villages.

Here, 14.7 ha of land is directly managed by the base: 74 families of farmers with a population of 398 persons on 4822.7 ha of land are directly associated with the station, where 46% of the total land is rangeland, 13% is woodland, 3.9% is cultivated land, with 57.8 ha of irrigated land and 130.7 ha of nonirrigated land.

The shifting and semi-shifting dunes, i.e., severe degradation, accounted for 30%. Moderate degradation affected 5.5% of the land. Salinization of the lowland represented 12.5% of the degraded land. The original fixed sandy land was reactivated in shifting dunes. According to the Chinese scientists, the causes of degradation were overcultivation, overgrazing, fire wood collection and cutting for fuel in a dry environment.

The treatment was based first on biological measures: planting shrubs and grasses. Then socio-economical measures were applied: regulation of land-use including the proportion of farmlands and expanding woodland, limitation of the number of animals according to the carrying capacity of grassland, rational development and utilization of the underground water with strict limitation of the irrigated land.

In 1990, after more than two decades of treatment, interesting changes have appeared:

1. In the climate of the area: the increase in the number of trees, shrubs and grasses resulted in an increase in the surface roughness and therefore in a decrease in wind velocity, which is in fixed and semi-fixed dunes only 25–45% of that in shifting dunes. The amount of migrating sand decreased several times. The severe degraded land is reduced by 10%. The air humidity rose by 6–15% and the evaporation of the water surface was reduced by 15–34%.
2. In the biological environment: an increase in the woodland from 13 to 22.7%, in the available rangeland from 46 to 64.8%, changes in the proportion of irrigated farmland, which reached 81.1 ha, and dry farmland, which decreased to 74.2 ha. Before the treatment, the vegetative cover and grass yield were close to zero. After 2 or 3 years of plantation of *Salix psammophila, Hedysarum scoparium, Hedysarum leave, Artemisia sphaerocephala, Agriophyllum squarrosum,* the vegetative cover reached 25–37% and the grass yield was 465 to 3660 kg ha^{-1}. The average yield of *Triticum aestivum* increased from 354 kg in the past to 5080 kg. In small areas the yield can reach 7675 kg ha^{-1}.
3. In the socio-economical situation: from 1985 to 1990 the grain per person increased from 341.8 to 537.7 kg, sheep and goat from 1942 to 2752; livestock is stable at around 130 head, but the meat has increased. The annual gross income at the station has increased from 167 000 to 348 000 Chinese yuan and the average personal income from 391.8 to 893.8 Chinese yuan.

   A weather station was built with 11 lysimeters to survey evapotranspiration and 17 wells to survey the fluctuations of underground water. The station has 20 rooms for scientists and 15 permanent researchers who work hand-in-hand with the farmers.

This experience has a double advantage, since if focused on a small integrated area of being extended further and of establishing direct links between the scientists and the grass root level. There the farmers can apply the results of fundamental research in addition to their own experience and knowledge.

5.9.2 Africa Must Find its Own Specific Remedies

1. Communal land tenure cannot be combined with individual or family freedom to use the land vegetation and water at will, because direct responsibility for maintaining the quality of the land resources has been abdicated.

   Systems of land tenure which create a vested interest in the conservation of the quality of land by the land users are necessary.
2. Fodder reserves must be created. In North Africa certain varieties of *Atriplex* are successful with up to 150 mm of annual rainfall: certain varieties of *Opuntia* and also *Acacia* at increased altitude can also be cultivated.
3. To help the pastoralist to decrease the numbers of cattle, the meat market must be cleaned up. The price of meat is multiplied three or four times between the Sahelian areas of production and the Sudanian or Guinean areas of consumption.

IPAL in Kenya is trying to rationalize the meat market mainly by organizing transport of cattle from the Rendiles nomads grazing area, north of Marsabit, to Nairobi directly, without intermediaries between the nomads and the urban meat market (I have personal experience that this can produce some of the best and cheapest meat I have ever known).

The major mistakes which must be avoided in the future are:

1. Drilling of wells using advanced technology which may result in gathering nomads together, forcing them to concentrate and settle in order to receive the help which they need.
2. To intervene in the topography by leveling the undulating terrain. The top layer of fertile soil is mechanically eradicated on the formerly high areas and poorer layers are exposed. Near Kyzyl (Republic of Tuvan, former USSR), and area of fixed dunes was leveled so as to establish a mechanized strip cultivation of wheat. The team received congratulations from the Government for its project. Unfortunately, a couple of years after its initiation, wind erosion was already affecting the success of the project. Each heart of leveled dune, where loose sand is exposed, becomes an area of active wind erosion where blow-outs are dug and ripple marks formed in the loose sand. The golden rule in sandy areas is to follow the natural contours.
3. Introduction of unadapted high technology. The general cause of dryland degradation is the introduction in agriculture of a type of high technology unadapted to the ecological environment and aggravated by human and animal pressure, the result of the introduction of western medicine.

The final summary of this part of the book must be a sentence of El Baz (1988), which is applicable to desert and all temporary or seasonal drylands:

"Land degradation happens basically because of the lack of a deep understanding of the desert environment. Agriculture in the desert has to be different from agriculture in other places, and it requires a set of practices that vary from the norm.
We cannot establish the rules for such practices in a given arid area unless we fully understand the physical parameters that control its environment. Only through such understanding can we utilize parts of the desert for our benefit without degrading the terrain." (El Baz 1988).

Instead of adding new fields of study to the area of development and environment conservation, the aim should be to work on and improve the local existing roots; to make the old traditional branches more efficient by rediscovering the Earth, its soil, its water, and its vegetation; and also to remember that words are not actions and that diagnoses are not cures.

Chapter 6 General Conclusion: The Need for a New Realism

The concept of desertification was seen in this book as an extreme form of land degradation, irreversible on the human time scale.

"Desertification" is caused by human activities in which the carrying capacity of the land is exceeded. This becomes obvious in time of drought. The processes of desertification are exacerbated by both natural and man-induced causes and are manifested by interlinked mechanisms of soil degradation and vegetation deterioration which in human terms finally result in the irreversible diminution or even the destruction of the biological potential of the land and its capacity to support man and his livestock.

In fact, drylands are affected by gradual degrees of damage, until irreversible destruction takes place, which is called desertification.

Irreversible destruction means a deterioration of vegetation and soil which has reached a level impossible to rehabilitate within a reasonable time scale of one or two generations, even with human support, or when the cost of rehabilitation in poor communities exceeds the capital available, or will not attain the expected benefits for rich communities.

To accept as a well-established fact that the most degraded land could be rehabilitated with large amounts of capital and technology is not to deny the tragic level of the problems in semi-arid or semi-humid areas.

China offers good examples of reversing land degradation on a reasonable time scale (Table 50).

Is this book not appearing at the right moment, just at the end of the most severe drought of the 20th century, to demonstrate that the return of good rainfall in Africa can avoid the food crisis but cannot hide land and agricultural degradation and the increasing food dependency linked to overpopulation?

Record yields of cereal crops in 1988 have filled the granaries of the Sahel with reserves for half a year, but neither land and agricultural degradation nor economic difficulties have disappeared.

Table 50. Time necessary to reverse desertification in China. (Zhu Zhenda MAB 1984)

Annual rainfall (mm)	Time of reversing desertification (years)	
	On-going desertified land	Severe desertified land
400 – 550	3±	5±
300 – 400	3–5	5–8
200 – 300	5–7	>8

During the 4th CILSS ("Comité de Lutte contre la sécheresse au Sahel") meeting for "Promotion of Sahelian cereals", the ex-President of Chad Hissein Habré in his conference said: "Consommez Sahélien" ("consume Sahelian"). "The Sahel", said Hissein Habré, "already confronted with inclemency of nature, with degradation of the worldwide prices of raw materials, has in proportion to its income a heavy debt and cannot afford to utilize its meagre income for imports which are not necessary." In fact, says CILSS, the expenses of imported wheat and rice are as expensive as repaying of the debt. However, transformed local products based on local maize and millet are beginning to compete successfully with imported rice and wheat (Barrot and Seye 1989).

Focusing the world's attention on desertification has two advantages:

1. An emerging awareness of the vulnerability of our planet: this is the major progress of mankind in the 20th century
2. Emphasis on the necessity to augment food supplies for an increasing density of mankind particularly in the Sahelian and probably more in the Sudano-Sahelian zone and some other underdeveloped regions where crop production has not kept pace with population increase in the last decades.

In the African semi-humid Sudanian ecozone where, according to my assessment, environmental degradation is at a maximum, increasing inputs of water, fertilizers, pesticides, demographic explosion, and saturation of the occupation of land have made conventional agriculture unstable, with increasing risk of complete failure.

In 1986 in an introduction, entitled *The Vanishing Earth*, to an International Trust Viewer's Guide to a BBC TV program, the writer says: "The connection between soil erosion, deepening poverty, and, ultimately, famine is rarely made". Maybe an amendment to this sentence could be proposed: "The connection between land degradation (mainly soil erosion), deepening poverty, demographic explosion, and ultimately famine is rarely made." (Fig. 84).

In addition to the legend of the Fig. 84 the authors have written: "...the gains have not been evenly distributed...". In fact, the beauty and the tragedy of our planet is its heterogeneity, and precisely the heterogeneous distribution of good soils and natural wealth. There is a story told in the Central Africa Republic (RCA): "God was flying over Africa in an airship, throwing out of the window all kinds of wealth: good soils, good plants, good air, good rivers, good minerals. ... He was unfortunately asleep when he flew over the RCA..."

Reality is more complicated still. Harrison's (1987) description of the last Sahelian crisis in simple words demonstrates its complexity. In 1985 some 3 million people (50% of Niger's population) were hit by drought. Fulani and Tuareg pastoralists lost two-thirds of their herds. By June 1985 400 000 nomads had moved into the cities. North of Segou, when I arrived in July 1985, people were boiling roots, eating poisonous leaves and seeds.

Eight million people were affected in Ethiopia. In the highlands farmers sold their sheep, goats, young cattle, their cows, and even their draught oxen at half the normal prices, while the prices of grain increased. One 100-kg sack is enough

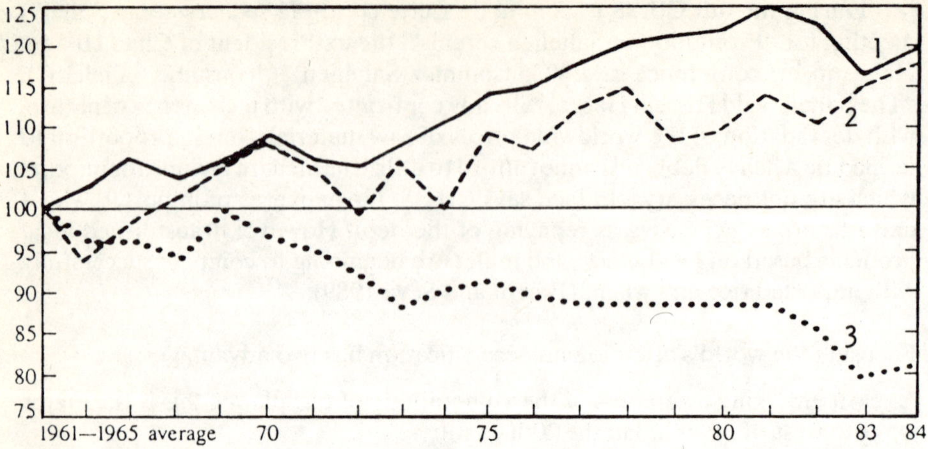

Fig. 84. Food production per person in sub-Saharan Africa, Latin America, and Asia 1965–83. Index of food production per person 1961–65 to 1983 (1961–65 average = 100). (US Department of Agriculture in International Trust viewer's guide to a BBC TV Programme). *1* Latin America; *2* Asia; *3* sub-Saharan Africa

to feed one person for 5 months. In 1982 an ox could buy six sacks, by 1984 one ox was necessary for one sack. As the food crisis continued, the farmers sold their tools: plows, hoes, and ate the seeds for the following year. All the fit young men had gone to town in search of work or help (Harrison 1987).

For Harrison (1987) the food crisis in Africa is fourfold:

1. Gradual decline of food and cash crop production per head. For one generation food production has been unable to support the demographic explosion. The evolving agrarian structures cannot adjust to the increasing population, and so result in the present crises.
2. The slow increase of nonagricultural production, too slow to compensate for the food deficit or to pay for financial loans.
3. The decrease in the value of African exports and the increase, on the other hand, of oil prices and costs of imports of manufactured products.
4. The acceleration of the environmental deterioration overlapping social and economic problems in terms of trade.

According to FAO, on the total surface of the soil of our planet Europe occupies 493 million ha, the former USSR 2240,North America 1968, Latin America 2057, the Middle East 1207, the Far East 11170, Africa 2503, and Oceania 850.

On account of demographic increase the surface available for each individual shrinks: the USA, for example, in 1985, with 935 million ha of cultivable land and 201 million people, had 4.7 ha per person. India had only 0.67 ha per inhabitant, Holland 0.27, and Burundi 0.8. In overpopulated countries, farmers have only

0.52 ha. In Java and Bangladesh, the figure declines to 0.16 and 0.18 ha per rural worker.

Nowadays the values are lower, as shown by Table 51:

Table 51. Demographic and economic statistics: Ethiopia, China, and the USA

	Ethiopia	China	USA
Population (millions)	43	1020	232
No. of children per family	6.7	2.5	1.9
GNP per person ($)	150	300	13, 160
Arable land per person (ha)	0.32	0.10	0.81
Average cereal yield per hectare of arable land (1984) kg ha^{-1}	1043	3894	4378
Calories available as % of requirements	92%	106%	138%
Military expenditure as % of GNP	8.8%	8.3%	6.4%

Egypt and Libya have only 3% arable land. Egypt, which is one of the countries most affected by aridity, must irrigate and conquer more and more arid land to feed its increasing population.

The world population which has now in 1989 reached and perhaps surpassed 5.5 billion people, was in 1971, according to FAO, 3.7 billion, and the Food and Agricultural Organization declared at that moment: “it seems therefore that the soils are enough for everybody” but added “however, neither population nor rich soil favorable to food agriculture are equally distributed and, of course, all the soils are not of good quality.” According to the FAO, Iran, for example, has a total surface of 166 million ha but 130 million are unproductive soils of which only 4 million can be improved by minor ameliorations to become usable.

High population growth rates under conditions of a poorly diversified economy in the drylands result in an increased pressure on unstable ecosystems. Such socio-demographic processes coincide with cultivation of marginal lands, e.g., mountain slopes and drier flat areas.

That the drylands environment is not the most suitable for human activities cannot be denied. Unfortunately, the long discussions on the causes of desertification have often masked the human responsibilities, both individual and collective, and attracted attention to the natural responsibilities, mainly drought. It must be admitted that man is the initiator of desertification and its victim; that drought is merely the revealer of desertification.

The failure to assign blame for degradation to human mismanagement has often led to the belief that nothing could be done to stop it.

Even if to diagnose is not to heal, a clear diagnosis is nevertheless indispensible for useful therapy.

An appropriate treatment, but not an easy one, has to take into account at one and the same time:

1. The causes; where they are human, they must be stopped; where natural, they must be avoided.

2. The mechanisms, which are always major natural processes, physical or biochemical, are open to technical solutions.
3. The manifestations, results, and consequences, are often only the symptoms of the disease. Treatment of the symptoms and the underlying causes will not result in a cure.

These three levels require twofold action:

1. A short-term remedial program for combating the mechanisms and manifestations: water erosion and gullies, wind erosion, deflation and active dunes, salinization, famine, etc.
2. A long-term solution to attack the underlying causes. In practice the solutions cannot succeed without considering a fourfold background: grass-root level action, rigorous formulation of the problems, ecological efficiency, and maintenance of soil quality.

1. *Grass root level action.* Three main obstacles appear:

a) Most projects run for a maximum of 5 years and are not always long enough for the initial difficulties to be sorted out or to win the confidence of the local people.
b) The level of investment is often too high for the receiver country to maintain after the aid finishes. Very little financial help is directed towards small-scale local agriculture. Most bilateral aid is still spent on prestigious schemes that do little to help the poor who are in most immediate need.
c) How to introduce new technologies at the village level where individual farmers or groups require technical guidance through appropriate extension activities. These include: low-cost techniques of soil and water conservation which have been developed to increase food production. Improved varieties of plants exist, a promising variety of sorghum has been identified, a breakthrough in millet is on the way, improved varieties of maize, cassava, and cowpea have been produced.

For a poor farmer, soil conservation is a luxury that he may not be able to afford, except through extended bush fallows. But even these are breaking down under pressure of increasing population. In addition, little money has been invested by governments and aid agencies in soil conservation. Incentives to help villagers to sustain conservation schemes is a new idea. Combating land degradation through technical solutions can be achieved on a short time scale: these technical solutions are the most efficient in the immediate future, but they are only palliative. To transform these palliative solutions into curative ones, changes of socio-economical structures are necessary. Such a long-term approach requires political will, integrity, and stability.

Improvement of international technology transfer so that it will reach the grass root level is essential. The village level is the level of survival; the majority of the success tends to be small-scale and localized. At this level a new philosophy is needed: farmers and stock breeders are very impressive in their knowledge of the land, but this knowledge is turned toward land exploitation rather than conserving resources; but until land users have satisfied their needs for food and energy resources, conservation has little meaning for them.

Production and conservation are linked: the government subsidies for conservation must be as high as those for production.

2. *More rigorous formulation of the problem.* An efficient organization is required to consider in greater detail:

a) the natural resources potential,
b) the main natural processes leading to degradation,
c) the man-induced factors leading to land degradation.

Research consists mainly in monitoring a number of physical, biological, and social factors and their combinations for identification of adverse mechanisms occurring in drylands.

We know how to acquire these indicators: they should be investigated locally for each program. A reasonable pilot phase will help the central governments, the aid agencies, and the individual farmers:

1. to understand the vulnerability of ecosystems vis à vis land degradation,
2. to predict the risks before the process becomes irreversible,
3. to control the process,
4. to choose the most suitable solutions.

A dryland soil takes several centuries to form; it can be spoiled in less than a decade. Vulnerability depends mainly on climatic background but also on man's poverty. FAO in 1971 estimates that 70% of the world's soils are poor, vulnerable, and unsuitable for agriculture because they are in regions which are too cold, or too steep, or too arid (meaning that the arable layer is too thin). The drylands are precisely those where arable land vulnerability is at a maximum and soil fertility minimum. "Low soil fertility, especially in west Africa, is an example of a problem which requires sustained basic research". (Eicher 1986).

If soil erosion is a natural worldwide phenomenon balanced by regeneration, it is at a maximum in the tropics where soils are not very fertile. The richest countries can largely mask all their deficits by the application of fertilizers; but it is tempting to adhere to the idea that fertilizers, pesticides, and imported fossil fuel-based production are not suitable for poor countries. Already much of the food presently consumed by the inhabitants of arid and semi-arid ecosystems is imported from other climatic zones. Drylands will serve as the home for about 850 million people by the year 2000. At the same time, our ability to continue to expand high-input agriculture is severely limited (Nabhan 1987).

Other solutions exist which consist basically in taking a long hard look at the design of drylands agrosystems. A return to a better knowledge of the Earth itself at the ground level, a return to a proper identification of the priority action areas and to the collection of information with a clear notion of practical use are the first steps of action.

3. *Ecological efficiency.* In modern agriculture, efficiency is often formulated in economic terms such as productivity of labor or cash inputs. In fact ecological

efficiency should become the key factor, with emphasis on the knowledge of the natural ecosystems to help to optimize the agro-ecosystems.

4. *Maintenance of soil quality.* The most efficient and modern agrosystems do not differ fundamentally from the most traditional ones and can be considered often as a systematic improvement of them. The empirical objectives of the most ancient practices and the scientific aims of the presentday ones converge in soil and water conservation. To achieve this goal the following techniques are required:

a) To keep the soil covered with growing plants or crop residues for all or most of the year. To return organic matter and mineral nutrients to the soil.
b) Multiple-cropping schemes with trees and shrubs conceived to increase productivity and conservation and to moderate the effects of drought, flooding, and range of temperature.
c) Maximum use of perennial crops to avoid disturbance of the soil.
d) Deep-rooted species to create an upward transfer of soil nutrients.
e) Perennials to protect the soil against wind and water erosion.

The set of conditions for sustainable agriculture is enumerated by Dover and Talbot (1987) as follows:

1. Replenishment of soil nutrients removed by crops.
2. Maintenance of the soil's physical condition.
3. Constant or increasing humus level in the soil.
4. No build-up of weeds, pests, or diseases.
5. No increase in soil acidity or toxic elements.
6. Control of soil erosion.
7. Minimization of off-farm contamination of the environment.
8. Maintenance of adequate habitat for wildlife.
9. Conservation of genetic resources."

Baker (1982) gives the most complete summary of "the main criteria for project identification, selection, and assessment in terms of the anti-desertification struggle:

1. All resource-based projects must aim at planning specifically for a sustained ecological balance between producer and resources. Anti-desertification/conservation should be built in from the outset, the environment impact statement being one way of doing this.
2. All land-use proposals should be within the context of a sound national resource policy and legislative framework, leading to systematic, integrated, land-use planning.
3. Full recognition must be made of the fact that all land-use proposals must deal explicitly with the impact of drought, which is a reasonable expectation in these areas.
4. Explicit attention must also be given to the management, training, and extension components at all levels for all projects envisaging changes in resource-use technology or land use.

5. Research and planning must operate within the framework of production systems.
6. Ecological problems require ecological solutions which, in turn, require the bringing together of social and physical thinking.
7. It is essential that local participation be involved in project work from the outset. Planning is a two-way flow. The previous orientation of taking projects from the planning office, or research from the research station, to the field is inappropriate and accounts for the failure of much technology transfer.
8. Many factors directly relevant to desertification lie outside the areas most directly affected, e.g., opportunities for migration, exchange and barter, symbiotic relations, opportunities for work or investment.
9. Effective monitoring is essential.
10. There is still a considerable amount of basic data gathering to be done, especially at the production-system level. Since this is often what is being changed by planning, it is essential that the workings of these systems be properly understood. This involves values, strategies, objectives, as well as techniques".

The attempt was made in this book to review on an integrated scale the nature of desertification and land degradation, to determine the processes involved in them, and to apply effectively the available resources of manpower, capital, skill, technology, research, and training *to halt* land degradation, *to conserve* the existing resources, *to regenerate* the degraded resources, and *to make more sustained use* of the threatened areas in the future to improve the quality of life of the inhabitants (Baker 1982).

Chapter 7 New Developments in Desertification (September 1993)

"The future of life on earth was once entirely beyond human control. We lived for most of our time on this planet as one species among many, depending on natural processes that functioned without our intervention. Today the future of life on earth depends on human action. At no time during the last hundred thousand years has the survival of the human species along with most other species been so imperiled. We have reached this state through an overwhelming increase in human numbers, made possible through our growing command of energy resources that we have used to modify the surface of the earth" Dasmann (1984).

1. Desertification, a Fluctuating Concept

The word desertification describes a fluctuating concept but may have a basic simple meaning: the degradation which introduces desert-like conditions or transforms an ecosystem into a desert-like landscape.

It acquired its patent of gentility when Aubreville (1949) used it – without defining it – for the first time (p. 6) and observed: *"Ce sont de vrais déserts qui naissent aujourd'hui sous nos yeux dans des pays où il tombe annuellement de 700 à plus de 1500 mm de pluies."* (Real deserts appear today under our eyes in areas where the annual rainfall is between 700 and 1500 mm.) This sentence of Aubreville is not premonitory or intuitive; it results directly from field observations. He, of course, thought of desertification as human-induced vegetation degradation and soil erosion following shifting cultivation in the humid tropics, and more precisely here the Sahelo-Sudanian ecosystem of the northern former Oubangui-Chari (at present Republic of Central Africa).

When at the end of the 1970s the ecological crisis reached south of the Sahara, in the Sahel, its first towering acme, UNCOD, proposed in 1977 a map of areas at risk of desertification from varying degrees and a comprehensive definition (p. 7, resolution number 7), which was propagated by the mass media: "Desertification is the diminution or destruction of the biological potential of the land, and can lead ultimately to desert-like conditions. It is an aspect of the widespread deterioration of ecosystems, and has diminished or destroyed the biological potential, i.e., plant and animal production, for multiple use purposes at a time when increased productivity is needed to support growing populations in quest of development."

In spite of this UN definition, the numerous observations made in different regions ended in the multiplication of the definitions which may all be classified in five categories, namely:

1. The spread of desert-like landscapes budding from the deserts with the wrong idea of encroaching desert.
2. The spread of exacerbated mechanisms of land degradation, characteristic of deserts (wind erosion, water erosion) in the semi-arid and dry subhumid ecosystems.
3. The decrease in the biological productivity of land. This is the sense given by Dregne (1983): "Desertification is the impoverishment of terrestrial ecosystems under the impact of man. It is the process of deterioration in these ecosystems that can be measured by reduced productivity of desirable plants, undesirable alteration in the biomass and the diversity of the micro and macro fauna and flora, accelerated soil deterioration, and increased hazards for human occupancy." Olsson (1985) adds a notion of time scale: "The long-term decrease of the lands biological productivity caused or accelerated by human activities in combination with the climate."
4. Degradation of the social and economic conditions. This was already introduced by Kates et al. (1977), who wrote: "It involves destructive processes in which the productive base deteriorates and the social system is imperilled. Unlike drought, which is usually a short-term diminution of a available moisture, the physical processes involved in desertification are long-term, chronic, and pervasive."

   Impoverishment of the social system has also been developed by Warren and Maizels (1977). They introduced in the definition the idea of a sustained economic impact, namely a decline in the yield: "A simple graphic meaning of the word desertification is the development of desert-like landscape in areas which were once green. Its practical meaning ... is a sustained decline in the yield of useful crops from a dry area accompanying certain kinds of environmental change, both natural and man induced."
5. The ultimate nonproductive desert-like and irreversible status of a deteriorated environment. "Desertification is the name given to the processes whereby such ecosystems lose this capacity to revive or to repair themselves" (Hare 1985). Hare introduces the idea of natural irreversible deterioration.

This dispersion is not satisfactory. In resolution 44/172 of December 1989 the United Nations General Assembly asked UNEP to undertake a general reevaluation of desertification to be discussed at the United Nations Conference on Environment and Development (UNCED) in Rio de Janeiro in 1992.

The *Third Meeting of the Technical Advisory Group on Desertification Assessment and Mapping*, covened by UNEP, 5–7 June 1991, proposed the following definition: **"Desertification is land degradation in arid, semi-arid and dry subhumid areas resulting mainly from adverse human impact."**

It is assumed that within the context of the above definition, **land** implies a declining of the *crop*, deteriorating *vegetative cover*, exacerbation of external dynamics at the *land surface*, qualitative and quantitative retrogression of *water resources*, degrading *soils* and pollution of the *air*. **Degradation** is a point of evolution which leads to reduction of the resource potential.

Taking into account this definition, the special session of the UNEP Governing Council, held in Nairobi (3–5 February 1992), summarized the impact of land degradation as follows: "... about 73 per cent of the rangelands, 47 per cent of the rainfed croplands and 30 per cent of the irrigated lands in the drylands, thus affecting more than 3.6 billion hectares of the total world area of arid, semi-arid and dry sub-humid lands, or about 25 per cent of the total world land area and about 900 million people, or one sixth of the world population" and "... the impact of desertification on Africa in particular where it is a serious contributory factor to famines, such as those which occurred in 1984 and 1985, affecting between thirty million and thirty-five million people, and in 1991, when some thirty million people were threatened by famine and needed urgent external food aid in order to survive." "Desertification is a global phenomenon directly affecting more than 60 per cent of the countries of the world."

The 1991 UNEP definition of desertification, which insists on man-induced causes of desertification, was discussed and amended by UNCED in Rio de Janeiro (Brazil) in June 1992. UNCED, after international negotiations, adopted the following definition: **"Desertification is land degradation in arid, semi-arid and dry subhumid areas resulting from various factors including climatic variations and human activities."** After Rio, the United Nations General Assembly resolution 47/188 decided to organize a followup and established an Intergovernmental Negotiating Committee to elaborate an International Convention to Combat Desertification (INCD) by June 1994. The first (of five) negotiating session of this Committee to combat desertification in those countries experiencing serious droughts and/or desertification, particularly in Africa, took place in Nairobi in June 1993.

The UNCED definition determined in Rio de Janeiro is justified. The meteorological data obtained at the beginning of the 1970s registered a climatic break with a decrease of rainfall in the southern latitudes, mainly in the tropical drylands but also in the monsoon and western winds ecosystems where rainfall became more irregular, and a noticeable warming in the higher latitudes. In Africa between 1951 and 1973 the 500 mm/year isohyet had shifted by 3° towards the south (Carbonnel 1993, unpubl.; Fig. 85). This phenomenon is considered as independent of human-kind but its environmental effects are aggravated by the demographic increase.

After the publication of the first edition of this book (1991), the greatest reserve which appeared was at the level of the statement (p. 16) that desertification means "irreversibly sterile land". Unfortunately, the latest UNCED definition does not answer this question. Neither the notion of degrees of land degradation nor the ultimate degree when land degradation becomes irreversible, which should be the usual meaning of desertification, are mentioned. In practice, the degree of land degradation may be considered as irreversible when seeds have disappeared from the soil or when, according to Dregne (1983), the soil is degraded to such a point that the seeds in the soil cannot germinate because the soil has lost its ability to conserve humidity.

We wish to state again that, according to field observations, the notion of irreversibility has a double content, of time and of economy:

- Of time: when introducing the idea of irreversibility in desertification we introduce simultaneously the idea of a time scale; irreversibility could correspond to one human generation, 25 years. The generation which stands desertification is not able to repair it. Rehabilitation, if possible, is the task of several future generations.
- Of economy: in Texas and New Mexico, in Tunisia, and in the Chinese deserts the degradation by wind results in the same hummock blow-out conditions. The degradation can be called "economically irreversible" because it would take large and expensive machinery to level the hummocks and restore the land to its original smooth conditions. And we are not sure that this will lead to successful results because it would disturb the sand structure and cause it to become a friable material that is easily deflated.

The expression "green desertification" was proposed and means the substitution, during all the stages leading to desertification, of a nonpalatable vegetation that is economically not interesting for a very useful edible one.

If the concept of desertification is synonymous with **land degradation**, it must be considered as the most important environmental problem of the world today. If the concept contains the idea of **irreversible land degradation**, i.e., the end of a chain of transformations which lead an area to become a sterile desert-like area with desert-like conditions, this status is almost nonexistent. According to Dregne (1983), only tiny areas of the planet are affected; Dregne's assessment is 0.2%.

In Africa, the semi-arid ecosystem of Khordofan was considered one of the most degraded areas. Olsson (1985) performed a satellite-based monitoring of the desertification in eastern Khordofan and observed that "signs of desertification can mainly be seen during extremely dry periods ... It has not been possible to find a consistent trend of degrading landscape".

In her study of fuelwood resources and land degradation in the same area, Olsson (1985) confirms this absence of desertification "no woody species seemed to have been eradicated from the area, no ecological zones had shifted southwards and the boundaries between different vegetation associations appear to be same as they were 80 years ago".

In China, drought and land degradation have been recognized by the Chinese as a problem for at least two millenia. The destructive process of environmental degradation is mainly contemporaneous with socio-economic disturbances and political upheavals aggravated by droughts (Zhu Zhenda and Liu Shu 1983).

Toulmin (1988), returning from the semi-arid regions of Mali, had the pleasure to describe recovering greening areas.

This shows the fundamental necessity of differentiating between land degradation, which can be solved, and desertification, which to be solved requires expansive or sophisticated techniques that are often economically too luxurious for the people inhabiting the area.

After obtaining a satisfactory definition, a second question arises:

2. *What is Really Going on?*

Three levels should be considered: causes, mechanisms, and impact on the environment and on socio-economic structures. The main controversy, however, concerns the causes, which are numerous and intricate. They can be classified as follows:

7.1 Natural Causes: Droughts Since the Beginning of the Century

- In the 20th century the Sahel has suffered four droughts in 1900–1903, 1911–1920, 1939–1944, and 1968–1985, with the maximum deficit in 1972–1973 and 1982–1984. In total there have been 47 deficit years between 1900 and 1990.
- The drylands in China had the same year-on-year variations and runs of dry years as did the Sahel from 1900 to 1909, 1913 to 1930 and 1965 to 1990.

Nevertheless, nobody has been able to demonstrate an upward or a downward trend for the number and severity of droughts in the drylands during contemporary times. But all the scientists in Africa have observed a slipping towards the south of the isohyets corresponding to the drylands (Fig. 85).

7.2 Human Causes

- Stagnation and even declining economies, mainly in the African drylands: only four countries in the Sudano-Sahelian region achieved more than 3% economic growth between 1980 and 1985; most experienced a decline (Warren and Khogali 1992).
- Civil wars and the associated train of disturbances: the three worst examples in 1993 are Angola, Somalia and Sudan.
- Bad agricultural choices: in the Sahel where drought was very severe in 1983 and 1984; in 1984 in five countries, namely Burkina Faso, Mali, Niger, Senegal, and Chad, the cotton yield beat all the records with 154 million tons (22.7 million tons in 1961–1962). But cotton is mainly produced by modest farmers and replaces food crops. The Sahel in the same year imported 1.7 million tons of cereals (200000 t/year in the early 1960s), which is also a record (Timberlake 1985).
- Demographic explosion common to all dry ecosystems, mainly in Asia and Africa. In the last half century, many drylands have in fact seen a rapid increase in their population at a rate of 2.5 to 3%/year, i.e., doubled at each generation. The *World Map of Soil Degradation* (Oldeman et al. 1990) highlights the correlation between soil degradation and demographic excesses, particularly in the drylands of western China, in the Sahel, in the Maghreb, in the Near and Middle East, and in eastern Africa, mainly Kenya where the population growth surpasses 4%/year.

The demographic explosion is responsible for: (1) increasing pressure on the land; (2) vegetation changes; (3) shortening of the fallow cycles and all the consequences, such as poorer soils and greater necessity of human labor for poorer results; (4) cultivation of marginal land; (5) soil degradation. Zhu Zhenda and Liu Shu (1983) have demonstrated in the sandy steppes of Inner Mongolia that wind erosion has truncated 6 to 20 mm/year in the previously fixed sands, i.e., 200 to 300 t/ha/year.

Demography means also the animal demographic explosion, with overgrazing, particularly in the drylands of Central Asia (Kazakstan, Kirghistan, Uzbekistan, Tadjikistan and Turkmenistan), of China (mainly in Sinkiang), in Latin America, in North America and less in Australia.

Uncontrolled demography has triggered overcultivation, overgrazing, overexploitation of marginal lands and shortening of fallow. Accompanied by bad management, environmental changes have appeared due to exacerbated physical and chemical mechanisms of degradation.

In the middle of 1985 the famine spread out through Africa from the Atlantic coast to the Horn of Africa and toward the south into Mozambique and the Bantustan lands of South Africa (Timberlake 1985). This situation is a temporary end of a long chain of events. The first apparent event was the drought which began at the end of the 1960s from 1968 to 1973 in the Sahel. After an interruption of a couple of years, the drought began again and culminated in 1984. The Horn of Africa, less affected than the Sahel during the first period of drought (1968–1973), was more dramatically affected in the second period of drought, mainly in 1984. Thirty-five million people experienced famine and 10 million were ecological refugees in the whole drought-affected countries of Africa.

Droughts are natural disasters, but famines are not caused simply by a deficit rainfall. Bad use of the land, mismanagement, and overexploitation, which became, because of demographic explosion, a necessity in the Third World, transformed drought into a major disaster which affected an increasing number of people. In the end of the 1960s drought affected 18.5 million people per years. The number increased to 24.4 million people per year in the years 1970s, and 30 million people in Africa alone for the year 1985 (Timberlake 1985).

The couple drought–famine exists only in a context of bad management, poverty, economic disorganization, underdevelopment, and, above all, illiteracy or lack of education. The drought of 1983 in the USA and in Australia did not end in death by famine; the victims were only economic victims and precisely those who before the drought were debt afflicted. Desertification is a variable phenomenon according to the degree of development. With financial resources, technical ability, skill, an analytical approach of the natural and human problems, political balance, and a balanced ratio between land surface and population, land degradation and even desertification are not problems.

Some Conclusions

1. Desertification is the ultimate step of land degradation: irreversibly sterile land, meaning irreversible in human terms and within practical economic limitations.
2. Where does the responsibility for land degradation lie? With human and/or natural causes? The discussions are ongoing. Man is both the creator and the victim of land degradation but drought is the revealer and the accelerator. Both man-made and natural causes are catalysts to land degradation. The proportion of each is difficult to assess.
3. The existence of desertification itself is also a debate. This debate is clarified by the distinction between land degradation, which may be considered reversible, and desertification, which should be considered as the irreversible degree of land degradation.
4. Another differentiation could help in the approach of the concept: (a) deterioration of the land quality; and (b) biological degradation of the productivity of land. The first may be seen as caused by man; the second has been caused by climatic factors.
5. The notions of carrying capacity and of the equilibrium of dry ecosystems are not quantitatively known. The level of rehabilitation of the equilibrium in dry ecosystems between two droughts is very difficult to assess.
6. The notion of sustainable development, which was launched by the Brundtland report, is still a theoretical notion and leads to the question: What is development, when development is in opposition with environmental conservation? The notion of limits of development should be taken into account in the framework of environmental preservation.
7. With the demographic growth, in advance of the growth of agricultural production, the bad use and overuse of land in the Third World has become an obligation and has transformed drought into an economic and social disaster.
8. We are able to distinguish three facets of desertification: **causes, processes**, and **effects**. Can we fight on a reasonable time scale against the causes? Certainly not, because we know nothing about the mechanisms of drought, and also because several generations will be necessary to transform the socio-economic structures of the affected countries. But immediate should be the battle against the physical processes and their effects of deterioration, and support for the needs as defined by the farmers and pastoralists themselves, who are the actual participants of the fight: "The front line troops in this battle against desertification are generally impoverished and illiterate, overworked peasant farmers or pastoralists and most often women. Their priority is to feed their families, now. The challenge, and the opportunity, for the Convention, is to provide real support to them, of the kind they need, so that land use becomes sustainable, not destructive" (Elisabeth Dowdeswell, Executive Director, UNEP, first session at the INCD Nairobi, May–June 1993).

9. At the level of strategies, the advantage of a global strategy is to bring about the possibility of thinking on a megascale, but land degradation cannot be combated globally because, even if it is a megascale phenomenon, it is in fact the juxtaposition of difficulties and problems at a local scale.

In this framework of global solutions are located the immense former USSR irrigation schemes and the greenbelts, among them the Algerian, Kazaks and Chinese ones: what are the objectives of such greenbelts? To contain the desert and mainly the aeolian material – essentially dust – which is exported out of the desert. But the desert is a continuum which has no fixed limits; its boundaries are fluctuating and so unstable that it is very difficult to fix a place for the greenbelts. On the other hand, a greenbelt requires large-scale collective work, most often achieved by a totalitarian political regime, in contradiction with the individualism of the rural people.

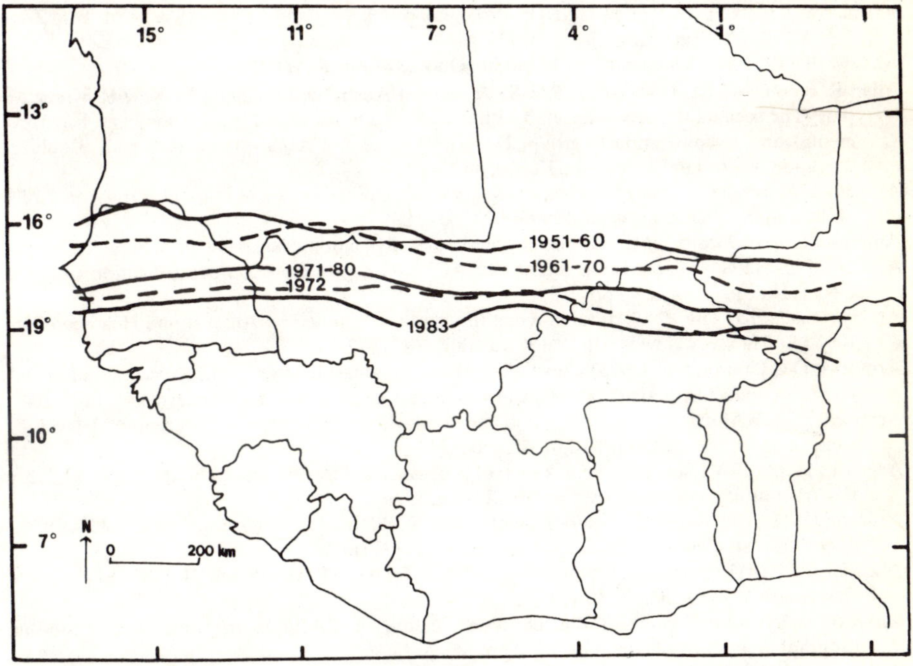

Fig. 85 Shifting of the isohyets in western Africa from 1951 to 1983 (after Carbonnel, unpublished).

References

Abbott EA (1884) Romance of many dimensions. Basil Blackwell, Oxford, 102 pp

Ahlcrona E (1988) The impact of climate and man on land transformation in Central Sudan. Applications of remote sensing. Meddelanden, Avhandlinger, 103, 140 pp

Ahmad MU (1988) Ground water resources: the key to combating drought in Africa. Desertification control bulletin UNEP Nairobi Kenya 16:2–6

Ahmad YJ, Kassas M (1987) Desertification: financial support for the biosphere. Hodder and Stoughton, London, 187 pp

Aladin NV, Khlebovich VV (1989) (eds) Hydrological problems of the Aral sea (8 papers). Trud Zool Inst Ak Nk T 199 (in Russian)

Alissow BP (1954) Die Klimate der Erde. Berlin, Quoted by Suzuki H (1981)

Allan R, Heathcote RL (1987) The 1982–83 drought in Australia. In: Glantz M, Katz R, Krenz M (eds) The societal impacts associated with the 1982–83 worldwide climate anomalies. Environmental and societal impacts group, National Center for Atmospheric Research. Boulder, Colorado, USA. UNEP, Nairobi, Kenya, pp 19–23

Almeida GA, Schutz L (1983) Number, mass and volume distributions of mineral aerosol and soils of the Sahara. J Climate Appl Meteorol 22:233–243

Amindo J (1986) Gazelle ranch is a splendid success. Daily Nation. Nairobi, Kenya, Sept 3, p 13

ANEN-UNEP (1988) How to stop the desert. An NGO action guide on desertification Control. UNEP, Nairobi, Kenya, 49 pp

Anonymous (1983) The 1980's drought-worst this century? Quoted by Allan R and Heathcote RL 1987 J Soil conserv New South Wales Australia 39(2):172–175

Armbrust DV, Paulsen GM (1973) Effect of wind and sandblast injury on nitrate accumulation and on nitrate reductase activity in soyabean seedlings. Commun Soil Sci Plant Anal 4(3):197–204

Armitage FB (1985) Irrigated forestry in arid and semi-arid lands: a synthesis. International development research centre. Ottawa, Canada, 160 pp

Arnoldus HJM (1974) Soil erosion: a review of processes and assessment techniques. FAO/UNEP Expert consultation on soil degradation. Rome, 53 pp

Aronson JA (1988) Haloph: salt-tolerant plants of the world. In: Arid lands newsletter, Vol 27. Office of arid land studies. Univ Arizona, Tucson, Arizona, 20 pp

Atkinson G (1984) Erosion damage following bushfires. In: Allan R, Heathcote RL 1987 J Soil conserv New South Wales Australis 40:4–9

Aubreville A (1949) Climats, forêts et désertification de l'Afrique tropicale. Soc d'éditions géographiques et coloniales, Paris, 352 pp

Babaev AG (1986a) Arid lands of the USSR and their productive forces. In: Arid land development and the combat against desertification -- an integrated approach -- UNEP, UNEPCOM, Moscow, pp 114–123

Babaev AG (1986b) (ed) Principles and methods of shifting sands fixation. Cent Int Proj GKNT, Moscow, 131 pp

Babaev AG, Freikin ZG (1977) Poostyni SSSR vchera, segodnya, zavtra (USSR deserts yesterday, today, tomorrow). Mysl' Publ., Moscow, 352 pp

Babaev AG, Zonn IS, Orlovsky NS (1984) The USSR experience in desert reclamation and desertification control. UNEP, UNEPCOM, Moscow, 104 pp

Babau MC (1983) The erosive capacity of rainfall. World Climate Programme 41, WMO, 28 pp

Bagnold RA (1941) The physics of blown sand and desert dunes. Methuen, London, 265 pp

Baker R (1982) Desertification: cause and control. A study of the U.N. Plan of Action and its possible applications. Sci Rev Arid Zone Res, Sci Publ, Jodhpur, 285 pp

Bardsley KL, Eraser AS, Heathcote RL (1983) The second Ash wednesday: 16 february 1983. Quoted by Allan R, Heathcote RL 1987 Austr Geogr Stud 21:129–141

Barrot P, Seye C (1989) "Consommez Sahélien!" Information, lettre d'information, novembre, CILSS, Club du Sahel, Paris, 6:14–15
Baumer M (1987) Agroforesterie et désertification. Le rôle possible de l'agro-foresterie dans la lutte contre la désertification et la dégradation de l'environnement. ICRAF, Cent Tech coopération agricole et rurale, 260 pp
Beaumont P (1986) Desertification in the Middle East. In: Arid land development and the combat against desertification: an integrated approach. UNEP, UNEPCOM, Cent Int Proj, Moscow, pp 79–84
Beckman GG, Coventry RJ (1987) Soil erosion losses: squandered withdrawals from a diminishing account. Search 18:21–26
Bell AE, Morrison S (1979) Analytical study of the Ogallala aquifer in Gaines County, Texas, Austin. Tex Dep Water Resour, 1 p
Benito GA (1974) Sand dune stabilisation at El Alaiun, West Sahara. Int J Biomet. 18:142–144
Billy B (1978) Navajo Indian culture and lands. In: Peterson DF, Berry Crawfords A (eds) Values and choices in the development of the Colorado river basin. Univ Arizona Press, Tucson, pp 269–243
Birot P (1981) Les processus d'érosion à la surface des continents. Masson, Paris, 579 pp
Bohrer V (1979) Eastern New Mexico University, Portales. Interview with David Sheridan, 19 Oct 79, in Sheridan 1981
Bovill E (1921) The encroachment of the Sahara on the Sudan. J R Afr Soc 20:175–185, 259–269
Brokensha D, Warren DM, Werner O (1980) Indigenous knowledge systems and development. Univ press of America, 460 pp
Brookfield N (1970) Dune trends and wind regimes in Central Australia. Inst Geomorph (Suppl) 10:121–153
Brooks CEP (1926) Climate through the ages. A study of the climatic factors and their variations. Ernest Benn Ltd, London, 439 pp
Brown J (1986) An ancient oil brings new hope to Chinese farmers. World food programme news Oct 86-March 1987:2–3
Brown LR, Wolf EC (1984) Soil erosion: quiet crisis in the world economy. Pap n° 60, Worldwatch Inst, Washington DC, 50 pp
Brunhes J (1956) La géographie humaine, 3ème édn. PUF, Paris, 393 pp
Bubenzer GD, Weis GG (1974) Effects of wind erosion on production of snap beans and peas. J Am Soc Hortic Sci 99(6):527–529
Buchanan F (1807) A Journey from Madras through the country of Mysore, Kanara and Malabar, 3 vol East India C°. London
Budyko MI (1958) The heat balance of the Earth's surface. Stepanova NA (trans) US Dep Comm, Washington, 259 pp
Buringh P (1981) An assessment of losses and degradation of productive agricultural land in the world. FAO, Roma Working group on soils policy. California Department of Conservation (undated) Quoted by Sheridan D 1981
Caborn JM (1957) In: Geiger R (ed) The climate near the ground. Harvard Univ Press, Cambridge, Massachusetts
Cana F (1915) The Sahara in 1915. Geogr. J XLVI: 33–357
Carter LJ (1977) Soil erosion: the problem persists despite billions spent on it. Science 196:409–411
Chelpanova OM (1963) Climate of the USSR. Issue 3, Central Asia, Gidrometeoizdat, Leningrad (in Russian)
Chembarisov EI (1974) Water salinity dynamics in some Central Asian rivers as a result of irrigation. Sci Cand Agric Diss, Moscow (in Russian)
Chemin MC (1985) Analyse des milieux éoliens dans le Sud-Est marocain et protection des cadres de vie. Primauté de la notion d'échelle. Thèse de 3ème cycle. Univ Reims, 199 pp
Chepil WS (1945) Dynamics of wind erosion. Soil Sci 60:305–320, 397–411, 475–480
Chisholm A, Dumsday R (1987) Land degradation: problems and policies. Cambridge Univ Press, London New York, 404 pp
Christiansson C (1988) Degradation and rehabilitation of agropastoral land. Perspectives on environmental change in semi-arid Tanzania. Ambio 17–2:144–152
CILSS - Club du Sahel (1989a) Récoltes céréalières, mieux connaître pour mieux gérer. Information, OCDE, Paris, 4:3–6
CILSS - Club du Sahel (1989b) Lettre d'information, Novembre, Paris, 6:24

Clark EH (1985) The off-site costs of soil erosion. J Soil Water Conserv 40:19–22
Clarke R (1986) The handbook of ecological monitoring. GEMS/UNEP Publ, Clarendon Press, Oxford, 298 pp
Colclough JD (1973) Salt. Tasmanian J Agric 44:171–180
Collectif (1989) Evolution des milieux sahélien et soudanien 1952–1987. Etude des critères accessibles par télédétection. Transect Mauritanie-Mali. Inst Géogr Natl (IGN FRANCE International), ORSTOM, Univ Reims (LGPZ), UNEP, Ministère des Affaires Etrangères, Ministère de la coopération fraçaise, Paris, 117 pp
Colwell JE (1974) Vegetation canopy reflectance. Remote Sens Environ 3:175–183
Cooke RU, Warren A (1973) Geomorphology in deserts. Batsford, Lond, 374 pp
Cooke RU, Brunsden D, Doornkamp JC, Jones DKC (1982) Urban geomorphology in drylands. United Nations Univ, Oxford Univ Press, New York, 324 pp
Curry W (1979) The desert blooms, but at an ever higher price. Wash Post June 25:A3
Dasmann RF (1984) Environmental conservation, 5 edn. John Wiley and Sons, New York, 486pp
Davies J (1987) Desalination – distillation or reverse osmosis? Developing world water. Grosvenor Press Internat, Hong-Kong, 588 p:222–224
Davy EG, Mattei F, Solomon SI (1976) An evaluation of climate and water resources for development of agriculture in the Sudano-Sahelian zone of West Africa. WMO 459:XV+289 pp
Delany AC, Zenchelsky S (1976) The organic component of wind erosion generated soil derived aerosol. Soil Sci 121–3:146–155
Derbyshire E (1978) The middle Huang He loess lands. Geogr J 144:191–194
Diderriksen RI (1981) Resource inventory — sheet, rill and wind erosion. Trans ASAE 24:1246–1252
Dissmeyer GE, Foster GR (1980) A guide for predicting sheet and rill erosion on forest land. Tech Publ SA-TP 11. For Serv, US Dep Agric, Washington DC, 40 pp
Di Xingmin (1988) The formation and development of desertification and the principles to combat it in the western part of the steppe zone (with Yanchi region as an example). Training package. International training course on desertification control. Lanzhou, China, 198:52–66
Dortignac EJ (1963) Rio Puerco: abused basin. In: Hodge C, Duisberg P (eds) Aridity and man. Am Assoc Adv Sci, Washington DC, BLM supranote 50:512–514
Douglass G (1984) The meanings of agricultural sustainability. In: Douglass G (ed) Agricultural sustainability in a changing world order. pp 3–29
Dover M, Talbot LM (1987) To feed the Earth: agro-ecology for sustainable development. World Resour Inst, a Centre for Policy Research, Washington DC, 88 pp
Downing TE (1986) Climatic change and variability in eastern Africa. Desertification Control Bulletin, UNEP, Nairobi, Kenya, 13:34–35
Dregne HE (1977) Desertification of arid lands. Econ Geogr 53–4:329
Dregne HE (1983) Desertification of arid lands. Advances in deserts and arid land. Technology and development 3. Harwood Academic Publications, New York, 242 pp
Dregne HE (1986) Magnitude and spread of the desertification process. Arid land development and the combat against desertification: an integrated approach. UNEP UNEPCOM, Moscow, pp 10–16
Dregne HE, Tucker CJ (1988) Desert encroachment. Desertification Control Bulletin, Nairobi, Kenya, 16:16–19
Dudal R (1981) An evaluation of conservation needs. In: Morgan RPC (ed) Soil conservation, problems and prospects. Wiley, Chichester, pp 3–12
Dzhanpeisov R, Dzhamalbekov Y (1978) Problems of soil conservation in Kazakhstan. Probl Desert Dev 4:63–69
Ebomo EG (1983) Shelterbelts: their establishment and effects, with particular reference to northern Nigeria. Ph D Thesis, Dep For Wood Sci, Bangor
Eckholm EP (1976) Losing ground-environmental stress and world food prospects. Norton, New York, 47 p
Eicher CK (1986) Transforming African agriculture. Hunger Proj Pap, San Francisco, 4:35
Ekern PC (1977) Turbidity and sediment, rating curves for streams on Oahu, Hawaii. In: Soil erosion: prediction and control. Soil Conserv Soc Am Spec Publ 21:242–254
El Amami S (1984) Les aménagements hydrauliques traditionnels en Tunisie. Cent Rech Génie rural, Tunis, République tunisienne
El-Baz F (1988) Origin and evolution of the desert. Interdiscip Sci Rev, Arrowsmith Ltd, 13/4:331–347

Elgabaly MM (1971) Reclamation and management of salt affected soils. In: Salinity seminar. Baghdad, FAO, Roma, Italy. Irrigation and drainage, 7:50–79
Ellison WD (1944) Studies of raindrop erosion. Agric Eng 25–4:131–136, 25–5:181–182
Ellison WD (1947) Soil erosion studies. Part 1. Agric Eng 28–4:145–146
El Swaifi SA, Dangler EW (1982) Rainfall erosion in the Tropics. A State-of-the-Art. In: Soil erosion and conservation in the Tropics. ASA Spec Publ. Reprinted 1986. Am Soc Agron Soil Sci Soc 43:1–25
Engsten HP, Hardie LA (1978) Saline lakes. In: Lerman A (ed) Lakes. Springer, Berlin Heidelberg New York, pp 237–293
FAO (Food and Agriculture Organisation) (1984) Spineless cactus. Filmstrip commentary, 11 pp
Farmer G, Wigley TML (1985) Climatic trends for tropical Africa. A research report for the overseas development administration. Climatic research unit. School of environmental sciences. Univ East Anglia, Norwich, UK, 136 pp
Faure H, Gac JY (1981) Will the Sahelian drought end in 1985? Nature (London) 291:475–478
Fedorovich BA, Babaev AG, Kes AS (1963) Natural conditions in deserts of Central Asia and Kazakhstan and ways of their development. In: Natural conditions, animal husbandry and forage base in deserts. Izd AN TSSR Ashkhabad: 7–22 (in Russian)
Finkel HJ (1986) Semi-arid soil and water conservation. CRC Press, Boca Raton, Florida, USA, 126 pp
Firmin R (1968) Forestry trials with highly saline or sea water in Kuwait. In: Boyko H (ed) Saline irrigation for agriculture and forestry. Junk, The Hague, The Netherlands, pp 107–132
Forse B (1989) The myth of the marching desert. New Sci 121–1650:31–32
Fryrear DW (1981) Long-term effect of erosion and cropping on soil productivity. In: Péwé TL (ed), Desert dust: origin, characteristics and effect on man. Geol Soc Am 186:253–259
Fryrear DW, Stubbendieck J, McCully WG (1973) Grass seedling response to wind and windblown sand. Crop Sci 113–6:622–625
Gans G de (1986) Taking indigenous knowledge seriously. The case of pastoral strategies among Turkana nomads of northwestern Kenya. ILEIA Newsl 5:14–16
Garduno MA (1977) Technology and desertification. UNCOD A/conf 74/6, Nairobi
Geiger R (1959) The climate near the ground. Harvard Univ Press, Cambridge, Massachusetts
Geraghty JJ et al. (1973) Water atlas of the United States. Water Inf Cent. Washington New York
Gerasimov IP, Kuznetsov NT, Kes' AS, Gorodetskaya ME (1983) The problem of the Aral Sea and of an anthropogenic desertification of the Aral Region. Probl Osv Pustyn' 6:22–33
Ghonaim OA, Gabriel B (1980) Desertification in Siwa Oasis, Egypt. Symptoms and Causes. In: Meckelein W (ed) Desertification in extremely arid environments. Stuttgarter Geographische Studien, Stuttgart, pp 157–172
Glantz MH, Katz R (1987) Africa drought and its impacts. Desertification control bulletin, Nairobi, Kenya, 14:22–30
Glantz M, Katz R, Krenz M (1987) The societal impacts associated with the 1982–83 worldwide climate anomalies. Environmental and societal impacts group. Natl Cent Atmos Res, UN Environ Progr, 105 pp
Glazovsky NF (1987) Recent salt accumulation in arid regions. Nauka, Moscow, p 182
Glazovsky NF (1990) The aral crisis: the source, the current situation and the ways to solving it. Int Symp The Aral crisis: origins and solutions, 2–5 October, Nukus, Karakalpak, Auton Sov Social Rep (Uzbek SSR), 55 pp
Gloyne RM (1954) Some effects of shelterbelts upon local and microclimate. Forestry 27–2:85–95
Gloyne RM (1955) Some effects of shelterbelts and windbreaks. Meteorol Magazine MO 590, 84–999:272–281
Graetz RD (1984) Monitoring desertification of arid rangelands using landsat: the Australian experience. Int Symp on integrated control of land desertification Proc, Nat Comm people's Republic of China for MAB, pp 4–18
Grave LM (1976) Technogenic mass exchange and complexes in the Kara Kum canal zone. Probl Osvo Pustyn 3–4:155–163
Grave MK, Grave LM (1986) Typical arid regions of the USSR and their integrated economic development; C. Kara Kum canal. Arid land development and the combat against desertification: an integrated approach. UNEP-UNEPCOM-Moscow, pp 129–132
Graves J (1971) "You ain't seen nothing yet". The water hustlers. San Francisco Sierra club, p 39
Grigoriev AA, Budyko MI (1959) Classification of climates in the USSR. IZV AN SSSR Geogr Ser 3

Gunn R, Kinzer GD (1949) Terminal velocity of water droplets in stagnant air. J Meteorol 6:243
Gupta JP, Aggarwal RK, Raikhy NP (1981) Soil erosion by wind from bare sandy plains in western Rajasthan, India. J Arid Environ 4:15–20
Gutentag ED, Weeks JB (1980) Water table in the high plains aquifer in 1978 in parts of Colorado, Kansas, Nebraska, New Mexico, Oklahoma, South Dakota, Texas and Wyoming. Hydrologic investigations Atlas HA 642 Reston, Va. US Geol Surv, Quoted by Sheridan D 1981
Gutteridge, Haskins, Davey (1970) Murray Valley salinity investigation. River Murray Commission, Canberra
Guyot G (1969) L'érosion éolienne. Bull Tech Inf 237:117–124
Guyot G (1987) Brise vent et rideaux abris avec référence particulière aux zones sèches. FAO Rome, Cah FAO Conserv 15:385
Hacker JB (1987) Genetic shift in *Setaria sphecelata* in response to grazing. Aust J Agric Res 38:907–916
Hagen LJ, Skidmore EL (1977) Wind erosion and visibility problems. Trans ASAE 20–5:898–903
Hall FF Jr (1981) Visibility reductions from soil dust in the western U.S. Atmos Environ 15:1929–1933
Hamilton GH, Lang RD (1978) Reclamation and control of dryland salt-affected soils. J Soil Conserv, New South Wales, Australia, 34:28–36
Hanley ML (1988) Des progrès décisifs dans la phytogénétique du maìs annoncent peut-être une nouvelle "révolution verte." Dev mondial PNUD, NY, 1–4:21–23
Hare FK (1983) Climate and desertification. Rep WMO Roma and UNEP
Hare FK (1984a) Recent climatic experience in the arid and semi-arid lands. Desertification control bulletin, Nairobi, Kenya, 10:15–22
Hare FK (1984b) Climate, drought and desertification. Nature Resour 20:2–8
Hare FK (1985a) Climatic variability and change. In: Kates RW et al. (eds) Climatic impact assessment. Wiley, Chichester, pp 7–68
Hare FK (1985b) Climate variations, drought and desertification. WMO 653:35
Harrison P (1987) The greening of Africa. Breaking through in the battle for land and food. International institute for environment and development. Earthscan Paladin. Grafton books. Collins, London, 380 pp
Heathcote RL (1983) The arid lands: their use and abuse. Longman, London
Heathcote RL (1986) Drought mitigation strategies in Australia: reducing the losses but not removing the hazard, Great Plains Quarterly 6:225–277
Hellden U (1984) Drought impact monitoring – a remote sensing study of desertification in Kordofan, Sudan. Lunds Univ Naturgeogr Inst, rapporter och notiser, Lund, Sweden, 61:61
Hellden U (1988) Desertification monitoring: is the desert encroaching? Desertification control bulletin. Stiles and Mainguet (eds) UNEP Nairobi, Kenya, 17:8–11
Higgins GM, Kassam AH, Naiken L (1982) Potential population supporting capacities of land in developing world. Proj INT/75.P13, FAO, Rome
Hogbom I (1923) Ancient inland dunes of north and middle Europe. Ann Geogr Stockholm V:113–243
Holburt MB (1978) International problems. Quoted by Peterson DF, Crawford AB, Values and choices in the development of the Colorado River Basin. The Univ of Ariz. Pr. Tucson, pp 269, 243
Horikava K, Shen HW (1960) Sand movement by wind. U S Army Corps of Engineers, Beach Erosion Board, Tech Mem 119:51
Hornburg CD (1987) Desalination for remote areas. In Pickford J (ed) Developing world water. Grosvenor Press International, London, pp 230–232
Hudson N (1963) Rainfall size distribution in high intensity storms. Rhodesian J Agric Res 1–1:6–11
Hudson N (1971) Soil conservation. Batsford Ltd, Lond, 330 pp
Hugo GJ (1984) The demographic impact of famine: a review. In: Currey B, Hugo G (eds) Famine as a geographical phenomenon. Reidel, Dordrecht, Boston, pp 7–31
Huibers FP (1985) Rainfed agriculture in a semi-arid tropical climate. Aspects of land and water management for red soils in India. Agric Univ Wageningen, The Netherlands
Hulme M (1990) The changing rainfall resources of Sudan. Trans Inst Br Geogr NS 15:21–34
Hurault J (1972) Phases climatiques tropicales sèches à Banyo (Cameroun, Hauts plateaux de l'Adamawa). In: Van Zinderen Bakker EM (ed) Paleoecology of Africa, the surrounding islands and Antarctica. Balkema, Cape Town, VI:93–102

Hurault J (1975) Surpâturage et transformation du milieu physique. Formations végétales, hydrologie de surface, géomorphologie. L'exemple des hauts plateaux de l'Adamaoua (Cameroun). Etude de photointerprétation, Inst Géogr Natl, Paris, 7:218

Hurault J, Lummaux JC (1979) Essai de cartographie de l'environnement d'un milieu tropical d'altitude: l'Adamaoua occidental (Cameroun-Nigeria). Symp Int sur la cartographie de l'environnement et de sa dynamique, Caen, pp 189–225, 5 cartes, 15 planches photographiques

Huszar PC, Piper SL (1986) Off-site economic costs of wind erosion in New Mexico. The off-site costs of soil erosion (edited by Waddel TE). Conserv Foundation, Washington DC, pp 143–166

Hyder DN (1978) Proceedings of the first international rangeland congress. Denver, Colorado, USA

ICIHI (Independent Commission on International Humanitarian Issues) (1986) The encroaching desert: the consequences of human failure. Zed Books, London, 132 pp

Institute of desert research, Academia Sinica Lanzhou and MAB UNESCO (1984) Int Symp on Integrated Control of Land Desertification Proc, 147 pp

Itard J (1986) Les glossines ou mouches tsé-tsé. Etudes et synthèse de l'IEMVT (Institut d'études médicales et vétérinaires tropicales) Paris no 15, 155 pp

Ivanov AP (1969) Wind erosion of sands, sandy and loamy-sandy soils in the USSR and combative measures. Moscow, p 58 (in Russian)

Jacks GV (1939) Soil erosion. The rape of the earth. A survey of soil erosion. In: Jacks GV, Whyte RO (eds) Soil erosion and its control. Faber, London, 313 pp

Jackson IJ (1989) Climate, water, and agriculture, Ind edn. Longman, London

Jaenicke R (1979) Monitoring and critical review of the estimated source strength of mineral dust from the Sahara. In: Morales C (ed) Saharan dust mobilization, transport, deposition. Wiley, New York pp 233–242

Johnson JD (1978) "Discussion". In: Reiming P (ed) Desertification papers. Wash DC Am Assoc Adv Sci, p 139

Jones LH (1983) The oil palm and its clonal propagation by tissue culture. Biologist 30–4:181–188

Judson S (1981) What's happening to our continents? In: Skinner BJ (ed) Use and misuse of earth's surface. Kaufman Inc, Los Altos, California, pp 12–139

Kates RW, Johnson DL, Johnson Haring K (1977) Population, society and desertification. Report of UNCOD A/CONF 74/8, Nairobi

Keetch RI (1981) Rangeland rehabilitation in central Australia. Land conservation unit. Conserv Comm North Territory, Alice Springs, NT 5750 - LG/79/732 pp

Khakimov FI (1989) Soil melioration conditions of desertification of deltas. Puschino, Moscow, 218 p

Kharin NG (1976) Remote sensing techniques and the monitoring of desertification processes in arid areas. Probl Osv. Pustyn' 3–4:45–54

Kharin NG, Kalenov GS (1978) The study of man-induced desertification from space imagery. Probl Osv. Pustyn' 4:25–28

Khatib AB (1971) Present and potential salt affected and waterlogged areas in the countries of the near East in relation to agriculture. In: Salinity seminar. Irrigation and drainage. FAO Roma, Baghdad, 7:13–38

Kiabou B (1989) Le maìs, la céréale africaine qui monte. Information, Lett Inf, CILLS, club du Sahel, Paris, nov 6:15

Kishk MA (1984) Present state of desertification in Egypt. In: Yang Y (ed) Proc Int Symp on integrated control of land desertification. National Committee of the People's Republic of China for MAB, pp 68–73

Kovda VA (1977) Land aridization and drought control. Nauka, Moscow, 272 pp (in Russian)

Kovda VA (1983) Loss of productive land due to salinization. Ambio 12:91–93

Kurochkina LY (1978) Psammophytes in the Kazakhstan deserts. Naoka, Alma-Ata (in Russian)

Kuznetsov NT (1980) Preliminary results and some trends of further research into the problem of the Aral sea. Probl Osv. Pustyn' 5:10–17

Lal R (ed) (1988) Soil erosion research methods. Soil Water Conserv Soc, Ankeny, Iowa, USA, 244 pp

Lal R, Hall GF, Miller EP (1989) Soil degradation: I Basic processes. Land Degradation and Rehabilitation. Wiley 1–1:51–69

Lamprey HF (1975) Report on the desert encroachment reconnaissance in northern Sudan. 21st Oct-10th Nov, UNESCO, UNEP, Paris/Nairobi, mimeo, 16 pp

Lancaster N (1982) Linear dunes. Progr Lin Geogr 6:475–504
Land T (1986) Desert tree with promise. Co-operation South, pp 17–18
Larousse du xx° siècle en six volumes (1929) Paris
Larson WE, Pierce FJ, Dowdy RH (1983) The threat of soil erosion to long-term crop production. Science 219:458–465
Laws JO (1941) Measurements of fall-velocity of water drops and rain drops. Trans. Am Geophys Union 22:709
Le Floch E, Floret C (1972) Désertisation, dégradation et régénération de la végétation pastorale dans la Tunisie présaharienne. IRAT, Projet Parcours Sud Mimeo
Le Houérou HN (1958) Ecologie, phytosociologie et productivité de l'olivier en Tunisie méridionale. Bull Carte Phytogéogr B, CNRS, Paris, IV-1:7–72
Le Houérou HN (1968) La désertisation du Sahara septentrional et des steppes limitrophes. PBI Coll Hammamet, London, 26 pp, et Ann Alg Géogr, Alger, 6:2–27
Le Houérou HN (1975) Science power and desertization. Meeting on desertisation, Dept Geogr, Univ Cambridge 22–28 sept, 26 pp
Le Houérou HN, Gillet H (1985) Conservation versus desertization in African arid lands. In: Soulé M (ed) Conservation biology, the science of scarcity and diversity. Sinauer associates, Publish Sunderland, Massachusetts, pp 444–461
Létolle R, Mainguet M (1993) Aral. Springer, Paris, 357 pp
Lettau H (1969) Evapotranspiration climatonomy: 1 A new approach to numerical prediction of monthly evapotranspiration, run-off and soil moisture storage. Mon Weather Rev 97:691–699
Liebman M, Robichaux RH (1988) Competition by barley and pea against weedy mustard: effects on resource acquisition, photosynthesis and yield. Agriculture, ecosystems and environment, (in press) Quoted by Dover M, Talbot LM (1987) To feed the Earth: Agro-ecology for sustainable development. World Resources Inst. Holmes, USA
Liu Tungsheng (1988) Loess in China, 2nd edn. Springer, Berlin Heidelberg New York Tokyo, pp 131–140
Lowdermilk WC (1935) Man-made deserts. Pacific affairs. Univ British Columbia 8–4:409–419. In: Morgan RPC (ed) Soil erosion and its control. Hutchinson Ross Publ, Van Nostrand Reinhold C° New York, pp 9–19
Lyles L (1977) Wind erosion: processes and effect on soil productivity. Trans ASAE 20–5:880–884
Lyles L, Takarko J (1986) Wind erosion effects on soil texture and organic matter. J Soil Water Conserv 41–3:191–193
Mabbutt JA (1977) Aspects climatiques et écologiques de la désertification. Nature et Ressources XIII-2:3–10
Mabbut JA (1978a) The impact of desertification as revealed by mapping. Environ Conserv 5–1:45
Mabbutt JA (1978b) Desertification in Australia. Water Res Found Aust 54:129
Mabbutt JA (1986) Desertification in Australia. In: Arid land development and the combat against desertification: an integrated approach. UNEP-UNEPCOM, Moscow, pp 101–112
Mainguet M (1972) Le Modelé des Grès Problèmes généraux. Inst Geog Nat Paris in Etudes de photo-interpret. 657 pp. 2 vols. 72 tableaux, 105 stéréogr., 333 fig et phot., 4 cartes ht en coul., 38 planches, cartes, coupes
Mainguet M (1982) Les marques géographies de l'aridité dans le Sahara et sur ses marges au Plio-Quaternaire. Bull AGF 483–484:25–29, 64–68
Mainguet M (1984a) Le vent, mécanisme d'érosion, de dégradation, de désertification. Trav Inst Géogr Reims, 59–60:142 pp
Mainguet M (1984b) Cordons longitudinaux (sand ridges), dunes allongées à ne plus confondre avec les sifs, autres dunes linéaires. Trav Inst Géogr Reims 59–60:61–83
Mainguet M (1984c) Un mécanisme de désertification méconnu: le "wala-wala". Trav Inst Géogr Reims 59–60;131–135
Mainguet M (1984d) Inventory of sand movement in Somalia. Natl Range Agency, UNSO/DES/SOM/80/001, 100 pp
Mainguet M (1985) Le Sahel, un laboratoire naturel pour l'étude du vent, mécanisme principal de la désertification. Univ Aarhus, Dep Theor Stat, Proc Int Workshop on the Physics of Blown Sand 3–8:545–561
Mainguet M, Chemin MC (1978) Dunes et autres édifices sableux éoliens — actions éoliennes (déflation, transport, corrasion), recherches bibliographiques. LGPZ, Univ Reims, 344 pp

Mainguet M, Chemin MC (1979) Lutte contre l'ensablement des palmeraies et des oasis dans le Sud marocain. Rapport technique: étude préliminaire de l'avancement du sable. Apport de la photointerprétation, de la télédétection et des critères géomorphologiques de terrain. FAO, Roma TCP/MOR/6702, 32 pp, 1 carte couleur hors-texte

Mainguet M, Chemin MC (1982) Formulation d'un programme de lutte contre l'ensablement des palmeraies de Bilma (Niger). FAO, Bureau des Nations Unies pour les régions soudano-sahéliennes UNSO, 106 pp

Mainguet M, Chemin MC (1983) Sand seas of the Sahara and Sahel: an explanation of their thickness and sand dune type by the sand budget principle. Int Assoc Sedimentol, 11ème Congr Int Sédiment, Hamilton, Ontario, aout 1982 In: Brookfield ME, Ahlbrandt TS (eds) Eolian sediments and processes. Elsevier Sci Publ BV, Amsterdam, pp 353–363

Malcolm CV, Stoneman TC (1976) Salt encroachment. The 1974 salt land survey. West Aust J Agric 17:42–49

Maley J (1973) Mécanisme des changements climatiques aux basses latitudes. Palaeogeogr Palaeoclimatol Palaeoecol 14:193–227

Maley J (1981) Etudes palynologiques dans le bassin du Tchad et paléoclimatologie de l'Afrique nord-tropicale de 30 000 ans à l'époque actuelle. Trav Doc ORSTOM 129:586

Martonne E de, Aufrère L (1928) L'extension des régions privées d'écoulement vers l'Océan. UGI Paris, 3:1–24

Mather JR (1974) Climatology: fundamentals and applications. McGraw-Hill, New York, 412 pp

Matheson WE (1968) When salt takes over. South Aust J Agric 71:226–272

McGuahey M (1986) Impact of forestry initiatives in the Sahel. Quoted by Dover M, Talbot LM 1987 Chemonics international. Washington, DC

Meckelein W (1986) Zu physischer Geographie und agraren Nutzungsproblemen in den innerasiatischen Wüsten Chinas. Geo-öko-dynamik, Darmstadt, RFA, VII, 1/2:27

Meigs P (1953) World distribution of arid and semi-arid homoclimates. In: Reviews of research on arid zone hydrology. UNESCO, Paris, France, Arid zone programme, 1:203–210

Meigs P (1961) Map of arid zone. UNESCO, Paris

Merva GE, Peterson G (1983) Wind erosion sampling in the North central region. Am Soc Agric Eng Pap 83:2133

Meyer LD, Wischmeier WH (1969) Mathematical simulation of the process of soil erosion by water. Am Soc Agric Eng Trans 12:754–758

Middleton NJ (1985) Effect of drought on dust production in the Sahel. Nature (Lond) 316, 6027:431–434

Milas SL (1983) The years of drought. Desert Control Bull 9:10–14

Molosnova TI, Subbotina OI, Chanysheva SG (1987) Climatic consequences of economic activity in the zone of the Aral Sea. Gidrometeoizdat, Moscow, 119p (in Russian)

Moore PD (1987) Mobile resources for survival. Nature (Lond) 325:198

Morozova OI (1959) Rangelands in a desert and a piedmont desert. Selkhozgiz, Moscow, 302 pp (in Russian)

Mou J, Meng Q (1980) Sediment delivery ratio as used in the computation of the watershed sediment yield. Beijing, China

Murray GE, Lowry GR (1976) S.C.S. and water quality: an overview. Soil Conserv p 97

Mutchler CK, Murphree CE, McGregor KC (1988) Laboratory and field plots for soil erosion studies. In: Lal R (ed) Soil erosion research methods. Soil Water Conserv Soc, Ankeny, Iowa, 244:9–36

Nabhan GP (1987) Using natural desert ecosystems as models for agroforestry: the gray, spiney revolution. In: Aldon EF, Gonzales Vicente CE, Moir WH (eds) Strategies for classification and management of native vegetation for food production in arid zones. USDA Forest service/SARH instituto nacional de investigaciones forestales y agropecuarias, general technical report, RM 150, Fort Collins, Colorado, pp 232–234

Nabhan GP, Robichaux R, Slater M, Nelson S (1988) Useful desert plants for low-input polycultures. In: Arid lands newsletter. Office of arid land studies, Univ Arizona, Tucson, 27:11–14

Nair PKR (1982) Some considerations on soil productivity under agroforestry land-use systems. In: Dover M, Talbot LM (eds) 12th Int Congr Soil Sci, New Delhi, India

Natarajan M, Willey RW (1986) The effects of water stress on yield advantages of intercropping systems. Field Crops Res 13:117–131

Nechaeva NT (1946) Effect of grazing on rangeland vegetation in the Kara Kum. Izv Turkmen FAN SSSR, Issues 3–4:81–89
Nechaeva NT (1954) Effect of grazing on Kara Kum rangelands. In: USSR deserts and their development. Izv AN SSSR, Moscow, Leningrad, II:370–391 (in Russian)
Nechaeva NT (1977) Problems of research stemming from the use of USSR desert ecosystems in range-based animal husbandry. Probl Desert Dev 2:21–32
Nechaeva NT (1979a) Problems of the desertification indicators development. Probl Osv. Pustyn' 4:18–24
Nechaeva NT (1979b) Effect of the management regime on the vegetation productivity in the Kara Kum. Probl Osvo Pustyn' 6:8–18
Nechaeva NT, Pelt NN (1963) Forage base of sheep farming in deserts of central Asia and Kazakhstan. In: Natural conditions for animal husbandry and forage base of deserts. Izv AN TSSR, Ashkhabad (in Russian)
Nelson R (1988) Dryland management: the "desertification" problem. The world bank policy planning and research staff. Environ Dep 8:42
Nicholson SE (1976) A climatic chronology for Africa: synthesis of geological, historical and meteorological information and data. PhD Thesis, Univ Wisconsin, Madison
Nicholson SE (1978) Climatic variations in the Sahel and other African regions during the past five centuries. J Arid Environ 1:3–24
Nicholson SE (1980) The nature of rainfall fluctuations in sub-tropical west Africa. Mon Weather Rev 108:473–487
Nicholson SE (1981) The historical climatology of Africa. In: Wigley TMI, Ingram MJ, Farmer G (eds) Climate and history. Cambridge Univ Press, Cambridge, pp 249–270
Nicholson SE (1983) Sub-Saharan rainfall in years 1976–80: evidence of continued drought. Mon Weather Rev 111:1646–1654
Nicholson SE, Flohn H (1980) African environmental and climatic changes and the general atmospheric circulation in late Pleistocene and Holocene. Clim Change 2:313–348
Nikolaev VN (1982) Experience of development and rational management of desert rangelands. In: Combating desertification in the USSR: problems and experience. Cent Int Proj, UNEPCOM, Moscow, 120 pp
Nikolaev VN, Amangeldyev AA, Smetankina VA (1977) Desert rangelands: forage evaluation. Nauka, Moscow (in Russian)
Northcote KH, Skene JKM (1972) Australian soils with saline and sodic properties. CSIRO Aust Soil Publ 27
Novikova NM (1990) Problems of conservation of the ecosystems of river deltas in the Central Asia. Prepared for the International Conference: "The Aral Crisis: Causes, Consequences and Ways of Solution". Nukus Symp, Karakalpak, U.S.S.R. Oct. 2–5, 1990
Noy-Meier I, Seligman NG (1979) Management of semi-arid ecosystems in Israel. In: Walker BH (ed) Management of semi-arid ecosystems. Elsevier, Amsterdam, pp 113–160
Oldeman LR, Hakkeling RTA, Sombroek WG (1990) World map of human-induced soil degradation. Colour map, 3 sheets 1/13,000th, ISRIC, Wageningen and UNEP, Nairobi
Oldham RD (1903) A note on the sandhills of Clifton near Karachi. M Geol Surv India XXXIV-1
Oliver J, Britton NR, James MK (1984) The Ash wednesday bushfires in Victoria 16 February 1983. Disaster investigation report No. 7. Quoted by Allan R, Heathcote RL 1987 Centre for disaster studies, James Cook Univ, Townsville, Australia
Ollier CA, Gallaway RW (1990) The laterite profile, ferricrete and unconformity. Catena 17:97–109
Olsson L (1985) An integrated study of desertification: applications of remote sensing, GIS and spatial models in semi-arid Sudan. Medd Lunds Univ Geogr Inst, Avhandlingar 98:170
Orev Y (1986) A practical handbook on desert range improvement techniques. Interagency group on agricultural meteorology. FAO-UNEP-UNESCO-WMO, Geneva, 165 pp
Orev Y (1988) Some agricultural considerations in the planning of runoff farming. In: Stiles D, Mainguet M (eds) Desertification control bulletin. UNEP, Nairobi, Kenya, 16:13–16
Orlovsky NS (1982a) Desertification: its mechanism and implications. In: Combating desertification in the USSR: problems and experience. Cent Int Proj, UNEP UNEPCOM, Moscow, 120 pp
Orlovsky NS (1982b) Natural conditions of deserts in the USSR and desertification processes, 2nd edn. In: Combating desertification in the USSR: problems and experience. Cent Int Proj, UNEP UNEPCOM, Moscow, 120 pp
Osborne M (1975) An assessment of the past, present and future status of the college farm shelterbelts. B Sci thesis Univ College North Wales, Bangor

Pachur HJ, Kröpelin S (1987) Wadi Howar: paleoclimatic evidence from an extinct river system in the south eastern Sahara. Science 237:298–300

Palutikof JP, Farmer G, Wigley TML (1982) Strategies for the amelioration of agricultural drought in Africa. In: Proc Tech Conf Clim Afr, WMO Genevane 596, pp 222–248

Pauli HW (1972) Relationships between land use and water pollution. In Water pollution. Water Res Found Aust 38 5.1–5.5:13

Penning de Vries FWT, Djiteye MA (1982) L'élevage et l'exploitation des paturâges au Sahel. In: Penning de Vries FWT, Djiteye MA (eds) La productivité des paturages sahéliens. Cent Agric Publ Doc, Wageningen, pp 1–19

Perry RA (1984) Desertification in Australia. In: International symposium on integrated control of land desertification. Inst Desert Res, Acad Sin Lanzhou and MAB UNESCO, pp 1–3

Perry RA, McAlpine JR (1976) Impact of isolated towns on the environment. In: New towns in isolated settings. Semin Aust Natl Comm UNESCO, Kambalda WA. Aust Gov Publ Serv, Canberra, pp 1–76

Petit-Maire N (1989) Interglacial environments in presently hyperarid Sahara: paleoclimatic implications. In: Leinen M, Sarnthein M (eds) Paleoclimatology and paleometeorology: modern and past patterns of global atmospheric transport. pp 637–661

Petrov MP (1950) Drifting sands and their control. Moscow, 454 pp (in Russian)

Petrov MP (1972) Desertification processes in arid regions and combative measures. In: Dokl K XXII MGK, Leningrad, pp 69–88 (in Russian)

Petrov MP (1973) Deserts of the world. Nauka, Leningrad, 435 pp (in Russian)

Petrova AV (1982) Changes in humus and nitrogen contents in calcareous chernozems caused by wind erosion. Agrokhimiya 1:76–80

Péwé TL (1981) Desert dust: an overview in desert dust: origin, characteristics and effects on man. Geol Soc Am Spec Pap 186:1–10

Pimental D, Allen J, Beers A et al. (1987) World agriculture and soil erosion: erosion threatens world food production. Bioscience 37–4:277–283

Pratt DJ, Gwynne MD (1977) Rangeland management and ecology in east Africa. Hodder and Stoughton, London

Prescott JA, Pendleton RL (1952) Laterite and lateristic soils. Commonwealth Bureau of Soil Science Technical Committee 47. Commonwealth Agricultural Bureau Harpenden

Prospero JM, Glaccum RA, Nees RT (1981) Atmospheric transport of soil dust from Africa to South America. Nature (Lond) 289:570–572

Puterbaugh TL, Motha RP, Strommen ND, Decker WL, Le Duc SK (1983) A regional comparison of seasonal African rainfall anomalies. In: Proc 2nd Int Meet Stat Climatol, Lisbon, Inst Nacl Meteorol Geofis, Portugal, 8.5.1–8.5.8

Quinn F (1968) Water transfers: must the American west be won again? Geogr Rev 51–1:115

Rapp A (1974) A review of desertization in Africa -- water, vegetation and man. Secretariat Int Ecol, Stockholm, Sweden, 77 pp

Reij C (1983) L'évolution de la lutte anti-érosive en Haute-Volta depuis l'indépendance, vers une plus grande participation de la population. Inst Environ Stud, Free Univ Amsterdam, The Netherlands

Reijntjes G (1986) Water harvesting, a review of different techniques. ILEIA (Information centre for low external input agriculture) Newsl 5:7–8

Reijntjes G (1986) Runoff farming vital part of the Tunisian water management system. ILEIA (Information centre for low external input agriculture) Newsl 5:8–9

Romanov NN (1960) Dust storms in central Asia. Tr Tash GU, pp 174–198

Rowan JN (1971) Salting on dry-land farms in North-western Victoria. Quoted by Mabbutt JA 1986 Soil conservation authority of Victoria. Melbourne

Rozanov BG (1984) Principles of the doctrine on the environment. IGU, Moscow, p 273 (in Russian)

Russell B (1947) History of western philosophy. And its connections with political and social circumstances from the earliest times to the present day. 2nd impression. G. Allen and Unwin, Lond 916 pp

Sandford S (1976) Pastoralism under pressure. ODI Rev 2:45–68

Sapozhnikova SA (1970) Schematic map of a number of days with dust storms in the arid zone of the USSR and adjacent territories. Tr NIIAK 65:61–69

Schneider JL (1969) Evolution du dernier lacustre et peuplements préhistoriques aux pays bas du Tchad. Bull Inst Afr Noire A 31:259–263

Sdasyuk GV (1986) Desertification in India and problems of integrated regional development. In: Arid land development and the combat against desertification: an integrated approach. UNEP-UNEPCOM, Moscow, pp 87–93

Shah M (1978) Salt affected soils in Pakistan. In: Proc workshop/seminar on membrane biophysics and salt tolerance in plants. Univ of Agric, Faisalabad, Pakistan, pp 47–64

Sharp RP (1964) Wind-driven sand in Cachella valley, California. Bull Geol Soc Am 75:785–804

Shearer W (1986) In search of fresh water. Dev Forum, UN Univ, July-August, 1 p

Sheridan D (1981) Desertification of the United States. Counc Environ Qual, 142 pp

Sheridan D (1986) Problems of desertification of the United States. In: Arid land development and the combat against desertification: an integrated approach. UNEP-UNEPCOM, Moscow, 145 p, pp 96–100

Shlikhter SB (1986) The role of basic infrastructure in integrated development of arid regions. In: Arid land development and the combat against desertification: an integrated approach. UNEP-UNEPCOM, Moscow, pp 48–53

Shmueli E (1973) Efficient utilization of water in irrigation. In: Yaron B, Danfors E, Vaadia Y (eds) Arid zone irrigation. Springer, Berlin Heidelberg New York, pp 411–423

Sidhu PS (1977) Aeolian additions to the soils of north-west India. Pedology 27–3:323–336

Sinha SK (1987) The 1982–83 drought in India: magnitude and impact. In: Glantz M, Katz R, Krenz M (eds) The societal impacts associated with the 1982–83 worldwide climate anomalies. Environ Soc Impacts Group, Natl Cent Atmos Res, UNEP, Boulder Colorado Nairobi Kenya, pp 37–42

Sircoulon J (1976) Les données hydropluviométriques de la sécheresse récente en Afrique intertropicale. Comparaison avec les sécheresses 1913 et 1940. Cah ORSTOM Sér Hydrol 13:75–174

Skoupy J (1976) Afforestation in the arid Mediterranean and near East areas. Silvaecultura tropica and subtropica. For Sci Inst, Univ Agric, Prague, 19 pp

Smith RCG, Mason WK, Meyer WS, Barrs HD (1983) Irrigation in Australia: development and prospects. In: Hillel D (ed) Advances in irrigation. Academic Press, London New York, pp 99–153

Stebbing EP (1938) The advance of the Sahara. Geogr J 91:356–359

Stiles D (1988) Arid land plants for economic development and desertification control. Desertification control bull, UNEP, Nairobi Kenya, 17:18–21

Street P (1987) Land use policy. Butterworth, London, 4–4:362

Street-Perrott FA, Roberts N (1983) Fluctuations in closed-basin lakes as an indicator of past atmospheric circulation patterns. In: Street-Perrott A, Beran M, Ratcliffe R (eds) Variations in the global water budget. Reidel, Dordrecht, pp 331–345

Suliman MM (1988) Dynamics of range plants and desertification monitoring in the Sudan. Desertification control bulletin, UNEP, Nairobi Kenya, 16:27–31

Suzuki H (1981) The transcendent and environments. A historico-geographical study of world religions. Addis Abeba Sha Yokohama, Japan, 134 pp

Szabolcs I (1986) Agronomic and ecological impact of irrigation on soil and water quality. Adv Soil Sci 4:189–218

Tardy Y (1993) Petrologie des latérites et des sols tropicaux. Masson, Paris, 472 pp

Teakle LJH (1938) Soil salinity in western Australia. West Aust J Agric 15:434–452

Thompson BW, Mörth HT (1965) Notes from East Africa, no 1, Weather 20:226–227

Thompson CH, Moore AW, Northcote KH (1982) Soils and land use. Soils: an Australian viewpoint. Division of soils, CSIRO, Melbourne/Academic Press, Lond New York, pp 757–775

Thompson M, Warburton M, Hatley T (1986) Uncertainty on a Himalayan scale. Milton for Ethnografia Press, Lond, 162 pp

Timberlake L (1985) L'Afrique en crise; la banqueroute de l'environnement, L'Harmattan-Earthscan éd, Paris, 294pp

Thompson RD (1975) The climatology of the arid world. Geogr Pap 35, Univ Reading

Tilho J (1910–11) Documents scientifiques de la mission Tilho. 2 volumes. In 8°, Impr Nation, Paris, I:412 pp, II:631 pp

Toulmin C (1988) Smiling in the Sahel. New Sci 12 Nov:69

Tseo GKYT (1986) Longitudinal dunes: their genesis and ordering. Ph D thesis, Univ Adelaide, I 153 pp, II 46 plates 100 figures

Tsoar H (1983) Dynamic processes acting on a longitudinal (seif) sand dune. Sedimentology 30:567–578
Tsoar H (1989) Linear dunes -- forms and formation. Prog Lin Geogr 13:507–528
UNEP-UNCOD (1978) United Nations conference on desertification. 29 Aug-9 Sept 1977 Round-up, plan of action and resolutions, New York, 43 pp
UNEP (1980) Annual review. Nairobi Kenya
UNEP-UNEPCOM (1980) Shifting sands in the deserts of the USSR: stabilization and afforestation. In: Babaev AG (ed) Cent Int Proj, Moscow, 318 pp
UNEP-UNEPCOM (1982) Combating desertification in the USSR: problems and experience. Babaev AG (ed) Cent Int Proj, Moscow, 120 pp
UNEP-UNEPCOM (1984) The USSR experience in desert reclamation and desertification control. Babaev AG, Zonn IS, Orlovsky NS (eds) Moscow, 104 pp
UNEP-UNEPCOM (1986a) Principles and methods of shifting sands fixation. Babaev AG (ed) Cent Int Proj, Moscow, 132 pp
UNEP-UNEPCOM (1986b) Arid land development and the combat against desertification: an integrated approach. Cent Int Proj, Moscow, 146 pp
UNESCO (1974) Influence of man on the hydrological cycle: guidelines to policies for the safe development of land and water resources. In: Status and trends in research in hydrology 1965–74. Working Group Rep, Paris, pp 30–70
US Bureau of reclamation and California department of water resources (1979) Agricultural drainage and salt management in the San Joachin Valley, Fresno, California, 62 pp
US Department of agriculture, soil conservation service (1977) A guide for draining agricultural land along the westside of the San Joachin Valley, Portland, Oregon 4 pp
US General accounting office (1977a) Ground water: an overview. CED-77–69:1–4
US General accounting office (1977b) To protect tomorrow's food supply, soil conservation needs priority attention. Washington DC, US government printing office, 5 pp
US Geological survey (1974) Hydrologic unit map, State of California. Quoted by Sheridan D 1981
US Geological survey (undated) Field observations of the December 1977 windstorm. San Joachin Valley, California. Quoted by Sheridan D 1981, 1–8–10–18–19
Utkin GN (1986) Problems of industrial development of arid lands. In: Arid land development and the combat against desertification: an integrated approach. UNEP-UNEPCOM, Moscow, 146 pp
Van Dijk DC (1969) Relict salt, a major cause of recent land damage in the Yass Valley, Southern Tablelands, NSW, Australian Geographer, 11, 13–21
Van Noordwijk M (1984) Ecology text book for the Sudan-Karthoum Univ Press, 280 pp
Vasil'yev GI, Bulgakov DS, Gaavrilenko LN, Kalinichenko AS (1978) Conditions under which dust storms develop in the northern Caucasus. Sov Soil Sci 10–1:60–69
Vergara NT (1982) New directions in agroforestry: the potential of tropical legume trees. Honolulu, Hawaii, Environment and policy institute, east-west centre. Quoted by Dover M, Talbot LM 1987, 88 pp
Vietmeyer ND (1986) Forgotten crops of the Incas. Bostid Dev 6–2:16–17
Vietmeyer ND (1987) Forgotten crops of the Incas. Bostid Dev 7:8–12
Viktorov SV (1970) Protection of deserts as a man's habitat. In: Problems of geography. Mysl, Moscow, Issue 82:95–102 (in Russian)
Wadsworth RM (1964) Wind speed and plant growth. 2nd Symp Shelter Res, Edinburgh, Sept 1964. Ministry of Agriculture, Fishery and Food
Wang Kangfu (1988) Fixation of shifting sands on both sides of the Baolan railway in the Shapotou region. In: Training package. International training course on desertification control, IDRAS, Lanzhou, China, pp 52–66
Warren A (1984) Productivity, variability and sustainability as criteria of desertification. In: Fantechi R, Maragris NS (eds) Desertification in Europe. Proc Inf Symp EEC Progr Climatol Mytelene, Greece. Reidel, Dordrecht, pp 83–94
Warren A, Agnew C (1988a) An assessment of desertification and land degradation in arid and semi-arid areas. Int Inst Environ Dev. Drylands programme. Univ college, Lond, 72 pp
Warren A, Agnew C (1988b) An assessment of desertification and land degradation in arid and semi-arid areas. Int Inst Environ Dev (IIED), Drylands Pap 2:30
Warren A, Maizels JK (1977) Ecological change and desertification. UNCOD A/CONF 74–7:1
Wasson RJ, Nanninga PM (1986) Estimating wind transport of sand on vegetated surfaces. Earth surface processes and land forms 11:505–514

Weber FR, Stoney C (1987) Reforestation in arid lands. Volunteers in technical assistance, Arlington, Virginia, USA, 345 pp

Wellton-Mohawk irrigation and drainage district. Crop Prod Rep (1978) US Bureau of reclamation, Corp Prod Rep 1955. Quoted by Sheridan D 1981, 142 pp

Wigley TML, Atkinson TC (1977) Dry years in south-east England since 1698. Nature (Lond) 265:431–434

Wilshire HG (1981) Human causes of accelerated wind erosion in California's deserts. In: Coates DR, Vitete JD (eds) Thresholds in geomorphology. Allen and Unwin, Lond

Wippermann FK, Gross G (1986) The wind-induced shaping and migration of an isolated dune: a numerical experiment Boundary-Layer Meteorology 36:319–334

Wischmeier WH (1959) A rainfall erosion index for a universal soil loss equation. Soil Sci Soc Am Proc 23–3:246–249

Wischmeier WH, Smith DD (1978) Predicting rainfall erosion losses. Agric handbook 537, US Dep Agric, Washington DC 58 p

WMO (1978) An evaluation of climate and water resources for development of agriculture in the Sudan-Sahelian zone of West Africa. A summary of WMO Spec Environ Rep 9, prepared for UNEP, 45 pp

WMO (1983a) Meteorological aspects of certain processes affecting soil degradation -- especially erosion. Tech Note No 178, WMO N° 591:149

WMO (1983b) Climatic situation and drought in Africa. Rep Expert Group Meet, Geneva, 6–7 Oct, WCP 61:32

Wollny E (1877) Untersuchungen über den Einfluss der Pflanzen und der Beschattung auf die physikalischen Eigenschaften des Bodens. Berlin

Wolstad A, Botts F (1984) Pigeon pea, Rhodes grass, and napier for soil conservation. FAO, 21 pp

Woods LE (1983) Land degradation in Australia. Quoted by Allan R, Heathcote RL 1987 Aust Gov Printing Serv, Canberra, Australia

Yensen NP (1988) Plants for salty soil. In: Arid lands newsletter. Office of arid land studies, Univ Arizona, 27:1–10

Young K (1981) The impact of erosion on the productivity of soils in the USA, pp 295–303. In: De Boadt M, Gabriels D (eds) Assessment of soil erosion. John Wiley & Sons, Chichester

Zakirov RS (1982) Prevention of sand drifts on railways, roads and irrigation systems. USSR/UNEP Proj, UNEP Publ Inf Support Progr USSR, published by the centre of international projects GKNT, Moscow, 182 pp

Zaletaev VS, Novikova NM (1990) Changes in biota of the Aral region as result of anthropogenic impacts in the period between 1950 and 1990. Prepared for the International Conference: "The Aral Crisis": Causes, Consequences and Ways of Solution". Nukus, Karakalpak, U.S.S.R. Oct. 2–5, 1990

Zhao Songqiao (1986) Physical geography of China. Sci press, Beijing, China, Wiley, New York, 209 pp

Zhu Zhenda (1984) Desertification processes and their reversing in arid and semi-arid lands in China. Yang Youlin (ed) MAB:19–26

Zhu Zhenda, Liu Shu (1983) Combating desertification in arid and semi-arid zones in China. Inst Desert Res, Academia sinica, Lanzhou, China, 69 pp

Zhu Zhenda, Liu Shu (1988) Desertification processes and their control in the north of China. In: Training package. International training course on desertification control. Academia Sinica, Lanzhou, China, pp 1–15

Zhu Zhenda, Liu Shu, Wu Zhen, Di Xinmin (1986) Deserts in China. Inst Desert Res, Academia Sinica, Lanzhou, China, 132 pp

Zonn IS (1981) Irrigation development of deserts and amelioration of irrigated lands. In: Combating desertification in the USSR: problems and experience. Cent Int Proj, UNEP-UNEPCOM, Moscow, pp 75–91

Zonn IS (1986) Land use and water resources of arid areas. In: Arid land development and the combat against desertification: an integrated approach. UNEP-UNEPCOM, Moscow, pp 23–33

Zonn IS, Orlovsky NS Anthropogenic Factors of Desertification 1986 pp 16–23 in Arid land development and the combat against desertification, an integrated approach. UNEP-UNEPCOM, Moscow 146 p

Author Index

Subject Index

Page numbers in bold refer to the principal place in the text.